Dynamics of
Biological Membranes

Dynamics of Biological Membranes

Influence on Synthesis, Structure and Function

MILES D. HOUSLAY
Department of Biochemistry
The University of Manchester Institute of Science and Technology
P. O. Box 88, Manchester M60 1QD, UK

and

KEITH K. STANLEY
European Molecular Biology Laboratory,
Postfach 10.2209, Meyerhofstrasse 1,
D-6900 Heidelberg, W. Germany

A Wiley–Interscience Publication

175 YEARS OF PUBLISHING

1807 1982

JOHN WILEY & SONS
Chichester · New York · Brisbane · Toronto · Singapore

Library of Congress Cataloging in Publication Data:

Houslay, Miles D.
    Dynamics of biological membranes.

    'A Wiley–Interscience publication.'
    Includes index.
    1. Membranes (Biology)   I. Stanley, Keith K.
II. Title.
QH601.H686   574.87'5   81-14759
ISBN 0 471 10080 3 (cloth)   AACR2
ISBN 0 471 10095 1 (paper)

British Library Cataloguing in Publication Data:

Houslay, Miles D.
    Dynamics of biological membranes.
    1. Membranes (Biology)
    I. Title II. Stanley, Keith K.
    574.87'5   QH601

ISBN 0 471 10080 3 (cloth)
ISBN 0 471 10095 1 (paper)

Typeset by Preface Ltd, Salisbury, Wilts.
Printed in the United States of America

Contents

Preface

Books on biological membranes have, in the main, discussed this topic primarily in terms of the detailed composition of membranes and the reactions, such as transport and enzyme functions, that characterize them. Over the past few years, however, the functioning of biological membranes at the molecular level has begun to be understood. In particular, biological membranes have emerged, not as static inert boundaries to the cellular and intracellular compartments, but as dynamic structures intimately involved in many, if not most, of the biochemical processes of the cell. The emphasis of this book is therefore on structural and functional aspects which contribute to our concept of the dynamic membrane. In view of this we do not aim to provide a catalogue of reactions that have been observed in biological membranes. As such we give no detailed description of the kinetics of transport of material across membranes as details of these, like that of enzymes, are inappropriate to the aims of this book and can be found in other standard biochemical texts. Hopefully, this approach will give the reader an insight into our current understanding of the many aspects of biological membranes and the direction future research is likely to take.

Much of what we describe in this book is the fruit of biochemical and biophysical techniques that have been developed over the last decade or two. We are conscious, however, that the information regarding the structure and function of biological membranes is going to increase at an enormous rate with the application of genetic engineering techniques. Rarely does a month pass without the sequence of another membrane protein, deduced using these techniques, appearing in the literature. Not only do these techniques provide a means of deducing a protein sequence without having to purify the protein and analyse the amino-acid sequence directly (both difficult and tedious processes), but there is also scope for the investigation of structure and function of membrane protein using systems in which genetically engineered DNA molecules are expressed *in vivo*.

We have compiled this book with the needs of scientific and medical students in mind, as well as those of research workers wishing to be acquainted with current developments in this fast-moving field. A rapid appreciation of the contents of each chapter may be made by reading only the summary and figure legends of each section. For those wishing more details and explanation, the full text may be read. Finally the references at the end of each section and those cited in figure legends will lead the interested reader to some of the key reviews and published papers. We hope that the concepts which we have tried to lay down will be clear and might even stimulate some to further investigations into the fascinating properties of biological membranes.

We wish to thank all those who gave helpful comments and criticism on the manuscript and Rhian Houslay and Elizabeth Wright who helped in the production of the manuscript.

Chapter 1

Membrane components and their organization

1.1 THE IMPORTANCE OF MEMBRANES IN CELL STRUCTURE AND FUNCTION

1.1.1 Definition of the cell surface

A living cell, by definition, consists of a discrete mass of cytoplasm separated from its environment by a membrane. The most obvious function of this membrane, called the plasma membrane, is to define the boundaries of the cell. No single structure for this membrane can be described, however, since it

1

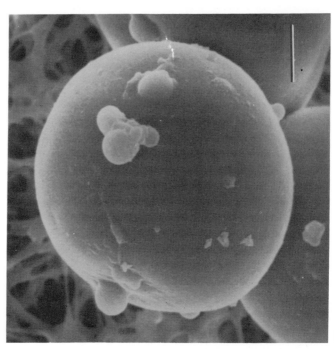

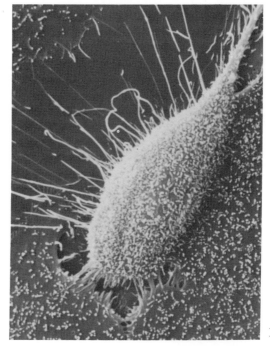

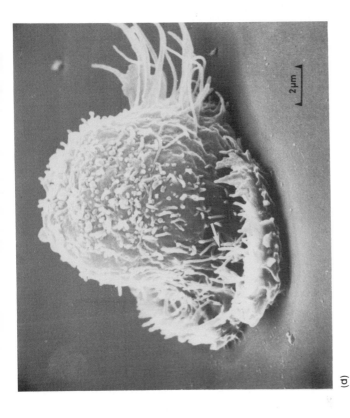

(d)

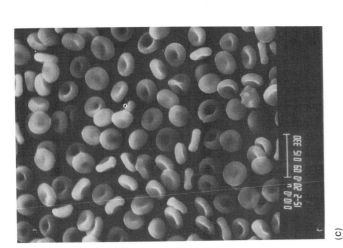

(c)

Figure 1.1 The morphology of some mammalian cells. Scanning electron micrographs reveal the different surface structures of four mammalian cell types: (a) a MDCK cell rounded up during division showing large number of microvilli on the cell surface, scale bar = 10 μm. (Photograph kindly given by Hubert Reggio.) (b) Epinephrine-stimulated adipocyte showing smooth surface with occasional blebs, scale bar = 10 μm. (Photograph kindly given by Robert Smith and Leonard Jarett.) (c) Human erythrocytes showing biconcave shape. Scale bar = 10 μm. (Photograph kindly given by Dr. S. W. Hui.) (d) Primary rabbit carcinoma cell with extending lamellipodia (arrow), bar = 2 μm. (Photograph kindly given by the division of Cancer Research, Institute of Pathology, University of Zurich, Switzerland)

can take on a number of forms according to the physiological role of the cell. A single cell can also have several different areas of plasma membrane with different morphology and function. For instance it might have domains specialized for interaction with a basement membrane, interaction with adjacent cells and for absorption from body fluids. A further complication is that some cells are able continually to change their shape. Some examples of the diversity of cell morphology are shown in Figure 1.1. The apical surface of the MDCK cell line, derived from dog kidney, is convoluted into narrow cylindrical structures called microvilli (Figure 1a), which greatly enhance the surface area of the cell and aid absorption. Shown here is a cell that has rounded up during cell division. In contrast the plasma membrane of the adipocyte is smooth and featureless (Figure 1b). The function of this cell is principally to store fat, so no specialized structures in the plasma membrane are required. The cell is very large (about 50 μm diameter) and contains a central droplet of triglyceride over which a thin layer of cytoplasm and the plasma membrane is stretched. On the other extreme of size is the erythrocyte (7.7 μm diameter $\times$ 1.9 μm in depth) which is the principal oxygen carrier in the blood. Erythrocytes are all very uniform in shape, having a characteristic biconcave appearance (Figure 1c). Even when the plasma membrane is isolated from the cell contents this shape is retained, showing that in this case the membrane structure itself is responsible for determining the cell shape. Despite its rigid appearance it is flexible enough to be squeezed through the narrow blood capillaries of the peripheral circulation. This may be demonstrated *in vitro* by the ease with which they may be drawn up into narrow capillaries several times smaller than the diameter of the cell. Flexibility of a different sort is seen in fibroblasts and some tumour cells (Figure 1.1d) which can flatten on to a solid surface, extend lamellipodia and crawl along at several μm per minute (see section 6.7.3). Here the plasma mambrane is distensible and irregular in shape, quite unlike the erythrocyte.

Such a diverse morphology in mammalian cells is a reflection of the complexity with which the membrane is organized. Close beneath the membrane and in the cytoplasm is a network of polymeric protein structures called the cytoskeleton (see section 6.7.3). Together with the membrane structure itself these determine the shape of the cell. Thus the shape of microvilli is largely governed by the actin filaments which line up along the axis of the structure. The regular shape of the erythrocyte membrane on the other hand is largely governed by cross-linking between membrane proteins. In the fibroblast, the ability to change shape and move is a combination of the fluid and flexible nature of the plasma membrane and the rapid polymerization and depolymerization of cytoskeletal proteins. This differentiation of cell shape to suit its function is unique to animal cells. Bacteria and higher plants also have a plasma membrane analagous to animals, but it is protected by an outer cell wall which is rigid and of uniform shape.

1.1.2 Control of the intracellular environment

Perhaps the most important property of the cell surface or plasma membrane is the formation of a continuous barrier around the cell which limits the diffusion of substances into and out of the cytoplasm, thus protecting the cell from fluctuating or adverse conditions in the environment. The origin of its impermeability lies in the lipid core of the membrane which, being hydrophobic, does not allow the passage of water or charged molecules. It is, however, punctuated with a wide variety of protein pores and transport systems whose function is to regulate the permeability of the plasma membrane according to the particular needs of the cell (see Chapter 7). Thus the lipids and proteins play complementary roles in the control of membrane permeability. Whilst some small molecules are able to gain access to the cell via these transport systems in the plasma membrane, no large particles, proteins or oligopeptides can enter directly in this way. For these a special mechanism exists. First the particle or macromolecule must bind to the cell, usually to specific receptor proteins embedded in the plasma membrane. The ligand–receptor complexes then aggregate into small areas of the plasma membrane which pinch off to form a vesicle in the cytoplasm. This process is called receptor-mediated or absorptive endocytosis. The endocytotic vesicles then usually fuse with the lysosomes where their contents are degraded, although it is also possible for them to be transported intact to other parts of the cell (see section 6.7). Defects in these transport systems severely affect the metabolism of the cell, usually with pathological consequences.

1.1.3 Communication between cells

While an isolated cell is capable of individual existence it does not normally function in this manner in a tissue. The activity of a cell is regulated by those around it, and also by cells in remote locations. For instance cells growing in a monolayer normally stop growing when they come into contact, a phenomenon known as 'contact inhibition'. Presumably this process is involved in the normal process of limiting cell growth in an organ, since cancer cells, which do not exhibit contact inhibition, grow over each other to produce a tumour. The nature of the intercellular communication involved in contact inhibition is not known, but presumably it takes place at the level of the plasma membrane.

Cells at remote locations may also be linked together by the action of hormones. These are trace substances produced in the endocrine glands in vertebrates which act as chemical messages between distant organs. In mammals a large number of different hormones exist which can evoke a wide variety of responses, ranging from long-term growth and maturation, to short-term regulation of metabolic activity. Hormonal control is organized in a hierarchical manner, secretion of hormones by the adrenal cortex, thyroid, gonads and pancreas being itself under the control of hormones released by the pituitary gland. This in turn is controlled by factors released from the hypothalamus. In each case specific targets are selected by the presence of an appropriate receptor protein on the cell surface which allows the hormone to bind. The hormone may then have a direct effect on the metabolism of the cell by activating a membrane-bound enzyme facing the cytoplasm or facilitating a movement of ions across the membrane. Alternatively, it may have secondary effects which follow absorptive endocytosis of the hormone–receptor complex (see section 7.7).

Another circulating molecule which has recently provoked intense interest is the small polypeptide called interferon, so named after its ability to protect cells from viral infection. Interferon induces two enzymatic activities in the cytoplasm. The first synthesizes unusual oligonucleotides from ATP containing $2'-5'$ phosphodiester bonds. These oligonucleotides activate an endonuclease that cleaves mRNA. The second enzyme activity is a protein kinase which phosphorylates the initiation factor, eIF-2, thus inhibiting protein synthesis. Although most cells are capable of making their own interferon, it appears that it can also be used to co-ordinate the defence of distant cells against viral infection, rather like a hormone. Similarly when interferon is added to some tumour cells in culture it inhibits their proliferation, a result which has led to widespread claims for its therapeutic use. It is not known how interferon acts on the cell, but presumably it binds to a membrane receptor.

Communication can also be achieved directly through small molecules. For instance through ion movements across membranes which control the excited state of nerve or muscle cells, or the nerve transmitters which couple adjacent synapses.

In each of these examples the communication between cells relies on structures contained in their plasma membranes. In the absence of these structures a cell would be isolated from its neighbours.

1.1.4 Cell adhesion and immunogenicity

Many of the proteins embedded in the plasma membrane have large carbohydrate structures covalently linked on the external face of the cell. These glycoproteins frequently terminate in sialic acid residues giving the cell a negative charge. The carbohydrate side chains on these glycoproteins are also important in the adhesion of the cell to its neighbours or to its substrate. Another important property of the cell surface is that it is very immunogenic, mainly as a result of the exposed glycoproteins and glycolipids. The structure of these cell surface antigens is under genetic control, and is used by the immune system to distinguish between 'self' and 'non-self'. Thus, all cells in one individual bear similar cell surface antigens, but differ from those in other individuals. It is the compatibility of these antigens which determines the success of a blood transfusion or graft transplantation.

The most important surface antigens determining the success of a blood transfusion are the blood group antigens. The origin of this antigenic behaviour is the precise structure of the carbohydrate moiety of glycoproteins and glycolipids on the erythrocyte cell surface. Many systems of blood-group specificity exist in man, but the one which is best studied and understood is that of the so-called ABH and Lewis (Le^a and Le^b) antigens. Genetic control of these five antigens is made by four separate groups of genes which control the synthesis of specific glycosyltransferases (see Table 1.1). These act sequentially to produce the characteristic A, B, O, Le^a and Le^b blood groups. Both glycoproteins and glycolipids in erythrocyte membranes can carry these blood-group antigens. Thus the major glycoproteins of the erythrocyte membrane (see section 4.4) can carry blood-group specific antigens. For example band III is amongst those proteins which carry blood-group A activity, and Glycophoryn may carry antigens of the MN blood group.

A second set of antigens determining the individuality of cell surfaces is the transplantation or histocompatibility antigens. Unlike blood-group antigens, which define (in most individuals) the antigenicity of body fluids as well as cell surfaces, histocompatibility antigens are restricted to the cell surface. The antigenic determinant in this case is the polypeptide chain of a group of transmembrane proteins coded for directly by the major histocompatibility complex (MHC) on the mammalian genome. In mice the MHC is known as the H-2 complex and is present on chromosome 17; in man it is called the

Table 1.1 Genetic control of ABH and Lewis blood group antigens. The four genes Le, H, A and B each code for a specific glycosyltransferase. Expression of different combinations of these genes in different individuals gives rise to five phenotypes with the blood groups A, B, O, Le^a, and Le^b. Sugar residues are as following: Gal = D-galactose, GlcNAc = N-acetyl-D-glucosamine, GalNAc = N-acetyl-D-galactosamine, Fuc = L-fucose. Sugar linkages are omitted for clarity

Gene	Gene enzyme	Product	Blood group
Le	α-4-fucosyltransferase	Gal—GlcNAc— | Fuc	Le^a
H	α-2-fucosyltransferase	Gal—GlcNAc— | Fuc	O
Le + H +	α-4-fucosyltransferase α-2-fucosyltransferase	Gal—GlcNAc— | | Fuc Fuc	Le^b
A + H +	α-GalNAc transferase α-2-fucosyltransferase	GalNAc—Gal—GlcNAc— | Fuc	A
B + H +	α-Gal transferase α-2-fucosyltransferase	Gal—Gal—GlcNAc— | Fuc	B

HLA complex which is found on chromosome 6. The MHC is made up of a number of loci for each of which up to 100 alleles exist in the population. This high degree of polymorphism explains the ability of the antigens to function as markers of the cell surface 'individuality'. The importance of these antigens lies not only in graft rejection but also in their role in the immune recognition of viral antigens, and the linkage of particular alleles with susceptibility to certain diseases.

The plasma membrane of mammalian cells thus expresses, through histocompatibility and blood-group antigens, the genetic individuality of the cell. This complex system, while not essential for the continued existence of a cell (or individual), is valuable in the defence of an organism against invasion by virus containing antigenic determinents picked up from their previous host. The modification or loss of these antigenic markers which occurs in transformed cells might also prove to be a useful way of screening for tumours.

1.1.5 Functions of intracellular membranes

In the electron microscope the most prominent feature of a mammalian cell is the abundance of subcellular organelles partitioned off by intracellular membranes (see section 1.2). All of these membranes conform to the same general pattern as the plasma membrane, having two layers of lipid molecules with a hydrophobic centre, but the metabolic functions differ quite radically according to the proteins which are embedded in the membrane.

One of the simplest functions of these intracellular membranes is to provide a support to which enzymes may bind. Thus, a number of glycolytic enzymes, which one might have assumed would be soluble enzymes, have been found in association with the plasma membrane in erythrocytes (see section 3.2). It is not entirely clear why this should occur. In other cases, where the substrates are lipid soluble a more efficient metabolic pathway is produced by restricting the reactions to the membrane surface.

A second, and very important, function of intracellular membranes is to define compartments within the cell which can maintain a different environment from the milieu so as to support their own specialized function. Thus the lysosomes maintain an acid pH for the function of acid hydrolase enzymes, the mitochondrial inner membrane maintains an ion and proton gradient which is used as an energy store in the synthesis of ATP, and the ratio of ATP to AMP and NADH to NAD can be independently regulated to suit the metabolic activities of the cytoplasm. Intracellular membranes, however, are not just permeability barriers defining intracellular compartments, they also contain many integral and spanning membrane proteins which perform many of the reactions characteristic of each organelle. Thus the mitochondrial inner membrane is composed of protein complexes which themselves act as the proton and electron carriers between the cytoplasmic and mitochondrial compartments. Other proteins exchange ATP or carboxylic acid ions which directly feed the catabolic pathways inside the mitochondria and generate most of the energy supply for an aerobic cell. Similarly, integral membrane proteins are responsible for recognizing extracellular molecules during hormone action or absorptive endocytosis at the plasma membrane and probably are also responsible for recognizing the leader sequences of secretory and membrane proteins, being synthesized in the endoplasmic reticulum. Thus intracellular membranes play a key role in the function of cell metabolism.

Summary
  Biological membranes are essential components of living cells. Few cellular activities could continue in their absence.

1.2 THE MORPHOLOGY OF MEMBRANES

In the electron microscope the most prominent feature of a mammalian cell is the abundance of subcellular organelles partitioned off by intracellular membranes. These organelles may be classified into a number of types (e.g. Golgi apparatus, endoplasmic reticulum, lysosomes) which are present in almost all eukaryotic cells. The appearance of any one type of organelle, however, is very variable depending on the cell type. Thus the endoplasmic reticulum is very prominent in secretory cells, and the Golgi apparatus which has only 1–3 cisternae in rat liver may have up to 30 cisternae in some plant cells. For simplicity we have chosen one cell type, the rat hepatocyte, in order to illustrate what the intracellular organelles and their membranes look like. It must be remembered, therefore, that other cell types may have similarly named structures with a rather different appearance.

Figure 1.2 shows a thin section through a rat hepatocyte viewed in the transmission electron microscope. In cross-section the hepatocyte has a polyhedral shape and is about 20–25 μm in diameter. Alternate faces of the cell interact with adjacent hepatocytes and contain the bile capillaries. In between these are the blood sinusoidal faces which are separated from the circulating blood by flattened endothelial cells. The adult mammalian liver is thus a sponge-like cellular mass perforated by two systems of communicating cavities.

The staining procedure used to prepare specimens for the electron microscope results in electron-dense osmium and lead compounds being deposited in the membranes (as well as some other structures) of the cell. The complexity of the intracellular structure of a hepatocyte, seen in Figure 1.2 is therefore largely a reflection of the large number of membranous organelles that fill the cytoplasm of mammalian cells. These organelles, which may be identified as morphologically distinct structures, are shown in Figures 1.3 and 1.4.

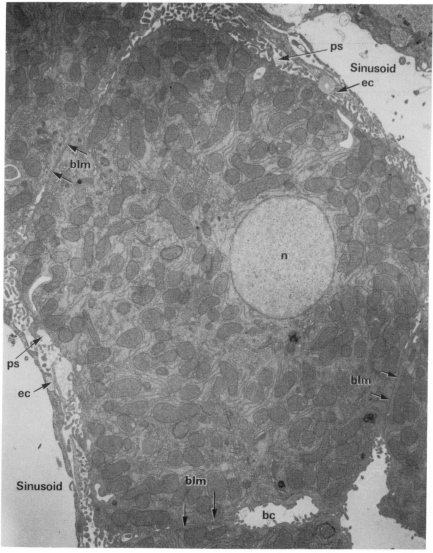

Figure 1.2 **Thin section of a whole rat hepatocyte.** In the electron microscope the most prominent feature of mammalian cells is the abundance of membranous organelles; ec = reticulo-endothelial cell, ps = pericapillary space (or space of Disse), bc = bile capillary, blm = basolateral membrane, n = nucleus. Magnification × 5780. (Micrograph kindly given by Stephen Massardo and Kathryn Howell)

1.2.1 The plasma membrane

Surrounding the cell is the plasma membrane which in hepatocytes is differentiated into three domains that may be distinguished on the basis of their morphology and composition. Where adjacent cells are apposed, the

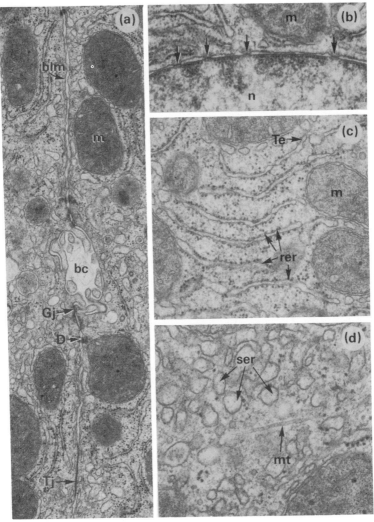

Figure 1.3 Structure of the plasma membrane, nuclear envelope, rough and smooth endoplasmic reticulum of the rat hepatocyte. (a) Adjacent hepatocytes come into contact at the basolateral membrane, which is characterized by junctional complexes; Tj = tight junction, D = desmosome, Gj Gap junction, bc = bile capillary, m = mitochondrion, blm = basolateral membrane. Magnification × 22,500. (b) The nuclear envelope contains two membranes which fuse at points to make nuclear pores (shown by arrows); n = nucleus, m = mitochondrion. Magnification × 22,000. (c) Secretory and membrane proteins are synthesized on ribosomes attached to the rough endoplasmic reticulum (rer). Smooth transitional elements (te) are often seen on the tips of each cisterna; m = mitochondrion. Magnification × 37,500. (d) Smooth endoplasmic reticulum (ser) contains no attached ribosomes; mt = microtubule. Magnification × 50,000. (Micrographs kindly given by Stephen Massardo and Kathryn Howell)

12

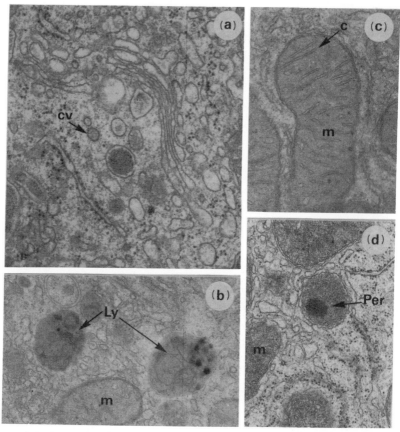

Figure 1.4 Structure of the Golgi apparatus, lysosomes, mitochondria and peroxisomes of the rat hepatocyte. (a) The Golgi apparatus consists of a small stack of dish-shaped cisternae. Coated vesicles (cv) are sometimes seen in its vicinity. Magnification × 37,500. (b) Lysosomes (Ly) contain degradative enzymes used to digest phagocytosed material; m = mitochondria. Magnification × 48,000. (c) Mitochondria (m) are recognizable by their double membrane and internal cristae. Magnification × 36,000 (d) Peroxisomes (Per) contain darkly staining crystals of urate oxidase; m = mitochondrion. Magnification × 26,400. (Micrographs kindly given by Stephen Massardo and Kathryn Howell)

plasma membrane is relatively smooth in outline. This area, called the baso-lateral membrane, is also rather sparse in marker enzymes (see section 1.3). It contains, however, a number of junctional complexes which allow the membrane to be identified in isolated preparations on morphological grounds. Three types of junctional complex have been described (Figure 1.3a). Close to the blood sinusoids are the tight junctions which seal the membrane of adjacent hepatocytes together and prevent the direct trans-fer of material between the bile and blood space. Using freeze–fracture microscopy it has been shown that intramembranous particles are aligned in his region to form an interdigitating meshwork of ridges. This probably controls the lateral diffusion of proteins in the bilayer as well as the trans-epithelial flux between cells. The second type of junctional complex has been called a Gap junction on account of the ability of some electron-dense small molecules to penetrate between the apposed membranes. These junctions are relatively stable in non-ionic detergents which has enabled them to be isolated. Structural studies on these isolated Gap junctions have shown that they contain a regular array of proteinaceous membrane pores aligned in the membranes of the two cells (see section 7.4). These allow a flux of ions and small molecules between the cytoplasm of each cell. The third type of junction is the desmosome, which is characterized by a local contact of adjacent cell membranes and a corresponding thickening in the surrounding cytoplasm. Desmosomes function as sites of adhesion between adjacent hepatocytes, and the origin of cytoskeletal attachment in the cytoplasm.

In contrast to the basolateral domain of the plasma membrane, the blood sinusoidal and bile cannalicular domains are highly convoluted with microvilli and are rich in marker enzymes. Concentrated at the blood sinusoidal face are the receptors for hormones, IgA and asialoglycoproteins present in the circulating blood. These materials are taken up into the cell by absorptive endocytosis. The blood sinusoidal membrane is also responsible for the uptake of various metabolites and the export of lipoprotein particles. At the bile cannalicular membrane bile salts synthesized by the hepatocyte and IgA absorbed from the blood are secreted. These transport processes are much facilitated by the large surface area of membrane presented by the microvilli.

1.2.2   The nucleus

In the centre of the hepatocyte is the nucleus (Figure 1.3b), which contains the DNA of the cell. Two forms may be distinguished morphologically. At the perimeter of the nucleus, in close proximity to the nuclear envelope, is the granular, condensed form of chromatin that is inactive in RNA transcription. In the centre of the nucleus the dispersed and transcriptionally active DNA is visible as a fine textured filamentous network.

Surrounding the nucleus, and partitioning the nucleoplasm and the cytoplasm, is the nuclear envelope. This contains an inner and an outer membrane separated by a cisterna of variable size. After cell division the nuclear envelope is formed around the dispersed chromosomes by elements of the endoplasmic reticulum, and certain features of this membrane are retained in the nuclear envelope. Most prominent are the polysomes attached to the outer nuclear membrane and the presence of physical continuities between the outer membrane and the endoplasmic reticulum. It has also been found that both membranes contain certain enzyme activities and antigenic markers in common. A distinctive feature of the nuclear envelope is the presence of pore structures with a characteristic morphology (Figure 1.3b). These consist of a circular orifice about 65–75 nm in diameter formed by the fusion of inner and outer nuclear membranes. Within the pore at high magnification an inner and outer rim of eight subunits is visible. Material is also frequently seen at the centre of the pore, although part of this may be due to ribonuclear protein fixed in the process of passing through the pore. The nuclear pore complexes allow the passage of mRNA molecules and regulator proteins between nucleus and cytoplasm as well as ions and small molecules of $M_r < 1000$.

1.2.3 The endoplasmic reticulum

Filling much of the cytoplasm and connected to the nuclear envelope is the endoplasmic reticulum (Figure 1.3c). This system of interconnected flattened cisternae is frequently covered by ribosomes engaged in the synthesis of membrane and secretory proteins. In the hepatocyte the tips of each cisternae lack attached ribosomes and are referred to as transitional elements of the endoplasmic reticulum. It is here that secretory and membrane proteins are packaged prior to transport to other parts of the cell.

Other areas of the endoplasmic reticulum have smooth membranes (Figure 1.3d). These domains of the endoplasmic reticulum are concerned with the fatty acid desaturation, detoxification reactions and microsomal electron transport functions of the hepatocyte. Close to the Golgi apparatus some smooth endoplasmic reticulum vesicles are found to stain with cytochemical stains for acid phosphatase. These vesicles, which are therefore believed to be on the pathway of lysosome biogenesis, have been called the GERL (Golgi-associated endoplasmic reticulum from which lysosomes form). It is possible that the GERL acts as an alternative route to the Golgi apparatus in the transport of lysosomal enzymes from the endoplasmic reticulum.

1.2.4 The Golgi apparatus

The central feature of the Golgi apparatus is a stack of disc-shaped flattened cisternae (Figure 1.4a). Each cisterna is usually dilated at the ends and may be fenestrated, giving rise to the vesicles seen at the rim of the cisternae in cross-section. A number of vesicles may also be seen on both sides of the Golgi apparatus which can occasionally be identified as coated vesicles or secretory vacuoles. In many cells the Golgi apparatus is situated near the nucleus, allowing the two faces to be defined as proximal and distal. This is difficult in the hepatocyte as they are usually situated further away from the nucleus, close to the bile cannaliculae. Examination of the function of the Golgi apparatus, however, suggests that it is still polarized with *cis*- and *trans*-faces. Whilst being involved in a variety of biochemical processes in different cells, the principal function of the Golgi apparatus is in the post-translational modification of the carbohydrate structure of glycoproteins. Secretory proteins packaged in the transitional elements of the endoplasmic reticulum are most likely delivered to the *cis*-face of the Golgi apparatus by coated vesicles. Various glycosidases and glycosyltransferases in the Golgi apparatus then trim and modify the core oligosaccharides previously synthesized in the endoplasmic reticulum (see section 6.3). Finally, the mature proteins are transported in vesicles to the plasma membrane. The details of this pathway of secretion are still under investigation.

1.2.5 The lysosome

Two lysosomes may be seen in Figure 1.4b. These membrane vesicles contain a wide variety of degradative enzymes which are used in the breakdown of material engulfed during phagocytosis and absorptive endocytosis. Most cellular components are also continually turned over by breakdown in lysosomes and then resynthesis. Breakdown of the cells' own constituents in lysosomes is referred to as autophagocytosis. Some lipoprotein particles and membranous material is visible in the lumen of the lysosomes in Figure 1.4b.

1.2.6 The mitochondrion

In Figure 1.2 it may be seen that the hepatocyte cytoplasm is largely occupied by the mitochondria. These are ovoid structures some 3 μm in length which may be recognized by the presence of internal membranes called cristae. Under higher power (Figure 1.4c) this organelle can be seen to contain two membranes, the inner membrane being invaginated to form incomplete transverse septa through the matrix known as the cristae. It is in the membranes of these cristae that mitochondrial electron transport occurs (see section 7.6.5).

1.2.7 The peroxisome

Another type of specialized oxidative organelle is the peroxisome (Figure 1.4d). This organelle may be recognized by its single surrounding membrane and the presence of electron-dense crystals of urate oxidase which diffract the electron beam in a regular pattern. Peroxisomes contain enzymes which can oxidize amino acids, xanthine and possibly fatty acids. They earn their name from the catalase that they contain which breaks down the hydrogen peroxide generated by these oxidative reactions.

Clearly, this brief account of the membranous organelles of the hepatocyte does not include all the specialized structures seen in other cells. Enough has been said, however, to convince the reader of the immense diversity of membrane structures in the eukaryotic cell. Strangely, when examined under very high power, the membranes of each organelle appear remarkably similar. After osmium fixation the usual appearance is of a trilamellar sandwich (visible in Figure 1.3d). It is difficult, however, to correlate this appearance directly with the arrangement of lipids and proteins in the membrane, both because little is known of the chemistry of staining, and also because similar profiles may be obtained after serious perturbation of the original structure.

Summary

Eukaryotic cells contain a large number of membranous organelles with specialized functions.

Each membrane appears similar in the electron microscope.

1.3 PREPARATION OF MEMBRANE FRACTIONS

At some point in most biological investigations a purified preparation is required to carry out structural studies or *in vitro* reconstitution of function. In order to make reliable interpretations of these experiments the membranes (or other structure) must be in a high state of purity so that it is clear that the results are attributed to the correct (rather than some contaminating) membrane fraction. They should also be obtained in high yield so that a preparation of a minor part of a membrane will not be taken as representing the whole structure. Thus the plasma membrane of rat liver for example contains three major domains, the blood sinusoidal, basolateral and bile cannicular faces, as well as specialized structures like the gap junctions and desmosomes. These structures differ morphologically and biochemically, although they are all genuine subfractions of the same membrane (see section 1.2). The endoplasmic reticulum also has two major domains, the rough and the smooth endoplasmic reticulum, and it is probable that there are more subtle differences within these major groupings. In general, any membrane with multiple biological activities is likely to have a mosaic of different functional domains. Most fractionation procedures enrich the membranes in one of these domains at the expense of the others and so do not represent the overall structure. It is recommended, therefore, that more than one 'marker enzyme' is used for measuring purity, and that both purity and yield are recorded during any preparation. Unfortunately, the dual requirements of high purity and high yield are not simultaneously possible as one is usually achieved at the expense of the other. Consequently, some caution must be exercised when extrapolating biochemical data on purified fractions back to the original cells.

The various organelles were originally named and defined on the basis of their morphology in thin sections of whole cells (see section 1.2). During the purification of these organelles, however, much of the characteristic morphology of the organelles is lost as a consequence of the homogenization procedure. Thus, smooth vesicles derived from the plasma membrane look similar to smooth vesicles derived from the endoplasmic reticulum or Golgi apparatus. It is therefore necessary to be able to define the various intracellular membranes on the basis of enzymes that they contain. For this purpose integral membrane proteins (see section 1.4) are chosen since these are not washed away in the media used for isolation. Underlying this biochemical definition of intracellular organelles is the assumption that each membrane possesses a unique and different set of enzyme markers. Some caution must, however, be exercised in the use of marker enzymes. Firstly, it is clear that not all the activity of plasma membrane enzymes can reside at the cell surface since a certain percentage must be present inside the cell where the enzymes are synthesized (see section 6.2). Secondly, it has been shown that some enzymes are able to move into intracellular vesicles during endocytosis of the surface membrane and during membrane circulation (see section 6.7). Structural proteins may make better marker enzymes in this respect since they are usually locked into a specific membrane by interaction with peripheral proteins and sometimes cytoskeletal elements (see section 2.14). The isolation of monoclonal antibodies to these structural proteins has shown that they can have unique locations in the cell as determined by immune fluorescence localization, and can even discriminate between different domains of individual organelles, e.g. the individual stacks of the Golgi apparatus. Evidently these immunological markers will become increasingly important in the biochemical characterization of membrane fractions.

Antibodies can also be used to isolate membranes, either by perturbing their normal density, or by direct absorption onto immunoadsorbants. These techniques offer a means of purification quite distinct from the traditional rate and equilibrium density centrifugation techniques, and promise to be of great use in the future characterisation of biological membranes.

Summary

Biological membranes from different organelles may be isolated by making use of their different biochemical, physical and immunological properties.

1.4   THE STRUCTURE OF MEMBRANE COMPONENTS

Biological membranes contain an immense diversity of protein and lipid molecules and to describe them all would take a whole book. Furthermore, such a laborious description would give the impression that membrane structure itself is concerned with slight variations in the structure of the components. Quite on the contrary, it is the common features upon which the structure of membranes rests, and it is therefore these features on which we shall place most emphasis. Those interested in the diversity of possible structures, especially those of lipid molecules, may consult other texts.

Most of the mass of a biological membrane is composed of lipid and protein, although the exact proportion varies according to the source. As a general rule the more metabolically active membranes contain a higher proportion of protein. Thus the myelin sheath contains about 75 per cent of the dry weight as lipid, while the mitochondrial inner membrane contains about 25 per cent lipid, the remainder in each case being chiefly the protein. This agrees with the contention, mentioned above, that the lipid provides the structural backbone punctuated by the proteins which have a functional role. In addition to the lipid and protein, biological membranes can contain up to 10 per cent of their dry weight as carbohydrate. This is found principally on the external surface of the plasma membrane where it is covalently attached to both protein and lipid molecules. Owing to the abundance of negatively charged terminal sialic acid residues this 'glycocalyx' confers on the cell a net negative charge. Also associated with biological membranes is a considerable amount of water (about 20 per cent of the total mass) and a number of chelated calcium and magnesium ions.

1.4.1 Lipids

Almost all of the lipid molecules in a biological membrane share one important feature—they are amphipathic. This means that they have two different natures, one which is charged or polar and is stable in aqueous solution, the other is a non-polar aliphatic or aromatic hydrocarbon which is most stable in a non-aqueous environment. Figure 1.5 illustrates how representative lipids from all the major lipid classes fulfil this design.

The most abundant lipids are the phospholipids (Figure 1.5a). The hydrophobic moiety of these lipids is derived from long-chain fatty acid molecules which are esterified, via a glycerol molecule and phosphate ester, to a hydrophilic base. Most polar lipids in biological membranes have a similar overall structure, but with minor variations in hydrophobic, linker or hydrophilic regions. We have tried to illustrate the similarities in Figure 1.5 and noted some of the possible variations in Tables 1.2–1.4. The fatty acyl chains are subject to a similar variation in each class of lipid. Shown in Figure 1.5a are two isomers of an 18-carbon atom fatty acid known as oleic and stearic acid. The former has a *cis*-double bond at the ninth carbon atom causing a bend of about 30° in the hydrocarbon chain. Evidently such a *cis*-unsaturated molecule cannot pack so tightly as a saturated hydrocarbon resulting in a more fluid membrane (see Section 2.6). This example serves to illustrate how minor modifications of membrane components can change the detailed properties of a membrane without changing the overall structure. In general the fatty acyl chains can be between 14 and 24 carbons long and may be unsaturated to varying degrees. The most common are shown in Table 1.3.

The two fatty acid residues of a phospholipid are esterified to a molecule of glycerophosphate to form phosphatidic acid, which in turn condenses with a base to form either phosphatidyl choline (PC), phosphatidyl serine (PS), phosphatidyl ethanolamine (PE), or phosphatidyl inositol (PI). The synthesis of these compounds is outlined in Figure 6.8. The headgroup of these phospholipids is very polar, containing the negative charge of the phosphate hydroxyl (pK 1–2) together with the charges and polar groups on the bases. Whilst PC and PE are both neutral at pH 7.0, PS and PI carry a net negative charge. The phospholipid headgroups are shown in Table 1.3.

In Figure 1.5 the changes in structure relative to a phospholipid molecule

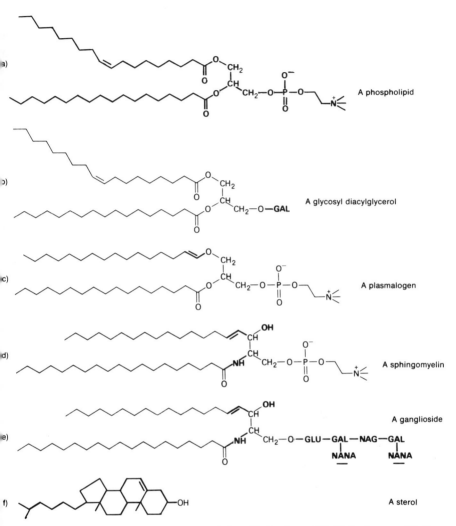

Figure 1.5 The structure of membrane lipids. Each class of lipid is shown with its hydrophobic area on the left-hand side. In order to emphasize the similarity between lipid structures we have shown the areas of each molecule which differ from a phospholipid structure in bold type. Symbols for sugar residues are explained in Table 1.5

are indicated in bold type. Thus, it may be seen that the mono-, and di-saccharide diacylglycerols (Figure 1.5b) differ only in the structure of the polar region. In these glycolipids the glycerol is esterified directly to the sugar units. Glucose, mannose and galactose containing diacylglycerols are widely distributed in bacterial and plant membranes with different permutations of sugar moieties and linkages giving an immense number of structures.

The plasmalogens (Figure 1.5c) are also closely related in structure to the

Table 1.2   Common fatty acid moieties occurring in phospholipids. Phospholipids c higher animals and plants contain hydrocarbon tails derived from fatty acids wit 14–24 carbon atoms. Most common are those with 16 and 18 carbons. Unsaturate hydrocarbons have *cis*-double bonds occurring between the ninth carbon atom and th end of the chain. Each double bond is separated by a methylene group so that con jugation does not occur. Bacterial phospholipids contain shorter hydrocarbons chain with less unsaturations, but with occasional branched structures

Structure	Notation	Trivial name
$CH_3(CH_2)_{14}CO_2H$	16 : 0	Palmitic acid
$CH_3(CH_2)_{16}CO_2H$	18 : 0	Stearic acid
$CH_3(CH_2)_5CH\!=\!CH(CH_2)_7CO_2H$	16 : 1	Palmitoleic acid
$CH_3(CH_2)_7CH\!=\!CH(CH_2)_7CO_2H$	18 : 1	Oleic acid
$CH_3(CH_2)_4(CH\!=\!CHCH_2)_2(CH_2)_6CO_2H$	18 : 2	Linoleic acid
$CH_3CH_2(CH\!=\!CHCH_2)_3(CH_2)_6CO_2H$	18 : 3	Linolenic acid
$CH_3(CH_2)_4(CH\!=\!CHCH_2)_4(CH_2)_2CO_2H$	20 : 4	Arachidonic acid

phospholipids. Here the difference lies in the linkage of one of the fatty acyl chains to the glycerol molecule. A similar variation of fatty acid chains and headgroups can occur as in the phospholipids.

Although synthesized by a rather different route, the sphingomyelins (Figure 1.5d) also have close structural resemblance to the phospholipids. These lipids are condensed from a sphingosine long-chain base (which comprises both the hydrocarbon tail and linker moiety) with a fatty acid derivative. The sphingomyelins are the most abundant of a family of lipids with a similar structure. They are found in plant and animal membranes but are most abundant in nervous tissue membranes.

Glycosphingolipids and gangliosides (Figure 1.5e) are related to sphingomyelin, having a short carbohydrate chain in place of the phosphate and base moieties. The gangliosides contain sialic acid and therefore have a net negative charge. They have an important function on the cell surface where they act as receptors for some hormones and toxins. Table 1.4 shows the structure of some ganglioside carbohydrate moieties.

Lastly, cholesterol in Figure 1.5f represents the large class of sterol lipids. Cholesterol is found almost entirely in the plasma membrane of mammalian cells where it constitutes about 45 per cent of the total lipid. This molecule is quite different from the other lipids, having a large planar steroid apolar region with one hydroxyl group on the A ring constituting the polar region. Slight structural variations of cholesterol are found in micro-organisms and higher plants.

Focusing on the differences we see an enormous complexity of structures with slightly different charge, packing and hydrophobicity. There are unfortunately few general rules why particular lipids are present in a particular membrane and it is not clear why such a diversity is required (see Chapters 2 and 3).

Table 1.3 Phospholipid headgroups. The polar region of phospholipids is composed of a base esterified via a phosphate ester. Phosphatidyl choline (PC) and phosphatidyl ethanolamine (PE) are zwitterionic, but phosphatidyl serine (PS) and phosphatidyl inositol (PI) carry a net negative charge at neutral pH

Structure of base	Phospholipid

Looking at the similarities we see a simple strategy with a number of solutions. The molecules of Figure 1.5 are all amphipathic and all are of a similar dimension. How this is used in building biological membranes is the subject of section 1.5.

1.4.2 Proteins

The proteins of membranes are classified by the ease with which they may be removed, although in fact this property varies continuously and the two categories of 'peripheral' and 'integral' are rather arbitrary. This classification

Table 1.4 Commonly occurring ganglioside headgroups. Gangliosides are
notated according to the number of sialic
acid (NANA) groups that they contain.
M = mono, D = di-, and T = trisialo
ganglioside. Sugar residues are abbre-
viated: Glc = glucose, Gal = galactose,
GlcNAc = N-acetyl-D-glucosamine

Sugar residues	Name
—Glc—Gal—NANA	G_{M3}

—Glc—Gal—GlcNAc
        |
      NANA
	G_{M2}

—Glc—Gal—GlcNAc—Gal
        |
      NANA
	G_{M1}

—Glc—Gal—GlcNAc—Glc
        | |
      NANA   NANA
	G_{D1}

—Glc—Gal—GlcNac—Gal
        | |
      NANA   NANA
      |
      NANA
	G_{T1}

revolves around an operative definition: those which may be washed off the membrane using low or high ionic strength buffers are called peripheral proteins while those requiring detergents or organic solvents for extraction are called integral proteins. Care should be taken not to be confused by the alternative terms 'extrinsic' and 'intrinsic' membrane proteins. These are synonymous with integral and peripheral, but are somewhat less descriptive. From the treatments required to dissociate a protein we may deduce something about its interaction with the bilayer. Since salt ions compete with the interactions of peripheral proteins with the membrane the linkage must be of an electrostatic nature, either salt bridges or hydrogen bonds. Cytochrome c, which is a peripheral protein of the mitochondrial inner membrane, has a structure similar to soluble proteins with the charged or polar side chains near the surface of the protein and the hydrophobic residues buried in the apolar core. Such a structure is described as micellar, since it resembles the structure of a detergent micelle (see Figure 1.9). The insertion of integral membrane proteins is clearly quite different. Whilst it is still non-covalent, the requirement of detergent for solubilization suggests that these proteins contain considerable areas of their surface which are apolar in character. If the detergent is dialysed away after solubilization, these proteins usually form a denatured precipitate showing that the protein is unstable without bound detergent or lipid molecules. The number of bound detergent molecules, and hence the area of hydrophobic surface on the protein, can be measured by using radioactive detergents to solubilize the protein. Unbound detergent is then dialysed away and the radioactivity associated with the protein is measured. The presence of large hydrophobic regions in integral membrane proteins may also be inferred in the overall amino acid composition which usually contains at least 50 per cent hydrophobic amino-acid residues.

The two categories of membrane protein are illustrated in Figure 1.6. The peripheral protein (a) is shown on the cytoplasmic face of the membrane interacting with two types of integral proteins. Altogether five different types of integral protein are shown, all of which have some counterpart in real membranes (these will be described in later chapters). Integral membrane proteins which span the membrane may be of the fibrous type (b) or globular as in (c). Fibrous proteins usually have only a single strand of the polypeptide chain spanning the membrane and are often heavily glycosylated on the extracytoplasmic surface. Globular proteins have many loops of the polypeptide chain spanning the membrane and a much higher percentage of hydrophobic surface. Other proteins are simply anchored to the bilayer with hydrophobic pedicles and may be situated with the major hydrophilic mass on either side of the bilayer. These have been called ectoproteins and

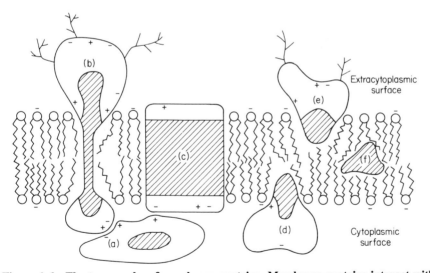

Figure 1.6 **The topography of membrane proteins. Membrane proteins interact with biological membranes in a variety of ways. Peripheral proteins (a) are hydrophilic and bind to the polar surfaces of other membrane proteins and lipids. All the integral membrane proteins (b–f) have apolar surfaces and intercalate in different ways with the tightly packed lipid bilayer. Proteins on the extracytoplasmic surface (b and e) may have carbohydrate attached. Hatched areas represent apolar regions**

24

endoproteins. It has been argued that ectoproteins (e) always span the membrane since it is difficult to see otherwise how they could be synthesized (see Chapter 6). Proteins completely within the apolar core of the bilayer (f) are extremely rare, the only possible example being the myelin proteolipid which is so hydrophobic that it partitions into organic solvents. The principal structural characteristics of integral membrane proteins to note at this stage are that they are amphipathic and globular. Much evidence suggests that integral membrane proteins retain a compact globular structure even when they interact with the apolar environment of the core of the membrane. Lipid molecules do not penetrate and interact with the proteins in an extended polypeptide chain form, rather, the hydrophobic environment and exclusion of water molecules favours the formation of intra- and interchain hydrogen bonds. As a result integral membrane proteins often display more secondary structure than their soluble counterparts. Even the fibrous integral membrane proteins usually have an α-helix conformation of the spanning hydrophobic segment.

Summary

Most membrane lipids are amphipathic molecules of similar overall dimensions, allowing them to pack closely in a lipid bilayer.

Integral membrane proteins are also amphipathic molecules with the apolar regions of their structure buried in the lipid bilayer.

Peripheral membrane proteins have a micellar structure similar to soluble proteins.

1.5 A MODEL FOR BIOLOGICAL MEMBRANE STRUCTURE

No single published model of membrane structure embodies all the features of a biological membrane. The very nature of a model is to simplify the design and emphasize the important features. The exercise of model-building is, however, very important since some physical relevance (or otherwise) must be assigned to experimental observations. Historically, three observations have been particularly influential in our present concept of biological membranes. The first was the demonstration that the plasma membrane was a discrete structure which could be separated by microdissection from the surface of amoeba. Previously the cell surface had been regarded simply as an interface between cytoplasm and medium. The observation that membranes could exist on their own paved the way towards their isolation and scientific investigation. A second leap was made when the lipid from a known number of erythrocytes was extracted and spread in a monolayer on a trough of water and shown to occupy about twice the surface area of the original cells. It was concluded that biological membranes consisted of a bilayer of lipid. Low-angle X-ray diffraction studies have since corroborated these observations.

The remaining problem was the location of the membrane proteins. Initially it was suggested that they lined the surface of the lipid bilayer or were combined in proteolipid complexes to form units for membrane construction.

The key to our present concept of the structure of membranes was the discovery that membrane proteins retain their globular structure and could actually span the membrane in an asymmetric fashion. This idea was taken up in the 'Fluid Mosaic' model which attempted to explain this structure in thermodynamic terms. Since this model still forms the basis of our present concept of biological membranes—and indeed is the basis for pictorial representations of membranes in this book—we shall describe it in a little detail. As its title suggests, two conclusions may be made from a thermodynamic consideration of membrane structure. Firstly the lipid and protein molecules must be arranged in a tightly packed, water-excluding mosaic in such a manner that apolar regions of proteins and lipids can interact in the core of the membrane whilst the hydrophilic regions of each amphipathic molecule are solvated in the aqueous environment. The second conclusion is that the individual components are free to move about in the plane of the membrane.

This model was proposed as a consequence of a clearer understanding of the forces operating between hydrophobic molecules. It had earlier been recognized that weak interactions could occur between apolar residues as a result of a synchronous oscillation of their electron clouds (van der Waals forces). Far more important than these direct interactions, however, is the indirect energetic advantage of excluding water from the membrane structure. For when an apolar group is solvated in an aqueous solvent a large energetic penalty has to be paid. This energy is referred to as the hydrophobic force.

In order to understand the nature of hydrophobic forces we must consider what would happen if a phospholipid molecule slipped out into the aqueous medium. At the interface of the hydrocarbon tail and the aqueous medium the water molecules cannot satisfy their full hydrogen-bonding potential since no polar groups are available. Little compensating attraction can occur so the water molecules in the first shell around the hydrocarbon are forced into a cage-like structure. Furthermore, the mobility of the hydrocarbon chains is reduced giving rise to a large overall decrease in entropy. The energies involved in this process may be measured for small hydrophobic molecules like the inert gases or methane. When these are taken from a lipophilic solvent into water there is an enthalpy increase (ΔH) of 2–6 kcal per mol and a decrease in entropy (ΔS) of 10–20 entropy units at 25 °C. Both terms make the free energy given by the equation:

$$\Delta G = \Delta H - T \cdot \Delta S$$

positive and the reaction is not spontaneous. Summed over the many hydrophobic residues of a biological membrane, hydrophobic forces provide a powerful deterrent to disaggregation. Conversely, when amphipathic lipids and proteins condense into a bilayer the structured water molecules are excluded and the process proceeds spontaneously.

The movement of charged residues through the apolar core of the membrane is as unlikely as the dissociation of phospholipids into the surrounding medium. The reason for this is the unfavourable energy term required to solvate charged groups in an apolar environment. For example the movement of 1 mol of glycine at 25 °C from water to acetone involves a free energy change of +6 kcal. Consequently, the exchange of phospholipid molecules between the two halves of the bilayer (called flip-flop) is an extremely rare event. For proteins with large hydrophilic regions often containing many carbohydrate moieties the process is virtually impossible. Biological membranes are therefore an interesting thermodynamic paradox. On the one hand, they are thermodynamically stabilized structures which form spontaneously from amphipathic molecules in aqueous solution. On the other hand, the same thermodynamic forces prevent asymmetric distributions of membrane components from coming to thermodynamic equilibrium. We shall discuss the implications of this paradox more fully in Chapters 4 and 6.

It is clear from this thermodynamic view of membrane structure that the protein must be contained in a mosaic with lipid molecules rather than be spread in a monolayer at the surface. For in the absence of water at the surface of the membrane the restrictions on the rearrangement of amphipathic molecules have vanished. This is the reason why organic solvents are effective at solubilizing and denaturing membrane components. They destabilize the bilayer structure by removing its energetic origin.

A further consequence of the entropic stabilization of membrane structure is that the $T \Delta S$ term decreases as the temperature is lowered. Thus at very low temperatures (-110 °C) when frozen cells are fractured with a microtome knife, the fracture plane is propagated along the hydrophobic core of the lipid bilayer. This technique of freeze–fracture gives a unique view of the inside surfaces of the plasma membrane, for after shadowing with carbon and platinum the replica surfaces may be examined in the electron microscope (see Figure 1.7). Integral membrane proteins with a large bulk in the centre of the membrane appear as particles or pits in these images.

Because of the asymmetric localization of these particles in the bilayer each half of the membrane has a different density of pits and may be distinguished as originating from either the cytoplasmic face (PF) or extracytoplasmic face (EF) of the bilayer. If the freeze–fracture specimen is warmed up very slightly (e.g. to -100 °C) before shadowing, some of the ice around the specimen sublimes, thus exposing the external surfaces (ES) of the membrane also. In the erythrocyte plasma membrane (Figure 1.7a) a random array of intramembranous particles may be seen which largely correspond to the integral membrane protein called band 3. The P face appears to contain more particles than the E face. After depleting the membrane of the peripheral protein, spectrin, and then causing the remaining spectrin to aggregate, the intramembranous particles are also seen to aggregate (Figure 1.7b). This demonstrates the association between band 3 and spectrin in the erythrocyte membrane (see section 4.4).

The second principle of the Fluid Mosaic model is that the individual molecules in the lipid–protein mosaic are able to diffuse with respect to each

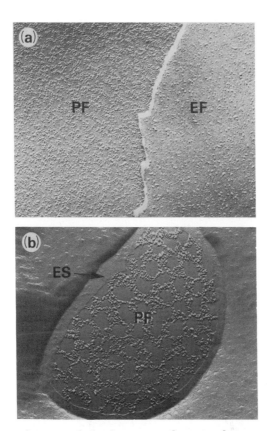

Figure 1.7 Freeze–fracture of the human erythrocyte plasma membrane. When frozen cells are fractured at low temperature the fracture plane progresses along the centre of the lipid bilayer, revealing details of the internal structure of biological membranes. (a) Erythrocytes fractured at $-110\,^{\circ}C$ and then shadowed with platinum and carbon ($\times\,53{,}600$). On the left is the cytoplasmic face (PF) observed from the exterior of the cell. On the right is the extracytoplasmic half of the bilayer (EF) of an overlying cell, as observed from the cytoplasm. (b) Erythrocyte ghosts fractured at $-110\,^{\circ}C$ and then etched for 60 s at $-100\,^{\circ}C$ before shadowing ($\times\,39{,}000$). The central region shows the P face of the erythrocyte membrane. Surrounding this is a narrow rim of the outer surface of the plasma membrane (ES) exposed by etching, and beyond this ice. The erythrocyte ghosts in (b) were pretreated for 30 min at $55\,^{\circ}C$ in 20 mM phosphate pH 7.6 to deplete the membrane of spectrin, and then for 30 min at $0\,^{\circ}C$ in 20 mM phosphate pH 5.5 to cause the spectrin to aggregate at its isoelectric point. (Photographs kindly given by David Shotton)

other. Above the melting temperature for the bilayer lipid the hydrocarbon chains in the core of the membrane are in a fluid-like state approaching that of a liquid hydrocarbon. In this respect the bilayer acts as a two-dimensional liquid through which the lipids and proteins may move laterally. We will discuss fluidity in more detail in Chapter 2. Suffice it to say here that the fluidity of a biological membrane is a variable property which depends on the packing of the hydrocarbon chains of the phospholipids. The more tightly packed they are the less fluid is the membrane.

Although the Fluid Mosaic model is frequently presented as an accurate description of membrane structure this is not really the case. The Fluid Mosaic model depicts biological membranes as a 'sea' of lipid in which globular proteins float. Whilst this might be true for myelin membranes which contain only 25 per cent protein by weight, the plasma membrane with 50 per cent protein and the mitochondrial inner membrane with 75 per cent protein have a quite different structure. In these membranes (which one is more entitled to describe as 'typical') protein molecules must come into close proximity. Thus, assuming diameters for an average protein and lipid to be approximately 4 and 1.7 nm respectively, one can calculate that 42 lipid molecules are required to form a single shell of bilayer lipid around the protein. The actual ratio of lipid to protein in the mitochondrial inner membrane is only 28, which means that for globular proteins (e.g. Figure 1.6c) there is insufficient lipid to form a complete lipid shell. For fibrous proteins (e.g. Figure 1.6b) with a single spanning portion of polypeptide chain about two shells of lipid bilayer are possible, but the hydrophilic domains on each side of the membrane still come into contact. In the plasma membrane about two shells are possible on average, giving a distance between proteins of 2.8 nm, while in the myelin membrane the distance is about 11 nm. These calculations do not take into account the mass of peripheral proteins which are included in the overall protein content of these membranes; in fact it turns out that only 30–40 per cent of the protein of the mitochondrial inner membrane is embedded in the bilayer. Thus, lateral diffusion of mitochondrial membrane proteins is probably important in the function of this membrane (see section 7.6), which actually has a rather low microviscosity coefficient of 0.9 P. Nevertheless, the idea of a 'sea' of lipid has turned out to be incorrect.

A second tenet of the original Fluid Mosaic model was the absence of long-range order in the protein molecules. This derives directly from the concept that the lipid is the major structural entity. Several exceptions to this are now known, however, including the quasi-crystalline arrays of bacteriorhodopsin (see section 7.6). More generally it is necessary to modify the concept of free mobile integral membrane proteins to include the possibility that they might be cross-linked to each other by peripheral membrane proteins or attached to cytoskeletal elements (see section 2.13). Both interactions give the protein networks in biological membranes an important structural role. Finally, the thermodynamic argument of the Fluid Mosaic model, whilst quite valid as an explanation of the stability of a mosaic model of membrane structure, does not in fact rule out a number of other structures which are equally thermodynamically likely.

Summary

   Biological membranes are held together by non-covalent bonds allowing a high degree of flexibility and fluidity.

The strength of these bonds relies on the exclusion of water from the core of the membrane.
Lipids and proteins are mobile in the plane of the membrane.

1.6 SOLUBILIZATION AND PURIFICATION OF MEMBRANE PROTEINS

A classical approach to the study of biological membranes is to extract and purify the individual components so that their structure and properties may be studied *in vitro*. The problems when trying to do this with membrane proteins, however, are much greater than those with soluble proteins. Integral membrane proteins are not soluble in aqueous solution and will denature and precipitate if solubilizing agents are removed. Some membrane-bound enzymes also require specific lipids as co-factors or require a shell or boundary layer of a particular lipid in order to express enzyme activity. Even if solubilization is achieved while retaining a native structure it is not always possible subsequently to separate the proteins using the techniques for soluble protein preparation. This is partly due to the presence of bound detergent around the protein which can change the size and charge of the molecule, partly because few solubilization procedures achieve a monodisperse solution of proteins and partly because many integral membrane proteins by virtue of their specialized structure tend to have similar molecular weights, hydrophobicities and isoelectric points.

The earliest methods used for solubilizing membrane proteins employed organic solvents. Acetone powders, made by homogenizing the membrane material in a large excess of acetone at $-20\,^{\circ}\text{C}$ and then filtering off the precipitate, have frequently been used as a starting material for membrane enzyme purification. In the hydrophobic solvent the thermodynamic stability of the lipid bilayer is destroyed and most lipids dissolve in the acetone and are removed. The remaining proteins usually retain some tightly bound lipids which allow them to be subsequently dissolved in aqueous solution. Unfortunately, many proteins are irreversibly denatured by this procedure. Less hydrophobic organic solvents like tertiary-butanol have also been used to effect in certain cases.

An alternative method of destabilizing the lipid bilayer is to use chaotropic ions. These are large ions with a low charge density like CCl_3OO^- and SCN^- which disrupt the hydrogen-bonded lattice of water molecules and so make the solvation of hydrophobic molecules less unfavourable. At concentrations of $1-2$ M a high proportion of the integral membrane proteins can be brought into solution. Furthermore, with a wide choice of chaotropic ions to choose from, some selectivity of solubilization may be achieved. The main disadvantage of chaotropic ions is that the conditions for destabilizing the lipid bilayer are also the conditions for denaturing proteins. In some cases it is possible to restore the native conformation by adding small highly charged ions like SO_4^{2-}. These have an anti-chaotropic effect. Care must also be taken

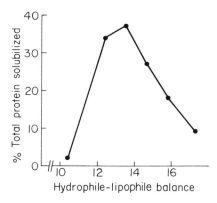

Figure 1.8 **Efficacy of detergent solubilization.** Detergents are amphipathic molecules like lipids, but are more hydrophilic, allowing them to exist in aqueous solution. By replacing the membrane lipids around an integral protein they can also effect the solubilization of membrane proteins. The ability to do this is related to the ratio of hydrophilic to lipophilic properties (called the hydrophile–lipophile balance). Of the Triton series of detergents with different length polyoxyethylene groups, Triton X-100 with an HLB of 13.5 is optimal at solubilizing membrane proteins. (Reprinted with permission from Collins, M. L. P., and Salton, M. R. J. (1979) *Biochim. Biophys. Acta,* 553, 40–53)

when using chaotropic ions to avoid auto-oxidation of membrane lipids. Reactive products of this process can modify both proteins and nucleic acids.

By far the most common method of solubilizing membrane protein is by the use of detergents. These may be considered as a class of lipid molecules having an amphipathic structure but with a higher ratio of hydrophilic to hydrophobic properties. This ratio is sometimes given an arbitrary theoretical value called the hydrophile–lipophile balance (HLB). Detergents with high HLB values are more hydrophilic and more powerful as solubilizing agents. The optimum value is about 13 (Figure 1.8).

The initial action of a detergent in solubilizing a biological membrane is to bind to the bilayer, intercalating between the lipid molecules. As the concentration of detergent increases, detergent-lipid micelles break off from the membrane into the surrounding medium. Finally, detergent molecules exchange with lipid molecules around the hydrophobic regions of the integral membrane protein. In this manner the proteins are able to retain their globular shape and normal distribution of hydrophobic and hydrophilic amino-acid residues. The detergent provides an apolar environment to stabilize the non-polar regions of the protein whilst being sufficiently hydrophilic to get them into solution (Figure 1.9). Clearly, it is necessary to add sufficient detergent to achieve this three-stage solubilization. Not only must the concentration (w/v) of the detergent be adjusted, but more importantly, the ratio of detergent to lipid in the membrane must be correct. On a weight-to-weight basis between one and ten times the amount of lipid is required.

(a)

(b)

(c)

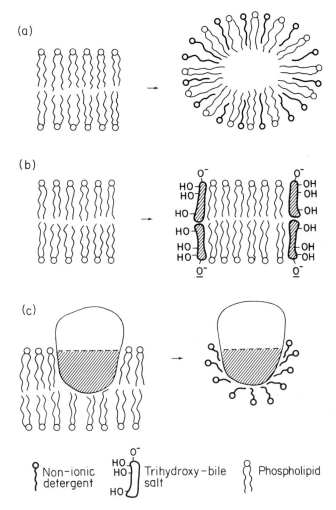

♀ Non-ionic
│ detergent

HO⌐ O⁻
HO┤│ Trihydroxy-bile
   │ salt
HO⌐

♀ Phospholipid

Figure 1.9 **The mechanism of detergent solubilization. When a detergent is added to a suspension of membranes the detergent intercalates with the lipid bilayer. Non-ionic detergents are wedge-shaped molecules and so cause surface curvature which allows lipids to break away from the membrane as lipid–detergent micelles with a small radius of curvature (a). Bile salts solubilize lipid in a different fashion. These molecules are planar and contain one polar and one apolar face. The result is that these detergents cut out disc-shaped pieces of bilayer, the hydrophobic edges of the lipids being protected by the detergent molecules (b). Integral membrane proteins are solubilized with a layer of detergent molecules maintaining an apolar environment around their hydrophobic surfaces (c). (Reprinted with permission from Yedgar, S., Barenholz, Y., and Cooper, V. G. (1974)** *Biochim. Biophys. Acta,* **363, 98–111; and Small, D. M., Penkett, S. A., and Chapman, D. (1969)** *Biochim. Biophys. Acta,* **179, 178–189)**

Table 1.5 The structure and properties of some detergents

Name	Structure	Type	CMC(mM)
Sodium dodecyl sulphate (SDS)		Anionic	8.2
Sulphobetain (SB14)		Zwitterionic	0.6
Sodium cholate		Bile salt	14.0
Triton X-100		Non-ionic	0.2
CHAPS		Complex	—

There are many different detergents available which fall into several categories. Some of the more popular are shown in Table 1.5. Most powerful of the commonly used detergents is the anionic detergent sodium dodecyl sulphate (SDS). When the concentration of this detergent is raised above that required to saturate high-affinity sites on a protein, a second co-operative phase of binding occurs. This is accompanied by the unfolding of the protein until about 1.4 g of SDS are bound per g of protein. Whilst not of great value for the purification of proteins in their native state, it does have great value in their molecular weight determination. For these rod-like detergent–protein particles have a uniform charge per unit length and so migrate in a cross-linked polyacrylamide gel according to their molecular size when an electric field is applied. This process is called SDS polyacrylamide gel electrophoresis (SDS-PAGE). Other alkyl ionic detergents are not as powerful as SDS, but most denature membrane proteins at the concentrations and temperatures required for complete solubilization.

At the other end of the scale are the non-ionic detergents. There are a very large number of these with polyoxyethylene or sugar head groups. Table 1.5 shows the structure of the popular detergent, Triton X-100. Most commercial preparations of non-ionic detergents are of heterogeneous chain length and contain other impurities. More consistent results may be obtained by using detergents with a defined structure, e.g. $C_{12}E_8$ which contains an alkyl chain of 12 carbon atoms with a headgroup of 8 polyoxyethylene moieties. Octylglucoside is also a popular non-ionic detergent. Non-ionic detergents have the advantage of being non-denaturing, but are also inefficient at breaking protein–protein interactions. This can be both an advantage and a disadvantage. For instance when removing the cell contents to examine the structure of the cytoskeleton or dissociating membrane-bound polysomes from the endoplasmic reticulum it is an advantage if the detergent does not dissociate protein–protein interactions. On the other hand, non-ionic detergents cannot usually be used for protein purification since the individual proteins are not adequately solubilized. Another disadvantage is their small critical micellar concentration (CMC). This is the highest concentration at which free detergent molecules can exist in solution at a given pH, salt concentration and temperature. Since it is the monomer which is active in solubilizing membranes, the low CMC effectively limits the maximum concentration that can be added. It also makes their subsequent removal difficult since the micellar form does not pass through dialysis bags.

When a detergent solution is heated a temperature is reached (the cloud point) at which the solution suddenly becomes turbid. This is a result of the increased size of detergent micelles at elevated temperatures. For Triton X-114 this cloud point occurs at about 20 °C. Above this temperature, therefore, two phases are produced, a detergent and an aqueous phase. Using this simple system it is possible in one step to separate peripheral and soluble proteins from integral membrane proteins. First, the proteins are dissolved in dilute Triton X-114 solution at 0 °C. They are then warmed to 30 °C and centrifuged to separate the phases. Integral membrane proteins are then found in the detergent phase.

The ability of non-ionic detergents to solubilize proteins is a result of their lipid dispersing power. By intercalating in the bilayer the wedge-shaped detergent molecules facilitate the formation of mixed micelles of lipid and detergent with small radii of curvature (Figure 1.9). Bile salts have a quite different method of solubilizing lipid. These molecules are planar with one face hydrophobic and one hydrophilic. They solubilize the bilayer by chopping out disc-shaped pieces of lipid with the hydrophobic edges protected by the bile detergent. *In vivo* the cholate detergents are found as the tauro- and glyco-cholate conjugates. These trihydroxy bile salts have a high CMC making them easy to remove and particularly suited to membrane reconstitution studies (see Chapter 5). More powerful than either the trihydroxy bile salts or the non-ionic detergents is the dihydroxy bile salt, deoxycholate. This detergent is effective at disrupting protein–protein interactions and rarely causes denaturation of membrane enzymes. It is therefore particularly useful for the purification of membrane proteins.

Bile salts, however, have the disadvantage of being charged and so usually change the properties of a protein when it is subsequently subjected to protein purification procedures. For this reason the zwitterionic sulphobetains were developed (Table 1.6). These detergents have no net charge between pH 2 and pH 12 and are therefore suitable for methods which separate proteins on the basis of their charge (like ion-exchange chromatography and isoelectric focusing). They are very effective at disaggregating proteins and have been used successfully to purify several membrane proteins. Some proteins, however, are denatured at detergent concentrations required to completely solubilize the membrane. This is presumably due to their short, flexible alkyl chains. Recently a complex detergent, CHAPS, has been synthesized which combines the bulky non-denaturing apolar region of a bile salt with the neutral but powerful zwitterionic headgroup of a sulphobetain (Table 1.5). Initial results suggest that this might be a very useful detergent. It is not possible, however, to recommend a single detergent as different experiments require detergents with different properties. It should also be noted that it is possible to exchange protein-bound detergents so that one detergent may be used for solubilization and another for characterization of the protein or reconstitution. Finally, a note of caution should be raised about the term 'solubilization'. This could be taken to mean: (1) the clarification of a turbid solution; (2) the reduction in sedimentation coefficient of the material so that it no longer sediments at say $100,000 \, g \times 60$ min; (3) the formation of protein-detergent micelles containing only one molecule of protein; (4) the formation of detergent- and lipid- free protein. Clearly, a criterion at least as stringent as (3) is required for attempts at protein purification.

Most schemes of purification, once the membrane has been solubilized, are similar to those used for soluble proteins, except that a small percentage of detergent is retained in all the buffers. Selective denaturation of proteins using heat or pH, ammonium sulphate precipitation, gel filtration, ion-exchange chromatography and affinity chromatogrphy have all been used with success. When membrane proteins are purified by precipitation with ammonium sulphate it is frequently advisable to sonicate well so that all lipid or detergent–protein micelles are of the minimum size and do not co-precipitate. In some cases it is possible to purify a protein as a detergent or lipd–protein micelle by sedimentation on density gradients. If there is a particularly strong association of the protein of interest with a specific lipid, it may also be possible to purify the protein by purifying the lipid in question. The ectoenzyme 5′-nucleotidase partitions in this manner into the sphingomyelin fraction when this is precipitated from detergent-solubilized rat liver. Although several of the major proteins of membrane have been purified to a high state of purity this is not usually true for the membrane enzymes which occur at a much lower percentage of the total membrane protein.

Summary
Membrane proteins are best solubilized using detergents.
No single detergent can be recommended for all applications.
Solubilized membrane proteins may be purified using conventional techniques.

1.7 EXTRACTION AND SEPARATION OF MEMBRANE LIPIDS

Membrane lipids, being small molecules, do not denature in the same manner as membrane proteins, thus allowing more vigorous methods to be used for their extraction. The most popular method shown in Figure 1.10 is to homogenize an aqueous suspension of the membranes in a mixture of chloroform and methanol which achieves the twofold function of denaturing the membrane proteins (which may be removed by filtration) and of solubiliz-ing all the lipid material. Two phases are produced. The lower phase, which is composed mainly of chloroform and methanol, contains the lipids with a low HLB value while the more hydrophilic lipids remain in the aqueous upper phase. By adding a controlled amount of salt to the aqueous phase and extracting the chloroform phase several times a fairly clean separation of phospholipids and neutral lipids from gangliosides may be achieved. The glycolipids are distributed between the phases according to their carbohydrate content. Further separation of lipid classes may be achieved by liquid chromatography of the two phases on silicic acid thin layer plates or using silicic acid columns for preparative separations.

Figure 1.11(a) shows how this method may be used for the analytical separation of phospholipid classes. This employs a two-dimensional separa-tion by silicic acid thin-layer chromatography. The mixture of lipids is spotted in one corner of the plate which is then placed in a tank containing the first solvent. The solvent moves up the plate by capillary action in the finely divided coating of silicic acid. When at the top of the plate it is removed and

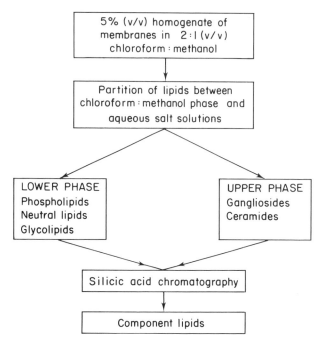

Figure 1.10   Extraction of membrane lipids

thoroughly dried before developing in the second solvent. The plate is arranged in the second solvent tank so that the solvent runs perpendicular to the first. During this process the lipids partition between the stationary silicic acid phase and the moving solvent phase. Each lipid has a unique partition coefficient in the two solvents of different pH and polarity and ends up in a discrete spot when the plate is developed with a staining reagent (Figure 1.11a). For a quantitative estimate of the distribution of lipids the spots may be scraped off the plate and assayed. Phospholipids for example may be estimated by their phosphorus content. In addition, the lipids eluted from the spots may be hydrolysed, methylated, and the volatile fatty acid methyl esters derived from the alkyl chains of the lipid separated by gas–liquid chromatogrphy. This is a very high-resolution chromatographic technique in which the volatilized substances being examined partition between a moving gaseous phase (usually argon or nitrogen) and a liquid phase coated on a finely divided solid support. Various greases and esters which melt in the temperature range 100–250 °C may be used as liquid phase. Each lipid may be identified by its characteristic retention time in each liquid phase. An example of the high resolution of this technique is shown in Figure 1.11(b).

Glycolipids, gangliosides and glycosphingolipids may also be separated by chromatography on silicic acid using solvents of different polarity and pH.

FID response (A × 10^7)

Solvent (1): Chloroform : methanol : NH$_3$ (65 : 25 : 5) ⟶

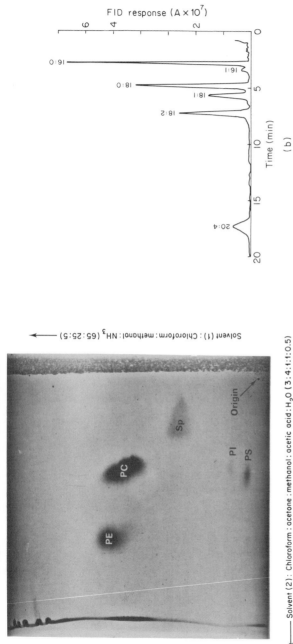

Solvent (2): Chloroform : acetone : methanol : acetic acid : H$_2$O (3 : 4 : 1 : 1 : 0.5) ⟵

(a)

(b)

Figure 1.11   Chromatographic separation of membrane lipids. Two applications of chromatographic separation are shown for membrane lipids. In (a) classes of phospholipids from an extract of rat liver plasma membranes are separated by two-dimensional thin-layer chromatography which uses silicic acid as a stationary phase and organic solvents as the moving phase. In (6) the alkyl chains from these lipids are separated by gas–liquid chromatography on a column of ethylene glycol succinate at 180 °C. PC = Phosphatidyl choline, PE = phosphatidyl ethanolamine, PI = phosphatidyl inositol, PS = phosphatidyl serine, Sp = sphingomyelin. For the key to fatty acid notations see Table 1.1. FID = flame ionization detector. (Reprinted with permission from Stanley, K. K., and Luzio, J. P. (1978) *Biochim. Biophys. Acta*, 514, 198–205)

After acid hydryolysis the sugar residues present in the headgroup may be separated by gas–liquid chromatography after trisilylation to render them volatile.

Summary

Lipids may be extracted in organic solvents and separated according to their partition in solvents of different hydrophobicity.

FURTHER READING

Evans, W. H. (1979) Preparation and characterisation of mammalian plasma membranes in *Laboratory Techniques in Biochemistry and Molecular Biology*, Part 1, Work, T. S., and Work, E., eds., North–Holland.

Evans, W. H. (1980) A biochemical dissection of the functional polarity of the plasma membrane of the hepatocyte, *Biochem. Biophys. Acta*, **604**, 27–64.

Folch, J., Lees, M., and Sloane Stanley, G. H. (1956) A simple method for the isolation and purification of total lipids from animal cells, *J. Biol. Chem.* **226**, 497–509.

Helenius, A., and Simons, K. (1975) Solubilization of membranes by detergents, *Biochim. Biophys. Acta*, **415**, 29–79.

Palade, G. (1981) in *Pathophysiology, the Biological Principles of Diseases*, Smith, L. H., and Thier, S. O., eds., W. B. Saunders & Co.

Ploegh, H. L., Orr, H. T., and Strominger, J. L. (1981) Major histocompatibility antigens: the human and murine class 1 molecule, *Cell*, **24**, 287–299.

Singer, S. J., and Nicolson, G. L. (1972) The Fluid Mosaic model of the structure of cell membranes, *Science*, **175**, 720–731.

Chapter 2

Mobility of the lipid and protein components of biological membranes

In Chapter 1 we emphasized the static arrangement of proteins and lipids in biological membranes. Biological membranes are not however static structures. In this Chapter we shall examine in detail how each type of molecule is able to move in the bilayer—this encompasses various translational, rotational, and vibrational modes of movement, and the special property which phospholipids possess of undergoing phase transitions between 'liquid crystalline' and 'crystalline' states. Later on in Chapter 6 we shall consider movements of the membrane as a whole. It is the importance of these dynamic properties of biological membranes which is the thesis of this book.

2.1 MODES OF MOBILITY OF MEMBRANE PHOSPHOLIPIDS

Individual phospholipid molecules have four permitted modes of mobility within the bilayer structure (Figure 2.1). These are:

(i) Fast *lateral diffusion* within the plane of the bilayer. This is diffusion constrained to the two dimensions of the horizontal plane of the bilayer. Indeed it would take only 1 s for a lipid molecule to travel the length of a bacterium.

(ii) *Phospholipid 'flip-flop'*. This is the movement of lipid molecules from one half of the bilayer to the other on a 'one-for-one' exchange basis. Such a process is extremely slow and it can take up to a day for a phospholipid molecule to traverse the 4 nm thickness of the bilayer compared with 2.5 μs for it to migrate that distance by lateral diffusion in the plane of the bilayer.

(iii) *Intra-chain motion* of fatty acyl chains attached to the glycerol backbone of the phospholipid molecule. In biological membranes at physiological temperatures the bilayer is in a fluid state and the fatty acyl chains are continually being distorted by rotation about the C–C bonds yielding kinks in the chain. These are extremely short lived states lasting for only about 10^{-9} s.

(iv) The fatty acyl chains and the lipid molecules can undergo a *fast axial rotation* when the bilayer is in the fluid state.

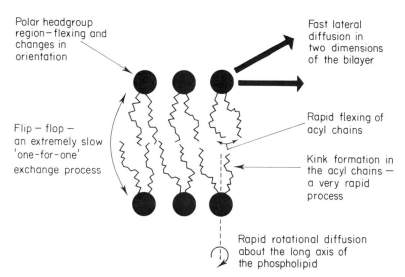

Figure 2.1 Modes of mobility of the lipid components of the bilayer

2.2 FAST LATERAL DIFFUSION OF PHOSPHOLIPIDS

Phospholipids, fatty acids and sterols can migrate laterally in the plane of the bilayer when the phospholipid is in the liquid-crystalline (L_α) state. The diffusion coefficients obtained for this mode of molecular motion are of the order of 10^{-8} cm^2 s^{-1} in both model and biological membranes (Table 2.1). Such rapid rates of diffusion can only be measured using sophisticated physical chemical techniques such as nmr, esr, and fluorescence photobleach recovery. Mechanistically, the diffusion measured by esr, nmr and fluorescence studies corresponds to a 'hopping' or 'jumping' event that occurs with a frequency of about 10^{-7} s. A variety of models have been put forward to explain the molecular mechanism whereby lipids undergo this movement within the bilayer which leads to their lateral diffusion. It is, however, difficult to assess the degree of long-range order present in both the solid and fluid states of the bilayer. Models have focused on analogies with diffusion in ordered crystals, which bear best comparison with the solid (crystalline) state of the bilayer where lateral diffusion is an extremely slow process. Three of the simpler mechanisms are outlined in Figure 2.2. For example two, or even more, molecules may exchange positions by rotation about a midpoint or, alternatively, lipids may exchange positions by migration through small 'gaps' or interstitial sites in the bilayer. Both of these processes are likely to involve considerable steric repulsion between adjacent molecules which could make them unfavourable. The third mechanism postulates the presence of

Table 2.1 Lateral diffusion coefficients (D_L) for phospholipids in model and biological membranes

Membrane	Detection method	T °C	D_L (cm^2 s^{-1})
Egg phosphatidyl choline	ESR (steroid)	50	10^{-8}
Egg PC	ESR (fatty acid)	20	10^{-9}
Egg PC	ESR (fatty acid)	25	10^{-8}
Egg PC	NMR (^{1}H)	20	10^{-8}
Egg PC + cholesterol (4 : 1)	NMR (^{1}H)	30	10^{-9}
DPPC	NMR (^{1}H)	60	10^{-8}
Sarcoplasmic reticulum vesicles	ESR (phospholipid)	40	6×10^{-8}
	NMR (^{1}H)	50	10^{-8}
Sarcoplasmic reticulum extracted lipids	ESR (PL)	40	10^{-7}
	NMR (^{1}H)	31	4×10^{-9}
Rat liver microsomes	ESR (fatty acid)	40	10^{-7}
Lymphocyte plasma membrane	Fluorescence (phospholipid analogue)	20	10^{-8}
Torpedo electroplax	NMR (^{1}H)	33	10^{-9}
Rabbit sciatic nerve	NMR (^{1}H)	31	10^{-8}
Escherichia coli	NMR (^{1}H)	31	10^{-8}
	ESR (fatty acid)	40	10^{-8}

42

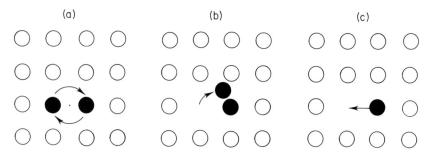

Figure 2.2   Mechanism of lateral diffusion. This could occur by a variety of processes, possibilities of which are (a) exchange about a midpoint, (b) migration through interstitial sites or small 'gaps' and (c) migration into vacant sites

vacancies in the bilayer into which an adjacent lipid can move, thus exchanging positions with the vacant site. Within a crystal structure there are likely to be a number of lattice defects giving rise to vacancies, or interstitial sites through which lipids can move. The rates of lateral diffusion in the crystalline state yield diffusion coefficients at least two orders of magnitude smaller than those observed in the fluid state, suggesting that this is a restricted process.

In the fluid (liquid-crystalline state; see section 2.6) there is far less long-range order. It is possible to envisage that diffusion in this state could occur by any one of the mechanisms outlined. Although the vacancy process may overestimate the degree of long-range order than can exist in this state, vacancies have been detected and quantified in liquid crystals of benzene and hexane. Indeed, positron annihilation techniques have been used successfully to detect the existence of vacancies in fluid bilayers. These vacancies could form 'transient' pores, providing not only a means for lipid lateral diffusion but also for the diffusion of water across lipid bilayers.

We might well expect that alterations in the properties of the bilayer would affect the lateral diffusion coefficient for lipids by changing the number of vacancies and by reducing their free energy of migration. One of the ways of altering the properties of the bilayer is to change the temperature. As the temperature is decreased the number of vacancies and the value of the diffusion coefficient both fall together in the same exponential fashion. Alternatively, cholesterol may be incorporated into the lipid bilayer. This molecule inserts itself into the bilayer in a highly specific fashion which leads to a condensing effect on the acyl chain region of the bilayer (see section 2.11). Cholesterol thus has an ordering effect on the liquid-crystalline (fluid) state of the bilayer. This would be expected to reduce the number of vacancies occurring, and is consistent with cholesterol reducing the passive permeability of lipid bilayers to water and small neutral molecules through such transient 'pores'. Such changes in the liquid-crystalline state of the bilayer may then explain why cholesterol has been shown to reduce considerably the rate of lipid self-diffusion in fluid bilayers and hence the value of the diffusion coefficient. Cholesterol appears to do this without altering the activation energy for the diffusion event. This implies that it exerts an entropy effect, either on the entropy of diffusion or the entropy of vacancy formation.

2.3 PHOSPHOLIPID FLIP-FLOP IN MEMBRANES

The exchange of phospholipid molecules between each half of the bilayer of a membrane has been termed 'flip-flop'. This is a 'one-for-one' exchange mechanism and thus the total number of molecules in each half remains constant.

Phospholipids form bilayers spontaneously in an aqueous environment and thus it is thermodynamically unfavourable to move the polar headgroup of a phospholipid through the apolar core of the bilayer. This is reflected in the extremely slow rates of flip-flop that have been observed, with half-times measured in the order of days. Indeed, a membrane potential of 1000 mV applied across a unilamellar liposome consisting of phosphatidyl ethanolamine and phosphatidyl serine failed to cause the flip-flop of the acidic phosphatidyl serine. These extremely slow rates of flip-flop mean that the lipid asymmetry of biological membranes can be preserved (see Chapter 4).

2.3.1 The assessment of phospholipid flip-flop

(i) *Esr methods*

Phospholipid flip-flop was first demonstrated in unilamellar phosphatidyl choline liposomes. The method, which is shown schematically in Figure 2.3, takes advantage of the ability of ascorbate to reduce the nitroxide spin label attached to the choline *N*-methyl group of a phospholipid. This destroys its paramagnetism. However, as liposomes are impermeable to ascorbate, only the label in the external half of the bilayer is susceptible to its action. Thus, by removing ascorbate and subsequently rechallenging the liposomes with the reductant, it is possible to assess the rate of appearance of the label in the external half of the bilayer achieved by the flip-flop process. This method indicated that flip-flop occurred with a half-life of about 6.5 h at 30 °C. For methodological reasons this is now believed to be a drastic overestimation of the true rate of flip-flop. The factors which contribute to this are the reoxidation of the spin label in the outer half of the bilayer by oxygen and the ability of ascorbate to gain access to lipids in the inner monolayer after the liposomes have been incubated for prolonged periods. Indeed, if ascorbate is immobilized on a high molecular weight support then this method raises the estimate for the half-life of the flip-flop process to >20 h Of course, the other criticism of this procedure is that the flip-flop of a spin-labelled lipid is being investigated rather than that of a natural lipid. Owing to the apolar nature of the nitroxide group, which is attached to the polar headgroup

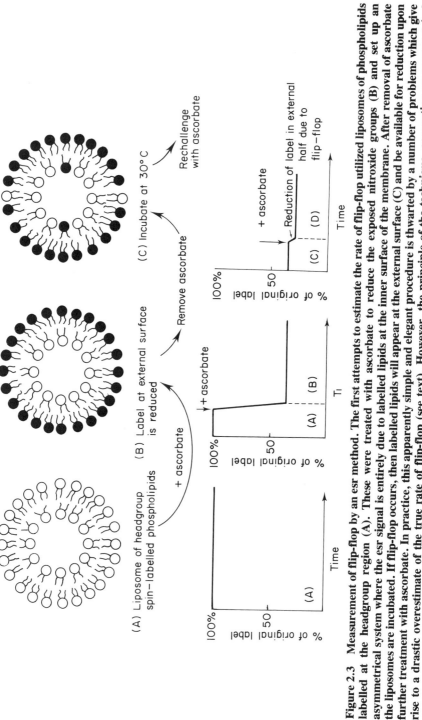

Figure 2.3 Measurement of flip-flop by an esr method. The first attempts to estimate the rate of flip-flop utilized liposomes of phospholipids labelled at the headgroup region (A). These were treated with ascorbate to reduce the exposed nitroxide groups (B) and set up an asymmetrical system where the esr signal is entirely due to labelled lipids at the inner surface of the membrane. After removal of ascorbate the liposomes are incubated. If flip-flop occurs, then labelled lipids will appear at the external surface (C) and be available for reduction upon further treatment with ascorbate. In practice, this apparently simple and elegant procedure is thwarted by a number of problems which give rise to a drastic overestimate of the true rate of flip-flop (see text). However, the principle of the technique, generating an asymmetrical system, is generally applicable

region, such a labelled lipid might well be expected to undergo an enhanced rate of flip-flop.

An interesting modification of this procedure has been used to assess flip-flop in intact erythrocytes. In this case the headgroup spin-labelled phospholipid was incorporated into the membrane by fusion which puts the label into both halves of the membrane bilayer. However, the label facing the cell cytosol is spontaneously reduced by agents within the cytosol, thus instantaneously setting up an asymmetric distribution of the label. Any further reduction in paramagnetism will presumably be due to flip-flop transferring oxidized species to the cytosol half of the bilayer. This procedure, which does not suffer from the problem of reoxidation of spin-label, yields a half-life for flip-flop in excess of 24 h. A similar result can be obtained for phospholipids in the mitochondrial inner membrane.

(ii) Phospholipid exchange protein methods

Phospholipid exchange proteins (PLEP) are small ($M_r = 14,000$) basic proteins that are found in the cytosol of eukaryotic cells. They catalyse the one-for-one transfer of phospholipids between the exposed surfaces of bilayer membranes or vesicles (see section 4.3).

Two different approaches have been made, using this exchange protein, to look at flip-flop rates in lipid vesicles (liposomes). In the first method, uniformly labelled ^{32}P-phosphatidyl choline, unilamellar vesicles were incubated with unlabelled mitochondrial membranes in the presence of PLEP. This depletes the vesicles of labelled phospholipid in the external half of the bilayer only. After removal of the PLEP, these vesicles can be incubated for various periods and rechallenged with fresh PLEP and unlabelled membranes in order to see if an increased amount of label is now available for exchange due to flip-flop. Over 5 d of incubation little or no change in the availability of the label occurred, which yields a lower limit of 11 d for the rate of flip-flop.

In the second approach unlabelled unilamellar phosphatidyl choline vesicles were incubated with PLEP and ^{32}P-labelled microsomes in order to introduce labelled phospholipids into the external half of the vesicles bilayer (Figure 2.4). After separation of these components, the vesicles were then incubated for various periods. They were then challenged with PLEP and unlabelled erythrocytes in order to assess the availability of labelled phospholipid for exchange. Over a period of several days of incubation it could be shown that all of the labelled lipid remained accessible, indicating that none had been transferred to the inner half of the bilayer of the unilamellar vesicles by flip-flop. Similarly then, a lower limit to the half-life of the flip-flop process could be set from this data at around 11 d.

Biological membranes with their lipids biosynthetically labelled with ^{32}P have also been probed with PLEP. From analysis of the rates of exchange of labelled species with exogenous unlabelled lipid it is apparent that there are at least two kinetically distinct pools present in the membrane. The rapidly

46

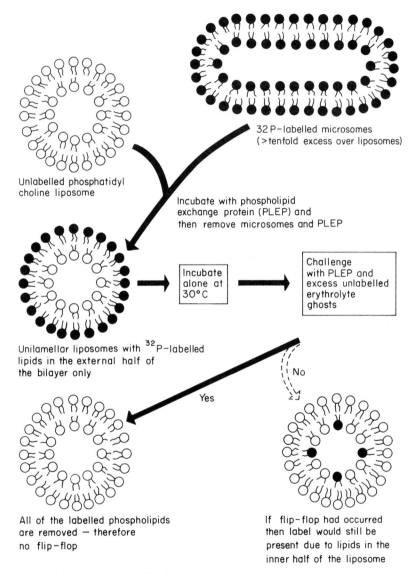

Unlabelled phosphatidyl
choline liposome

32 P-labelled microsomes
(>tenfold excess over liposomes)

Incubate with phospholipid
exchange protein (PLEP) and
then remove microsomes and PLEP

Incubate
alone at
30°C

Challenge
with PLEP and
excess unlabelled
erythrolyte
ghosts

Unilamellar liposomes with ^{32}P–labelled
lipids in the external half of
the bilayer only

No

Yes

All of the labelled phospholipids
are removed — therefore
no flip–flop

If flip–flop had occurred
then label would still be
present due to lipids in the
inner half of the liposome

Figure 2.4 **Measurement of flip-flop using phospholipid exchange proteins (PLEP).**
The flip-flop of phospholipids in liposomes can be assessed by initially generating an
asymmetric system. In this example PLEP is used to transfer phospholipids between the
external surfaces of lipid bilayers. Over the time courses of the experiments all of the
label could be demonstrated to be available for exchange. This suggests that flip-flop is a
very slow process with a half-life that can be measured in the order of days

exchanging pool can be shown to be provided by the lipids in the external half of membrane, and the much more slowly exchanging pool is presumed to reflect the rate of flip-flop of lipids from the inner half of the bilayer. However, there are alternative explanations for the presence of slowly exchanging pools. For example, certain endogenous lipids in the external half of the bilayer could exhibit a preferential interaction with cholesterol (see section 2.11) or proteins (see section 3.3) in the membrane which would reduce their availability for exchange. Indeed, the PLEP used in these studies cannot exchange phospholipids between the plasma membrane of intact erythrocytes and liposomes, yet it can carry out the exchange between resealed ghosts (see section 4.3) and liposomes. This demonstrates that PLEP is extremely sensitive to the presentation of the phospholipid.

In using PLEP for such studies a number of criteria must be, and indeed have been, met by systems under investigation. These are that the membrane vesicles should be impermeable to PLEP; that PLEP should not of itself catalyse flip-flop; that the changes in composition effected by PLEP should not catalyse flip-flop and that the conditions of the experiment, i.e. the media, should not catalyse flip-flop.

Phospholipids with a nitroxide spin label attached to a position on the acyl chain near to the headgroup have been incorporated into the external half of the bilayer of intact erythrocyte ghost membranes and mitochondrial inner membranes using PLEP. Under such conditions all of the label can be immediately reduced by ascorbate, even after the intact membranes have been subjected to prolonged periods of incubation prior to exposure to ascorbate. From such experiments the inferred rate of flip-flop has a half-life of the order of days.

(iii) *Chemical methods*

These rely on the ability of membrane impermeable chemical agents to modify primary amino groups on phospholipids exposed at the external surface only of sealed membrane vesicles and to generate an asymmetry. When isethionyl acetimidate (see section 4.3) is reacted with unilamellar liposomes formed from phosphatidylcholine and phosphatidyl ethanolamine then this reagent modifies >90 per cent of the externally facing phosphatidyl ethanolamine. The appearance of unmodified phosphatidyl ethanolamine in the external half of the bilayer, due to its flip-flop from the inner half, can be detected with another membrane impermeable probe, 2,4,6-trinitrobenzene sulphonate (see section 4.3). Using this method a lower limit of the rate of flip-flop can be put at about 80 d.

Summary

Phospholipid flip-flop in both model and biological membranes is normally an extremely slow process whose half-life can be measured in days.

Under certain conditions perturbation of the membrane structure can accelerate the rate of flip-flop.

2.4 CHOLESTEROL FLIP-FLOP IN MEMBRANES

On the basis of polarity alone one might have expected that cholesterol flip-flop should be a relatively rapid event. This, however, does not appear to be the case. Cholesterol is integrated in a specific fashion into membranes where it interacts strongly with adjacent phospholipids (see section 2.11). The slow rate of phospholipid flip-flop is thus conferred upon cholesterol.

Cholesterol flip-flop has been examined in erythrocyte ghosts and liposomes. When unilamellar liposomes with similar cholesterol/phospholipid ratios to the ghosts were incubated together there was a net exchange of cholesterol between the external halves of the bilayer of each of the membranes. However, this appeared to proceed without the occurrence of cholesterol flip-flop, putting a lower limit for this event at 6 d. Interestingly, if the vesicles were incubated with cholesterol-depleted erythrocytes so that a net transfer of cholesterol occurred from the vesicles to the membranes, then cholesterol in the vesicles that was previously unavailable for exchange now in fact became exchanged. This strongly suggests that under non-equilibrium conditions, where there is a net transfer of cholesterol from or even to the membrane, then cholesterol flip-flop can be induced. Indeed, if sealed membranes are treated with the enzyme cholesterol oxidase, which uses molecular oxygen to convert cholesterol to $\Delta 4$-cholestenone, then the kinetics of cholesterol depletion (conversion) are clearly a first-order process. This indicates that a relatively rapid rate of flip-flop can be induced if the size of the pool of cholesterol in one-half of the bilayer is altered.

Cholesterol exchange has also been studied in *Mycoplasma gallisepticum* by growing the cells on ^{14}C-cholesterol. Here cholesterol exchange was achieved using high-density lipoproteins (HDL) under conditions where the mycoplasma were not depleted of cholesterol. In broken cells only one rapidly exchanging pool was identified. However, using intact cells two kinetically distinct pools could be identified, each reflecting about 50 per cent of the total membrane cholesterol. The rapidly exchanging pool ($T_{1/2} = 4$ h) presumably reflects cholesterol in the outer half of the bilayer and the extremely slowly exchanging pool ($T_{1/2} > 18$ d) that of cholesterol flip-flop across the membrane.

Summary

Under equilibrium conditions cholesterol flip-flop is an extremely slow process whose half-life can be measured in days.

Under non-equilibrium conditions the rate of cholesterol flip-flop can be extremely rapid with half-lives measured in minutes.

Some methods used to assess cholesterol flip-flop may in fact promote its occurrence.

2.5 MOBILITY GRADIENTS WITHIN THE BILAYER

It is possible from ^{13}C nmr studies to obtain an idea of the degree of molecular motion occurring at different depths in the bilayer, ranging from the polar

Relaxation times T_1 in sec
in D_2O at 52 °C

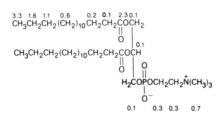

Figure 2.5   The mobility of regions of the dipalmitoyl phosphatidyl choline molecule in a fluid lipid bilayer. The mobility of particular carbon atoms in a molecule of DPPC can be measured by incorporating ^{13}C at particular locations in the molecule and studying its T_1-relaxation by nmr. The smaller the value of T_1, the more restricted is the motion. From this it is evident that the glycerol backbone of the molecule acts as a rigid anchor with mobilities increasing both towards the headgroup and along the fatty acyl chain. The most dramatic increases in motion are seen towards the terminal methyl group, indicating that the core of the bilayer is highly disordered. (Reprinted with permission from Levine, Y. K., Birdsall, N. J. M., Lee, A. G., and Metcalfe, J. C. (1972) *Biochemistry*, 11, 1416–1421. Copyright (1972) American Chemical Society)

headgroup to the terminal methyl groups of the fatty acyl chains. This has been achieved by synthesizing dipalmitoyl phosphatidyl choline with ^{13}C substituted at different positions in the molecule. Motion about specific C—C bonds can thus be assessed by observing the relaxation time, T_1 (Figure 2.5). In fully solvated lipid vesicles, motion is very restricted about the glycerol backbone of the molecule which acts as a relatively immobilized anchor for the flexible headgroup and acyl chain regions of the molecule. Indeed there is a significant increase in mobility about C—C bonds on moving away from the glycerol backbone. This increase is most dramatic towards the terminal end of the fatty acyl chains which are clearly undergoing extremely rapid motion. It is, however, not possible to ascertain whether or not the motion about individual C—C bonds is equal. This is because the resultant rate of motion of any particular C depends upon the motion of all the C—C bonds between it and the immobilized glycerol backbone. This, at least in part, explains why the resultant rate of motion of the terminal methyl group will be far larger than the motion of a C near the glycerol backbone.

The flexibility gradient within the bilayer core can be described in terms of an order parameter S, which measures the time average orientation of groups within the bilayer. This yields a quantitative estimation of the relative probabilities of *gauche* and *trans* conformers occurring in the lipid methylene chains. These are rotational isomers which are the result of a $120°$ rotation about C—C bonds. Both (i) esr methods with (m, n) phosphatidylcholine spin labels (Figure 2.6) where the spin-label nitroxide group is placed at various positions in the acyl chain, (ii) nmr methods with dimyristoyl phosphatidyl choline where selective carbons were made ^{13}C, and (iii) nmr methods with

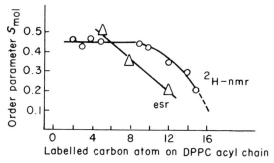

Figure 2.6 Mobility gradients within the bilayer. There is an increase in the frequency of *gauche* conformers and kink formation at C—C bonds towards the core of the bilayer (*S* decreases). This increase in mobility and disorder increases markedly towards the core of the bilayer and is detected by esr studies with nitroxide labels on the fatty acyl chain and also by nmr using selectively deuterated and ^{13}C labelled (Figure 2.5) carbons. The nmr studies use labels which effect a minimal perturbation of the bilayer and these show that the increase in mobility only suddenly increases over the last three or four carbon atoms. This is in contrast to the esr studies which show a linear increase, presumably because of the disruption caused by the bulky nitroxide group. (Reprinted with permission from Seelig, A., and Seelig, J. (1974) *Biochemistry*, 13, 4839–4845, copyright (1974) American Chemical Society; Hubbell, W. L., and McConnell, H. M. (1971) *J. Amer. Chem. Soc.*, 93, 314–326, copyright (1971) American Chemical Society)

dipalmitoyl phosphatidylcholine where selective methylene groups were made ^{2}H, have been used to monitor gradients within the bilayer.

These methods have detected the increase in disorder that occurs along the fatty acyl chains towards the core of the bilayer. However, the rate of decrease in order parameter (*S*) with increase in carbon number differs using the two experimental approaches. With the esr experiments there is a continuous decline in *S* until the terminal methyl group is reached, whereupon *S* decreases dramatically. In the nmr studies *S* is relatively constant over the first 8–10 carbons and then it falls very rapidly as the carbon number increases. This discrepancy may in part be due to the difference in time scale of the two methods, but the bulk of this effect can clearly be shown to be caused by the perturbing effect of the nitroxide group on the bilayer structure. The introduction of such a bulky constituent into the relatively ordered region next to the glycerol backbone of the bilayer will disrupt the structure and lead to these differences. This exemplifies the value of nmr studies which use lipid probes that so closely mimic the native ones.

Analysis of the nmr data allows us to assess the rotational diffusion coefficients due to the rate of motion about C—C bonds. These yield values of 10^9 s^{-1} over the first part of the chain and then rapidly increase towards the terminal methyl end with motion at the penultimate carbon yielding $D = 6 \times 10^9$ s^{-1} and at the terminal methyl $D = 8 \times 10^{10}$ s^{-1}. These data can be used to make an assessment of the 'microviscosity' at various depths in the bilayer and for dimyristoyl phosphatidyl choline this can vary from 210 mPa at the acyl chain next to the glycerol backbone to only 3 mPa at the terminal methyl group, indicating the considerable gradient that exists within the bilayer.

As well as motion about C—C bonds it is probable that the lipid molecules and their acyl chains undergo a fast axial rotation in the liquid-crystalline (fluid) phase (see section 2.6). This axial rotation has been demonstrated for steroid spin labels, of similar cross-sectional areas to phospholipids, which have been inserted into bilayers. The values for the diffusion coefficient for this process, $D_{axial} = 10^{-8}$–10^{-9} s, indicate that it is an extremely rapid process.

Summary

There is a mobility and microviscosity gradient within the bilayer in the fluid state.

The glycerol backbone acts as a rigid anchor restricting the motion of the acyl chain closest to it.

However, towards the core of the bilayer the end of the acyl chain undergoes extremely rapid flexing motion.

The headgroup region is also relatively mobile.

In the fluid state the entire lipid molecule can undergo a fast axial rotation as can its fatty acyl chains.

2.6 THE LIPID PHASE TRANSITION

Phospholipids can exist in several different physical forms: they are mesomorphic. In biological systems they are always found in aqueous environments where they form spontaneously fully solvated lipid bilayers (see Chapter 1). Under such conditions they can exist in two very distinct lamellar (L) physical states, known as the L_α and L_β forms. The L_β form is that found in the crystalline (solid) state where the acyl chains of the phospholipids are fully extended parallel to the bilayer normal with their C—C bonds in the all-*trans* position. The most detailed structural analyses of fully solvated phospholipid bilayers have been carried out with phosphatidylcholines which are major components of many biological membranes. In the crystalline state of this lipid the fatty acyl chains are not only fully extended but are tilted to the normal plane of the bilayer. This structure is denoted as the $L_{\beta'}$ form. However, with another major phospholipid species found in biological membranes, phosphatidyl ethanolamine, the acyl chains do not tilt in the solid state yielding an L_β form.

In the crystalline state these molecules do not undergo lateral diffusion but, pack together in a quasi-hexagonal array with their fatty acid chains lying parallel to each other. The transition from the $L_{\beta'}$ form to the L_α form found in the liquid-crystalline (fluid) state is most readily achieved by raising the temperature (Figure 2.7). This is accompanied by the occurrence of rotational isomers about the C—C bonds of the fatty acyl chains to form *gauche* isomers which are the result of a 120° rotation about the C—C bonds. These rotational isomers produce kinks with a *gauche–trans–gauche* conformation in the acyl chain. Such kinks appear to be highly mobile along the length of the

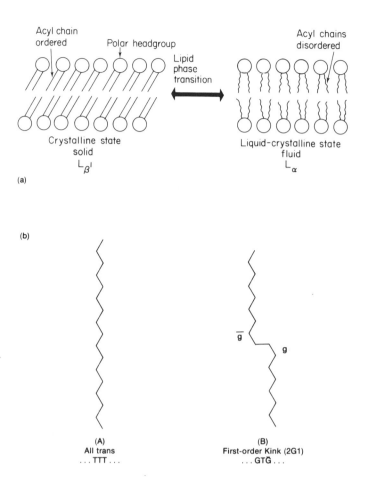

Figure 2.7   The lipid phase transition. (a) Fully hydrated phospholipids spontaneously form bilayers. These exist in two mesomorphic forms, a solid or gel form known as the crystalline (L_β) state and a fluid form known as the liquid-crystalline (L_α) state. In the crystalline state the fatty acid chains are fully extended in the all-*trans* position and the molecules pack together in a quasi-hexagonal array. With bilayers of phosphatidyl choline, the acyl chains pack tilted to the bilayer normal ($L_{\beta'}$, state). Conversion to the liquid-crystalline state can readily be achieved by increasing the temperature. As this happens then the molecules are able to undergo fast lateral and rotational diffusion, the intramolecular mobility about C—C bonds increases and the acyl chains can undergo a swinging motion. Most noticeable is the formation of rotational isomers about the C—C bonds, leading to kink formation. (b) This demonstrates the distortion of the fatty acyl chain in the all-*trans* state that occurs when *gauche* isomers are formed, producing kinks. Many phospholipids in biological membranes have acyl chains with a *cis*-double bond in them. This leads to the formation of a permanent kink, producing a highly fluid phospholipid bilayer with a low transition temperature. This reflects the difficulty in packing such a lipid into the crystalline state. (Reprinted with permission from Lagaly, G., Weiss, A., and Stuke, E. (1977) *Biochem. Biophys. Acta*, 470, 331–341)

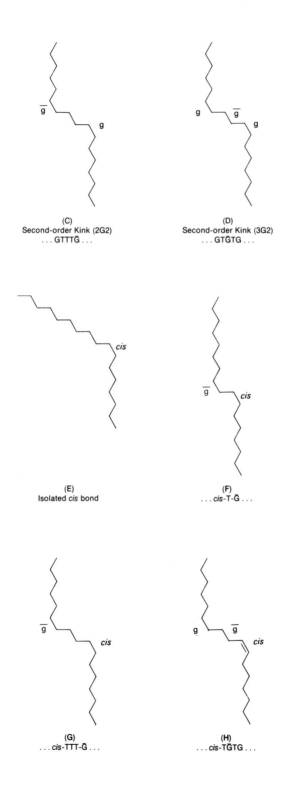

(C)
Second-order Kink (2G2)
. . . GTTTG . . .

(D)
Second-order Kink (3G2)
. . . GTGTG . . .

(E)
Isolated *cis* bond

(F)
. . . *cis*-T-G . . .

(G)
. . . *cis*-TTT-G . . .

(H)
. . . *cis*-TGTG . . .

chain and are extremely short-lived with correlation times of the order of 10^{-9} s. From esr studies the probability of kink formation can be shown to increase for C—C bonds closest to the terminal methyl group of the fatty acyl chain. This implies that there is a fluidity gradient, with the core of the bilayer being more fluid than the headgroup region (see section 2.5).

The transition of bilayer lipid from the crystalline to the liquid-crystalline state occurs at a characteristic temperature for a particular phospholipid species (Table 2.2). This has been termed the lipid phase transition temperature. The lipid phase transition is a highly co-operative event occurring over much less than a few degrees centigrade. At temperatures below the lipid phase transition the phospholipid fatty acyl chains are fully extended in the all-*trans* configuration, but as the temperature approaches the phase transition kink formation begins with an average of one kink existing per acyl chain. As the temperature increases above the lipid phase transition then the probability of kink formation rises and on average two kinks are found per acyl chain. These kinks affect the structure of the bilayer, causing chain shortening and an increase in the distance between individual molecules. The

Table 2.2 Lipid phase transition temperatures (T_t) of some defined, synthetic phospholipids

	Phospholipid	T_t °C	(pretransition, T_{pt} °C)
Same headgroup but	DMPC (1, 2 di 14 : 0)	24	(14)
different pairs of	DPPC (1, 2 di 16 : 0)	42	(35)
saturated acyl chains	DSPC (1, 2 di 18 : 0)	55	(49)
Same headgroup but	DMPE	51	
different pairs of	DPPE	63	
saturated acyl chains	DSPE	82	
Same headgroup but	DMPG	23	
different pairs of	DPPG	41	
saturated acyl chains	DSPG	54	
Same headgroup, same	DSPC (1, 2 di 18 : 0)	55	
length of acyl change but	DOPC (1, 2 di 18 : 1, 9-*cis*)	−22	
chains differ in saturation	DEPC (1, 2 di 18 : 1, 9-*trans*)	5	
Same headgroup, same	DSPE (1, 2 di 18 : 0)	82	
length of acyl chain but	DOPE (1, 2 di 18 : 1, 9-*cis*)	15	
chains differ in saturation	DEPE (1, 2 di 18 : 1, 9-*trans*)	41	
Same headgroup but	DPPA	67	
different pairs of acyl	DOPA	8	
chains			
Intramolecular mixtures	MPPC (1–14 : 0, 2–16 : 0)	35	(25)
	PMPC (2–14 : 0, 1–16 : 0)	27	(14)
	PSPC (1–16 : 0, 2–18 : 0)	47	(40)
	SPPC (2–16 : 0, 1–18 : 0)	44	(35)

Abbreviations: M–myristoyl; P—palmitoyl; S—stearoyl; O—oleoyl; E—elaidoyl; PC—phosphatidyl choline; PE—phosphatidyl ethanolamine; PG—phosphatidyl glycerol; PA—phosphatidic acid.

result of this is that the decrease in thickness of the bilayer is accompanied by a lateral expansion. It can be shown that the area taken up by individual phospholipid molecules increases from $0.48\ nm^2$ to $0.7\ nm^2$ and bilayer thickness decreases by $0.7\ nm$ at temperatures above the phase transition. This can readily be accounted for by the appearance of two kinks in each acyl chain. This phase transition is a highly endothermic process and the enthalpy change is in part the energy required for the sum of the *gauche* isomerizations and also the disruption of van der Waals forces between the neighbouring phospholipid headgroup region as the bilayer undergoes lateral expansion.

The temperature at which this phase transition takes place is influenced by the nature of the fatty acyl chains of the phospholipid molecule and by its headgroup. As a rule, the longer the fatty acyl chain in the molecule the higher will be the transition temperature (Figure 2.8), and the more

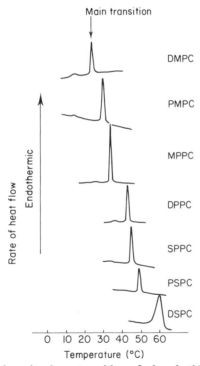

Figure 2.8 **The endothermic phase transition of phosphatidyl cholines. The phase transition is a large endothermic event that can be detected by differential scanning calorimetry (DSC). The longer the acyl chain the greater is the temperature of the transition, DMPC (di C_{14}) < DPPC (di C_{16}) < DSPC (di C_{18}). Intramolecular mixtures have phase transitions at temperatures midway between lipids with identical chains and this temperature is influenced by whether the longer chain is attached to the 1 or 2 position on the glycerol backbone. (Reprinted with permission from Keough, K. M. W., and Davis, P. J. (1979) *Biochemistry*, 18, 1453–1459. Copyright (1979) American Chemical Society)**

unsaturated bonds there are in the acyl chain the lower it will be (see Table 2.2). Unsaturated double bonds disrupt the hexagonal packing of the phospholipids in the crystalline state, with *cis*-bonds ($\diagdown\diagup$) being more effective than *trans* bonds ($\diagdown\diagdown$). This is because the form of the *cis*-double bond is more analogous to a permanent kink with a bond angle some 16° greater than that for a single saturated C—C bond, making it more difficult to pack in a crystalline array which requires an all *trans* configuration. In a detailed study of the position of the *cis*-double bond on the transition temperature of a phosphatidyl choline with two identical 18-carbon fatty acyl chains it has been shown that there is a minimum in the temperature and enthalpy of the phase transition when the double bond is in the 9,10 or 10,11 positions. Indeed, this agrees well with theoretical calculations that predict that with the double bonds in these positions the interaction potential energy (E_0) between adjacent chains will be at a minimum.

Naturally occurring phospholipids usually have one saturated and one unsaturated fatty acyl chain per molecule. Phospholipids with such an intramolecular mixture of fatty acyl chains exhibit sharp phase transitions at a temperature which is midway between the phase transition temperatures for the respective phospholipids with two identical acyl chains. This, however, is not true for mixtures (intermolecular) either of phospholipids with different acyl chains or with different headgroups but identical acyl chains (see section 2.10).

The other major factor influencing the lipid phase transition temperature is the nature of the headgroup of the phospholipid. Charge repulsion between adjacent negatively charged species can cause a lateral expansion of the lipid bilayer, favouring the liquid-crystalline state (see section 2.9). The conformation of the headgroup can also affect the phase transition by perturbing the packing of the bilayer in the crystalline state. For example, fully hydrated bilayers of phosphatidyl choline in the crystalline ($L_{\beta'}$) form have their acyl chains fully extended and tilted to the plane of the bilayer, whereas in phosphatidyl ethanolamine bilayers there is no such tilt exhibited by the acyl chains (L_{β}). The effect of this tilt is to lose the interchain interactions at the end of the molecule and reduce the effective chain length undergoing the phase transition, with a subsequent decrease in the phase transition temperature (T_t). This is amply demonstrated for dipalmitoyl phosphatidyl ethanolamine which has a T_t °C at 63 °C compared with dipalmitoyl phosphatidyl choline which has a T_t at 41 °C.

Summary

Phospholipid bilayers exist in two mesomorphic forms: a crystalline (solid) state and a liquid-crystalline (fluid) state.

At the lipid phase transition temperature the bilayer changes from the solid (L_{β}) to the fluid (L_{α}) state in a highly co-operative fashion

The nature of the headgroup, the length of the fatty acyl chain, and the number, position and type (*cis/trans*) of double bonds in the acyl chain of a phospholipid determine its phase transition temperature.

2.7 THE NATURE OF THE LIPID PHASE TRANSITION

The melting process of any organic compound can be shown to be associated with a variety of pre- and post-melting phenomena, and phospholipids are no exception to this. Indeed the melting of phospholipids cannot be described as a phase transition in the 'classical sense' as it does not occur at a single temperature. The lipid phase transition occurs over a narrow breadth of temperature, usually 1–2 °C, where the two phases co-exist. This change in state is continuous, being itself confined to a single phase, namely the phospholipid bilayer, and is believed to be first order as there are latent heat and volume changes associated with the transition.

The temperature at which the transition occurs and the width of the transition depend upon the individual phospholipid species. Indeed the form of the transition may be affected by low levels of chemical impurities in the phospholipid which might mask certain events such as a discontinuity exhibited by a true first-order transition. Lipid in the crystalline (solid) state has been shown by a variety of physical methods to exhibit a degree of deformation resulting from the occurrence of vacancies and defects or distortions in the lattice structure which will inevitably lead to a broadened phase transition. This is because defect formation is associated with changes in entropy, therefore its occurrence will increase with a rise in temperature, ensuring the co-existence of both solid and fluid lipid phases over a range of temperatures. Both esr and Raman spectroscopy have been used to quantitate these effects and it would appear that the fraction of disordered lipid within the crystalline state actually begins to increase some 15–20 °C below the temperature of the main transition. Indeed, with phosphatidylcholine which affords a change in the lattice structure of crystalline state lipid before the main transition ($L_{\beta'} \rightarrow P_{\beta'}$; see section 2.8), there is a marked reduction in the activation energy for the production of disordered lipid when the bilayer is in the crystalline, $P_{\beta'}$ state rather than the $L_{\beta'}$ state. This demonstrates that the probability of formation of disordered lipid is closely related to the packing of the bilayer in the solid state.

It is possible then to consider a structural model of the phase transition (Figure 2.9). As the temperature approaches that of the lipid phase transition, small pools of fluid lipid will form, centred at the points in the lattice where the lipid structure is more disorganized and exhibits a degree of rotational freedom. These pools of fluid lipid will be essentially contained by the overall order of the lattice of crystalline lipid, and hence both fluid and solid lipid will co-exist in the same phase over a relatively broad temperature range compared with the classical definition of a phase transition. The size of these pools will increase as the temperature rises, leading to their coalescence, which would provide the transformation from the crystalline to the liquid-crystalline state. It is implicit in such a description of the phase transition that after the initial melting of lipid at these defect sites, the subsequent melting of the rest of the bilayer would be a highly co-operative event. This would occur

58

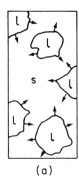

(a)   (b)

Figure 2.9   A structural model of the lipid phase transition. A hybrid single bilayer near the phase transition temperature, showing lipid in the crystalline phase under compression (a) and lipid in the gel phase under tension (b), since the molar volume of lipid in the fluid state is greater than in the gel state. This indicates the situations that might be encountered on melting (a) and solidifying (b). (Reprinted with permission from Lee, A. G. (1977) *Biochem. Biophys. Acta*, **472**, 237–281)

extremely rapidly, being propagated by the fast lateral diffusion of fluid lipid species.

The growth of these fluid lipid pools within the long-range order of the lattice will be subjected to energy barriers provided by the boundary surface of the crystalline-state lipid and presumably to the lateral diffusion of the lipid.

Approaching the lipid phase transition from a higher temperature, the 'freezing' phenomenon is characterized by the formation of clusters or islands of solid lipid within the bulk liquid-crystalline (fluid) phase. The packing of the lipid molecules in such clusters is likely to be variable because of the disorder of the liquid-crystalline phase. Some clusters will be too irregular to pack and crystallize into a regular lattice, whereas clusters of a regular array will seed the conversion of the whole of the bilayer from the liquid-crystalline to the crystalline state. As the phase transition is approached, the growth of lipid clusters and their aggregation takes the form of a co-operative change from fluid to solid phase lipid. However, within the solid phase lipid will be found the smaller 'non-crystallizable' clusters which will yield the nuclei of lattice defects which will form some of the centres for the pre-melting phenomena.

This structural model of the lipid phase transition (Figure 2.9) leads to the possibility of hysteresis occurring (Figure 2.10), since the pathway of the transition from the solid to the fluid state will not follow exactly the converse route. As one approaches the transition temperature from below, the fluid lipid pools that form will be under pressure as their molar volume is greater than that of the solid phase lipid in which they are contained. However, on cooling, fluid phase lipid clusters or islands of solid phase lipid will form within the bulk fluid phase, and as the solid lipid is more closely packed then

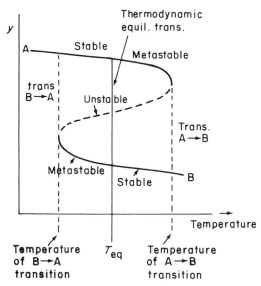

Figure 2.10 Hysteresis and the phase transition. This schematic demonstrates stable, metastable and unstable states of A and B in a hysteresis system. Hysteresis occurs if a process occurring in one direction fails to retrace its steps when going in the opposite direction. For example, if we follow the B to A transformation, then we would normally expect the transformation to occur at the temperature at which thermodynamic equilibrium occurs (T_{eq}). This process would show no hysteresis as the A to B transformation will follow an identical pathway. However, if B can attain a metastable state then an appreciably lower temperature has to be reached before the transformation occurs. Conversely, if A can attain a metastable state then a temperature greater than T_{eq} must be reached for the conversion. Consequently hysteresis occurs. (Reprinted with permission from Lee, A. G. (1977) *Biochem. Biophys. Acta,* **472,** 237–281)

it will tend to be under tension. In each of these cases, then, the free energy of the domains that are being formed within the bulk lipid phase will be different when one approaches the lipid phase transition from either above or below. In practice, hysteresis is slight for neutral phospholipid bilayers, but can be quite marked in bilayers consisting of charged lipids or mixtures of different lipids.

Essentially, three types of lipid domains exist over the transition: these are the solid (L_β) domain, the fluid (L_α) domain and a domain of so-called 'interfacial lipid' that forms the region juxtaposing the L_α and L_β domains (Figure 2.11). The lipid molecules in each of these three states will exhibit different free energies which will be derived from the rotational isomerization about C—C bonds in the acyl chains; van der Waals interactions between the acyl chains of neighbouring molecules; electrostatic interactions between adjacent polar headgroup regions and the overall molecular motion exhibited by the acyl chains as a whole. It is likely that the free energy of the lipid in the

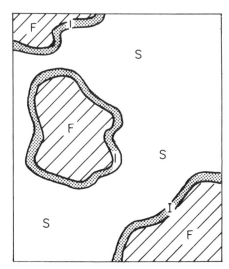

Figure 2.11 Interfacial lipid. At the phase transition both solid (L_β) and fluid (L_α) lipid will co-exist. This shows the lateral configuration of a lipid surface at this point. At the boundary between the solid, ordered lipid (S) and the fluid lipid (F) there will be areas of interfacial lipid (I) which are undergoing the transition. The areas of mismatch of molecular packing that occur at this interfacial region accounts for the dramatic increase in permeability to small molecules that occurs at the phase transition. (Reprinted with permission from Marsh, D., Watts, A., and Knowles, P. F. (1976) *Biochemistry*, 15, 3570–3578. Copyright (1976) American Chemical Society)

interfacial region will be the highest as there will be a contribution from the mismatch occurring between the molecular packing of the ordered and fluid domains. Thus the system will wish to reduce the number of lipid molecules occurring at such phase boundaries, favouring a rapid highly co-operative transition from the fluid to solid states.

The phenomenon of interfacial lipid arising from the mismatch of the solid and fluid lipid domains appears to account for the observation that the passive permeability of lipid bilayers to ions, e.g. K^+ and small molecules like sucrose and TEMPO-choline, reaches a maximum at a temperature corresponding to the lipid phase transition. Indeed, it appears that the degree of permeability closely follows the amount of interfacial lipid present, and it has been calculated that the permeability of dimyristoyl phosphatidylcholine vesicles to TEMPO-choline at the centre of the lipid phase transition is over 30 times that of the overall vesicle permeability. All this is presumably due to the mismatch in packing that occurs at the interface between the solid and fluid domains that co-exist over the phase transition. Indeed the susceptibility of vesicles made from a single type of phospholipid to exhibit a dramatic leakage close to the phase transition may explain why biological membranes contain complex mixtures of lipids in order to avoid the co-existence of solid and fluid phases under normal growth conditions (see sections 2.10, 11 and 3.5).

Summary

The lipid phase transition is a first-order process.

Solid and fluid lipid phases co-exist over the narrow breadth of temperature of the phase transition.

Various pre- and post-melting phenomena characterize the phase transition which is confined to the lipid bilayer.

2.8 THE PRE-TRANSITION OF PHOSPHATIDYL CHOLINES

Phosphatidyl choline and phosphatidyl glycerol, unlike other types of phospholipids, exhibit a rather broad first-order transition with a small endothermic event that precedes the large endothermic phase transition. This event, which exhibits considerable hysteresis, has been termed the pre-transition. No doubt part of the small enthalpy change that occurs prior to the main transition reflects the pre-melting phenomena discussed in section 2.7, but the pre-transition itself constitutes a defined structural change in the bilayer. Here we shall concern ourselves with studies on phosphatidyl choline as it is a major constituent of many biological membranes and has been studied in great detail.

At temperatures below the pre-transition, the lipids of the bilayer form a one-dimensional lamellar lattice with the acyl chains fully extended and tilted with respect to the plane of the bilayer ($L_{\beta'}$). This angle of tilt appears to be temperature dependent and falls to a minimum at the pre-transition temperature where a structural transformation to a two-dimensional mono-clinic lattice occurs ($P_{\beta'}$). This consists of a lipid lamellae distorted by a periodic ripple (Figure 2.12). The acyl chains remain tilted in this way until the main transition is reached where they assume a liquid-like conformation and the lattice reverts to a one-dimensional lamellar structure. These changes in state have been inferred from X-ray diffraction (XRD) data and have been detected by nmr and Raman spectroscopy and by their very different appearance in freeze–fracture studies. Thus, lipid in this intermediate temperature region forms a monoclinic structure (P) where all the molecules do not lie in a single plane as in the normal lamellar (L) form of the bilayer. The fatty acyl chains are, however, still fully extended in the all-*trans* state. Thus, on raising the temperature of crystalline, phosphatidyl choline bilayers a change, $L_{\beta'} \rightarrow P_{\beta'} \rightarrow L_{\alpha}$, is observed.

The $P_{\beta'}$, form requires that substantial changes in molecular positions be made and co-ordinated over fairly long distances. Such a long-range reorganization of an ordered matrix might be expected to be slow. Indeed the midpoint temperature of this event has been demonstrated to vary, dependent upon the rate of heating or cooling of the sample. Using multilamellar vesicles of dipalmitoyl phosphatidyl choline it is possible to demonstrate characteristic half-times for the pre-transition of the order of 3–11 min. This phenomenon may account for the marked hysteresis exhibited for the pre-transition peak, due to the formation of metastable states.

62

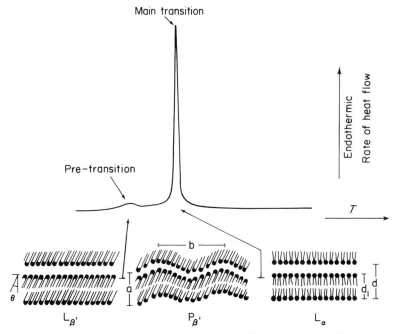

Figure 2.12   Pre-transition of phosphatidyl cholines. Phosphatidyl cholines exhibit a small rather broad endothermic event that precedes the main transition. This has been called the pre-transition. X-ray diffraction studies have been used to demonstrate that at temperatures below the pre-transition the acyl chains are fully extended and tilted with respect to the plane of the bilayer ($L_{\beta'}$), but packed in a distorted quasi-hexagonal lattice. As the temperature approaches the pre-transition this is associated with a transformation from a one- to a two-dimensional monoclinic lattice. This consists of lipid lamallae distorted by a periodic ripple with the acyl chains again tilted with respect to the plane of the bilayer ($P_{\beta'}$) and packed in a hexagonal array. Above the main transition the lattice reverts to a one-dimensional lamellar structure with disordered acyl chains (L_{α}). (Reprinted with permission from Janiak, M. J., Small, D. M., and Shipley, G. G. (1976) *Biochemistry*, 15, 4575–4580. Copyright (1976) American Chemical Society)

The midpoint temperature of the pre-transition obtained by fluorescence or esr studies is usually 1–3 °C lower than that obtained by differential scanning calorimetry (DSC). This is because the esr and fluorescence probes used to monitor this event actually perturb it, perhaps by partitioning more favourably into the $P_{\beta'}$ phase than the $L_{\beta'}$. Indeed, the fluorescence depolarization probe, 1,6-diphenyl-1,3,5-hexatriene, lowers the midpoint temperature of the pre-transition when present in DSC studies.

Summary

Phosphatidyl cholines exhibit a broad event of low enthalpy change that precedes the main transition: this is called the pre-transition.

This pre-transition is due to a structural reorganization of the bilayer into a two-dimensional monoclinic lattice that is stable at temperatures between the pre-and the main transition.

2.9 PHASE TRANSITIONS IN BILAYERS OF NEGATIVELY CHARGED PHOSPHOLIPIDS

Phospholipids bearing a net negative charge form a small but significant fraction of the lipid pool of most biological membranes. These lipids are phosphatidyl serine, phosphatidyl glycerol and phosphatidyl inositol which bear a net single negative charge, whereas phosphatidic acid and cardiolipin display two net negative charges.

There is a tendency for such negatively charged molecules to increase the area they occupy in order to decrease lateral repulsion between adjacent molecules and make the surface charge more diffuse. Increasing the pH and hence the degree of ionization results in an increase in the area occupied by individual molecules, favouring the liquid-crystalline state of the bilayer where there is a lateral expansion. This means that the lipid phase transition temperature of these bilayers is extremely sensitive to changes in pH, cation concentration and ionic strength. Thus, it is possible by manipulating these parameters to induce a lipid phase transition isothermally in a bilayer composed of these charged species. Increases in the pH will depress the lipid phase transition temperature, whereas increases in either ionic strength or cation concentration, both of which will reduce the bilayer surface charge, elevate the lipid phase transition temperature by promoting the crystalline state.

Recent XRD analysis of charged lipid bilayers has demonstrated that as surface charge increases and the lateral packing of the headgroups decreases leading to bilayer expansion, the acyl chains only partly follow their example due to the attractive van der Waals interaction between them. In order, then, to minimize a change in packing they tilt so as to maintain the interchain distance. In doing so they decrease the bilayer thickness and effective interacting length of the acyl chain undergoing the ordered to disordered transition. The decreases in T_t observed can be shown to be due both to the structural (tilt) changes in the form of the bilayer and the change in the electrostatic field due to lateral expansion of the bilayer. It would appear that these effects are of comparable magnitude in achieving a depression of T_t °C.

In aqueous solutions charged lipids will attract oppositely charged ions and repel similarly charged ones, forming an electrical double layer. This is because the region of the solution closest to the bilayer will contain more of the oppositely charged ion (Figure 2.13). The thickness of this layer will depend upon the ionic strength of the medium and can range from about 40 nm at 10^{-4} M salt to only 1 nm at 10^{-1} M salt. However, if the bilayer consists of a proportion of negatively charged phospholipids, then there will be an accumulation of H^+ at its surface. This results in a surface pH which is

64

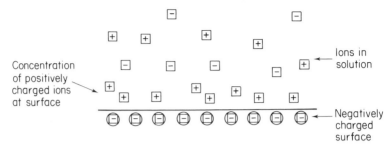

Figure 2.13 Electrical double layer. At a finite temperature a charged lipid bilayer in solution will attract ions of opposite charge and repel ions of a similar charge. This region of unequal negative and positive charge is referred to as an electrical double layer. Changes in the arrangement of surface charge, caused for example by a lipid phase transition, will affect the electrical double layer. This can not only modulate the phase transition by inducing hysteresis but in biological membranes may affect protein functioning by altering the surface pH

considerably lower than the bulk solution pH and can of course have a marked effect on the ionization of groups exposed at the surface. For example, the ionization of the first proton from a phosphate group is usually ascribed a pK of 2, but when associated with a negatively charged surface it exhibits a pK of 6. Changes in the electrical double layer may be expected to alter the thermodynamic properties of the phase transition and vice versa, but under physiological conditions by far the most important influence is that of divalent cations.

Here, Ca^{2+} and Mg^{2+} bind stoichiometrically to acidic phospholipids, reducing their surface charge and area, thus increasing the phase transition temperature. Their binding is influenced by surface charge, ionic strength and pH and is maximal when the lipids are in the close-packed crystalline state. Increase in Ca^{2+} concentration affects both the surface charge and the transition temperature in a linear fashion until the maximum level of one Ca^{2+} is bound per two lipid molecules. By binding to acidic phospholipids, Ca^{2+}, can lead to the production of solid 'clusters' of lipid at temperatures above the phase transition. These can co-exist within the bulk fluid (L_α) phase lipid. Their formation might explain the effect of Ca^{2+} in enhancing the hysteresis that is exhibited by the lipid phase transition of acidic phospholipids.

Acidic phospholipids are particularly sensitive to changes in their environment. The fluxes of cations, protons or changes in membrane potential that surround many cellular events may well lead to alterations in membrane structure and function by perturbing acidic phospholipid species. An interesting possibility that has yet to be experimentally explored is that the surface electrostatics on one side of the bilayer might affect those on the opposite side. This relies on the dependence of the surface potential on one side of the membrane being linked to the charge density on the other side. Thus, even in the absence of net charges on one side of the membrane, if there are net charges on the other then both sides will exhibit surface potentials that will be intrinsically coupled to each other. Changes in the charge density on one side, caused for instance by the binding of cation to a negatively charged group, will influence charge reactions on the other side of the bilayer. This in effect means that it is possible to transmit information vectorially across the bilayer, and implies that changes to the external surface of the plasma membrane may affect events happening internally. The properties of charged lipids and their interactions with cations and proteins (see Chapter 3) may prove to be of importance to our understanding of the functioning of biological membranes.

Summary

The phase transition of negatively charged phospholipids is markedly influenced by changes in pH, ionic strength and divalent cation concentration.

2.10 PHASE SEPARATIONS IN MIXTURES OF PURE PHOSPHOLIPIDS

Biological membranes are a complex mixture of phospholipid species which vary with respect to both their lipid headgroup and to their associated acyl chains. As a result they do not exhibit a single well-defined lipid phase transition where all of the lipid progresses from a solid to a fluid state at a well-defined temperature. Rather, they exhibit upon cooling what are termed lipid phase separations, which as one might expect are rather complex events. Effectively, a lipid phase separation in a biological membrane results in the co-existence, over a particular temperature range, of two or more distinct types of lipid phases with different chemical compositions. Most often the primary distinguishing feature between these two lipid phases is that one consists of solid-phase lipid and the other fluid-phase lipid. However, fluid–fluid and solid–solid phase immiscibility has been noted.

It is worth examining simple binary mixtures of pure phospholipids in order to appreciate some of the phenomena that are likely to be encountered in the much more complex situation found with biological membranes. In any study involving mixing phenomena it is important to appreciate that it can take a relatively long time for equilibrium to take place between solid phases where diffusion is slow and fluid phases where diffusion is rapid. Artefactual results can easily ensue if rapid changes in temperature are imposed on the system with little time taken for true equilibrium to be reached. As in most biological situations we assume a constant pressure and ignore water as a separate phase as it is in great excess.

If we consider a simple eutectic system of two components which exhibit no solution in the solid state but mix freely in the liquid state, yielding an ideal solution, then a typical phase diagram of the type shown in Figure 2.14 will be obtained. Three phases will be obtained on varying either the temperature or the relative proportions of X and Y. These will be pure solid X, pure solid Y

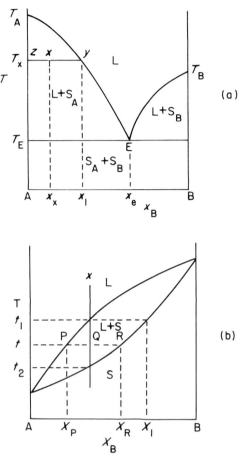

Figure 2.14 Two component mixtures. (a) Simple eutectic system of two components. Here the components form no solution in the solid state (S), but ideal mixing in the fluid state. Thus the only phases are a liquid and the two solid crystalline components, all of which are at equilibrium at a eutectic point. (b) A phase diagram for a two-component system with complete miscibility in the liquid and solid phases. As the proportions of A and B are varied then phases of a solid solution of A and B, a liquid of A and B and co-existing solid plus fluid phases are formed dependent upon the temperature. For a mixture of composition x at temperatures below t_1 then the liquid pool will be increasingly depleted of the higher melting point component (B) until a temperature t_2, is reached when a solid solution only exists. (Reprinted with permission from Lee, A. G. (1977) *Biochim. Biophys. Acta*, 472, 237–281)

and liquid, all of which will be in equilibrium at the eutectic point of fixed temperature and composition. The phase diagram shows an area, L, where liquid phase alone is stable, and an area X + Y where pure solid X and pure solid Y alone are stable and form a mechanical mixture of the two components. Also there will be two regions where the liquid phase will be in equilibrium with either pure solid X or pure solid Y, and in such regions the proportion of solid to liquid will vary on changing either the temperature or the composition. Furthermore, the composition of the liquid phase will vary with both temperature and relative amounts of X and Y. As one might expect, such a rigid state of affairs is unlikely to arise because no lipid preparations are going to be completely pure and one might always expect to see a solid solution of X in Y at extreme dilutions.

Lipid phase separations occurring in binary mixtures of phospholipids have been investigated using a variety of physical techniques. The correlation between the results that have been obtained is really very good, especially as these methods all have to face the inherent problem of defining the points at which all of the lipid attains either the solid or fluid state. These points correlate with, for example, the initiation of an endergonic event in DSC or an abrupt change in TEMPO solubility.

From a variety of studies that have been carried out it would seem possible to make certain predictions about the form of the eutectic diagrams that will be given by binary phospholipid mixtures. If the two lipids have identical headgroups and their fatty acyl chains are of similar length, for example dimyristoyl phosphatidyl choline (C 14 : 0) and dipalmitoyl phosphatidyl choline (C 16 : 0), then mixing is close to ideal in both the solid and liquid state. This gives rise to an extremely simple phase diagram (Figure 2.15) where there are only three phases: liquid, liquid plus solid and solid. No separation of either pure DMPC or pure DPPC lipid phases occurs as the lipids are structurally very closely related which allows them to pack together within crystals. Areas of lipid in the fluid and solid state that occur at various temperatures with different mixtures can also be identified using freeze–fracture electron microscopy (Figure 2.16).

The consequences of lipid phase separations can be appreciated in a relatively simplistic fashion by referral to Figure 2.14b. Consider a mixture of composition given at x. At all temperatures above t_1 there will be a completely miscible state of fluid lipids. However, as the temperature falls below t_1 then both solid and fluid lipid phases will co-exist until t_2 is reached when a homogeneous solid state is attained. One important consequence of the lipid phase separation is that when both solid and fluid lipids co-exist, the relative proportions of the different lipids within them will not be the same. At temperatures close to t_1 the solid phase will be enriched with the higher melting point lipid, but as the temperature falls towards t_2 those solidifying will contain an increasing proportion of the lower melting point lipid. Thus the occurrence of a lipid phase separation leads to temperature-dependent changes in the composition of the fluid and solid lipid pools.

With mixtures of lipids having the same headgroup but very different fatty acyl chains, then packing in the solid state will be different, leading to solid-phase immiscibility. This is characterized by the formation of heterogenous solid-phase lipid containing areas of pure lipid components as well as solid solutions. Solid-state immiscibility is also seen with mixtures of lipids having different headgroups due to their very different packing arrangements in the

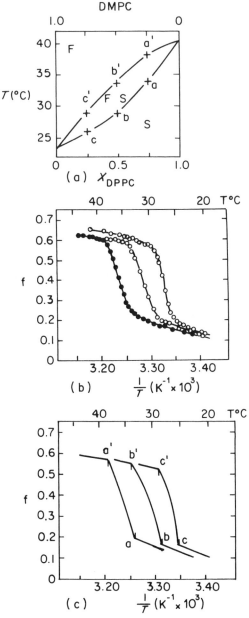

Figure 2.15 A phase diagram for a binary mixture of dipalmitoyl phosphatidyl choline and dimyristoyl phosphatidyl choline. DMPC and DPPC both have identical headgroups and possess acyl chains of similar length. They form ideal mixtures in both the solid (S) and fluid (F) states, yielding a simple phase diagram (diagram a). The onset (a, b, c) and completion (a′, b′, c′) of the phase separation is detected by the solubility (*f*) of TEMPO in fluid phase lipid (diagram b). This exhibits broadened transition curves from which the low-temperature onset (a, b, c) and high-temperature completion (a′, b′, c′) of the phase separation can be defined (diagram c). (Reprinted with permission from Shimshick, E. J., and McConnell, H. M. (1973) *Biochemistry*, 12, 2351–2360. Copyright (1973) American Chemical Society)

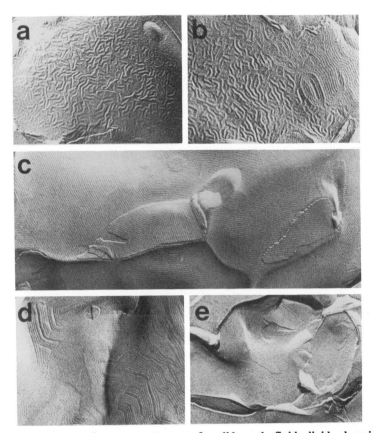

Figure 2.16 Freeze–fracture patterns of solid and fluid lipid domains in DMPC–DPPC mixtures. (a) Fluid lipid only. A mixture of DMPC : DPPC (4 : 1) quenched from 35 °C exhibits a jumbled texture characteristic of lipid in the liquid-crystalline state. (b) A mixture of DMPC : DPPC (1 : 1) quenched from 34 °C exhibits predominantly jumbled L_α lipid as would be expected from TEMPO analysis (Figure 2.15). However, on the right of this micrograph some areas of banded lipid are evident, suggesting the appearance of solid, $P_{\beta'}$ lipid. (c) DMPC : DPPC (1 : 4) from 35 °C. These banded arrays are characteristic of solid, $P_{\beta'}$ lipid. (d) DMPC : DPPC (1 : 4) from 27 °C. Contains a mixture of unmarked, solid ($L_{\beta'}$) lipid together with banded ($P_{\beta'}$) lipid. (e) DMPC : DPPC (1 : 4) from 19 °C. Unmarked fracture faces characteristic of solid ($L_{\beta'}$) lipid. This analysis implies that the solid lipid that co-exists with the fluid lipid domains in such mixtures (see Figure 2.15) is in the $P_{\beta'}$ state rather than the $L_{\beta'}$ state. (Reprinted with permission from Luna, E. J., and McConnell, H. M. (1978) *Biochem. Biophys. Acta*, 509, 462–473)

crystalline state. The one exception to this arises with mixtures of phosphatidyl choline and phosphatidyl glycerol where the acyl chains are identical. In this instance both types of lipid appear to pack identically in the solid state, producing ideal mixtures.

Whilst immiscibility in the solid state is a common phenomenon, immiscibility in the fluid state is comparatively rare. It has, however, been identified in mixtures of dielaidoyl phosphatidyl choline and dipalmitoyl phosphatidyl ethanolamine (Figure 2.15) which exhibit a limited fluid phase immiscibility as well as the expected solid-phase immiscibility. The occurrence of fluid-phase immiscibility may have consequences for the organization of membranes (see sections 2.13 and 3.3).

These simple analyses of mixtures containing phosphatidyl cholines take no account of the pre-transition and hence the formation of monoclinic, $P_{\beta'}$ lipid (see section 2.8). Unfortunately, only crude assessments of its contribution have been made using freeze–fracture electron microscopy to identify it. This is clearly unsatisfactory. However, in mixtures containing >20 mol per cent of phosphatidyl ethanolamines, there is no evidence for the formation of $P_{\beta'}$ lipid in the phosphatidyl choline. Indeed, $P_{\beta'}$ lipid is only found in pure phosphatidyl choline mixtures and not in those where any other headgroup phospholipid is present. This suggests that the pre-transition, which is associated with changes in headgroup orientation, is a highly co-operative event easily perturbed by alterations in long-range lipid order caused by the presence of lipids that do not respond similarly. Even though many biological membranes contain a high proportion of phosphatidyl cholines, the presence of a variety of other lipids would be expected to prevent the production of $P_{\beta'}$ phase lipid.

Rather interesting effects can be observed with mixtures containing negatively charged phospholipids in that treatment with divalent cations, in particular Ca^{2+}, can induce an isothermal lipid phase separation. Thus a domain of the Ca^{2+}–solid-phase complex of acidic phospholipids can be formed within the bulk fluid phase of neutral lipids due to the ability of Ca^{2+} to raise the transition temperature of acidic phospholipids (see section 2.9). Mixtures of neutral phospholipids with either phosphatidyl serine or phosphatidic acid are particularly susceptible to this effect and will also be influenced by changes in pH and ionic strength.

Phase separations are a common property of mixtures of phospholipids, and indicate that both fluid and solid domains of lipid can co-exist over a range of temperatures. The composition of the fluid and solid domains will be different and influenced by changes in temperature. Furthermore, if charged phospholipids are present the composition of the domains will also be determined by cations, pH and ionic strength. The wide variation of phospholipids found in biological membranes is likely to show non-ideal mixing. This will certainly be true of the solid state where the various headgroups exhibit different conformations and hence will not pack together in crystals but form various mixtures and solid solutions. It is also possible that non-ideal mixing will occur in the fluid state yielding fluid–fluid immiscibility (Figure 2.17). All of these phenomena may modulate the nature of the environment of protein species embedded in the membrane (see sections 2.13 and 3.3).

Summary

Mixtures of phospholipids will undergo phase separations that are influenced by temperature and composition, yielding solid, liquid and solid plus liquid states.

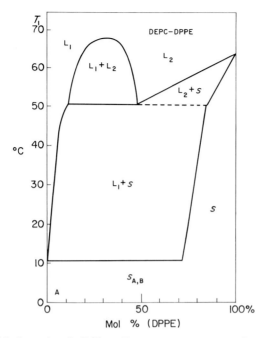

Figure 2.17 Fluid-phase immiscibility. Even two-component phase diagrams can become considerably more complex than that observed for DPPC : DMPC mixtures. This shows the result of mixing dielaidoyl phosphatidyl choline (DEPC) with dipalmitoyl phosphatidyl ethanolamine (DPPE). The feature of particular interest is that whereas these lipids are both miscible in the solid state, under certain conditions they are immiscible in the fluid state. This gives rise to co-existing domains of pure fluid DPPE and pure fluid DEPC, as well as conditions where either pure fluid DPPE or pure fluid DEPC are present with a solid mixture of DPPE and DEPC. (Reprinted with permission from Wu, S. H.-W., and McConnell, H. M. (1975) *Biochemistry,* **14, 847–854. Copyright (1975) American Chemical Society)**

Mixing in the solid state is only ideal if the component species pack in very similar fashions.

Non-ideal mixing in the solid state yields a mechanical mixture of the pure components together with the solid solution.

2.11 THE INTEGRATION OF CHOLESTEROL INTO PHOSPHOLIPID BILAYERS

Cholesterol is an extremely important constituent of many biological membranes where it often constitutes some 6–50 mol per cent of the total lipid pool.

The structure of cholesterol is discussed in Chapter 1. Briefly, it consists of a rigid sterol ring some 1.1 nm long with a 3β-hydroxyl group attached at one end. At the other end is a highly flexible hydrocarbon chain some 0.8 nm

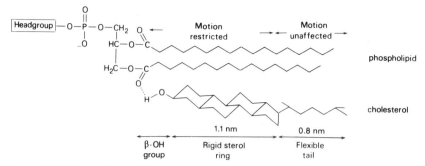

Figure 2.18 **Insertion of cholesterol into a phospholipid bilayer.** A schematic representation of the alignment of a cholesterol molecule with a phospholipid in a lipid bilayer

long. This molecule inserts into the bilayer with its hydroxyl group orientated towards the polar headgroup region and the hydrocarbon tail towards the apolar core of the bilayer (Figure 2.18). The long axis of the cholesterol molecule is thus orientated perpendicular to the bilayer plane. The precise orientation of the cholesterol molecule within the bilayer has been the subject of some controversy. However, it is now believed that the cholesterol molecule is inserted rather deeply into the bilayer. In this way its β-hydroxyl group can to some extent participate in hydrogen bonding with the carbonyl oxygen linking the phospholipid fatty acyl chain with its glycerol backbone (Figure 2.19). The rigid steroid nucleus will then interact closely with adjacent fatty acyl chains inhibiting the formation of kinks and reducing the number of *gauche* conformations. This leads to cholesterol having a condensing effect on the acyl chains of bilayers in the fluid state due to the reduction in the mean molecular area of the interacting components. This effect is severely reduced in bilayers containing lipid analogues which lack the carbonyl group, supporting the concept of a specific interaction of cholesterol with the phospholipids.

In direct contrast to the condensing effect cholesterol has on the acyl chain region of the bilayer, the interpolation of the hydroxyl group into the headgroup region of the bilayer leads to an increase in the separation between the polar headgroups until the intermolecular (electrostatic) interactions between them can no longer occur. This causes a marginal increase in the freedom of motion of the headgroup terminal moiety and is responsible for the subsequent increase in hydration of the bilayer. The extent of the increase in hydration observed depends on the chemical nature of the phospholipid constituting the bilayer. It would seem that the initial degree of hydration of the bilayer is inversely proportional to the strength of the intermolecular forces between the headgroups and thus to the extent at which these charged groups are accessible to water. Phosphatidyl ethanolamine bilayers, for example, exhibit a greater increase in hydration than do bilayers of phosphatidyl choline. This is because the headgroups of phosphatidyl

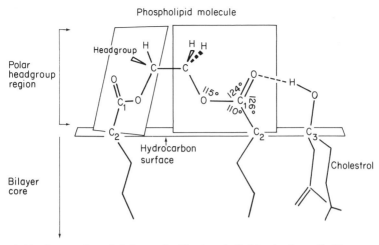

Figure 2.19 Interaction of cholesterol with phospholipids. A schematic diagram showing the interface region of the bilayer composed of PC and cholesterol molecules. The
planarity of the structural element $-COC_1OC_2-$ of the phospholipid molecule, and the
proposed hydrogen-bond formation between the cholesterol β-hydroxyl group and the
phospholipid carbonyl group are drawn diagramatically. (Reprinted with permission
from Huang, C.-H. (1977) *Lipids*, 12, 348–356)

ethanolamine bilayers are more tightly associated and exhibit a much reduced
motional freedom compared with phosphatidyl choline bilayers. The actual
conformation of both these headgroups appears to be little affected by the
incorporation of cholesterol into the bilayer as the zwitterionic groups remain
aligned parallel to the bilayer surface.

Cholesterol has then a condensing effect on the acyl chain region of fluid
lipid bilayers, producing a more ordered, rigid structure. However, when fluid
lipid bilayers containing cholesterol are cooled to temperatures below the
lipid phase transition temperature of the pure phospholipid, the phospholipids
do not pack together to form a solid phase excluding cholesterol. Instead,
cholesterol remains intimately associated with the phospholipids, preventing their acyl chains adopting the all-*trans* configuration which allows
them to pack together. This leads to a decrease in the fraction of acyl chains
found in the all-*trans* state at temperatures below the phase transition
temperature, and hence an increase in bilayer fluidity under these conditions.
Increasing amounts of cholesterol incorporated into the bilayer result in the
abolition of the highly co-operative phase transition. At 50 mol per cent
cholesterol, the normally sharp endergonic transition is completely smeared
out (Figure 2.20), being replaced by a gradual change from a relatively
ordered structure to a relatively disordered structure which occurs over a
wide temperature range. This gradual transition is not a first-order process
because the enthalpy change associated with the phase transition is progressively reduced to zero as the cholesterol content is increased. At the

74

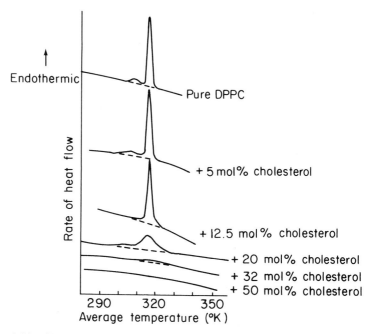

Figure 2.20 Cholesterol abolishes the endothermic lipid phase transition of phospholipids. By increasing the cholesterol content of phospholipid bilayers the endothermic lipid phase transition is gradually moved to lower temperatures, progressively broadened and eventually completely smeared out. Concentrations above 10 mol per cent suffice to abolish the pre-transition of phosphatidyl cholines and concentrations at or above 50 mol per cent completely abolish the endothermic event. These data show the DSC curves for dipalmitoyl phosphatidyl choline–cholesterol mixtures. Similar results can be observed using esr and other physical methods. Increase in temperature, at high cholesterol concentrations, cause a change from a 'relatively ordered' state to a 'relatively disordered' state of the phospholipids. (Reprinted with permission from Ladbroke, B. D., Williams, R. M., and Chapman, D. (1968). *Biochem. Biophys. Acta*, 150, 333–340)

same time cholesterol abolishes the weakly endergonic pre-transition of phosphatidyl cholines.

The observed effect of cholesterol on lipid bilayers is thus highly dependent on whether the temperature is above or below that at which the phase transition of the pure phospholipid would have occurred. At temperatures above this point, cholesterol increases bilayer rigidity by ordering the fluid-state lipid, whereas at temperatures below this point it increases bilayer fluidity by disordering the solid-state lipid. These apparently opposing effects lead to the smearing and eventual abolition of the lipid phase transition.

These properties of cholesterol are inherently related to its structure. The loss of its hydrocarbon chain at C17 to give androstan-3β-ol shows no condensing effect on phospholipid bilayers, and epicholesterol with a 3α-OH shows insignificant interaction. DSC studies confirm this as 20 mol per cent epicholesterol leaves the transition temperature of phosphatidyl choline and its associated enthalpy change unaffected. However, androstan-3β-ol does appear to broaden the shape of the transition and shift it to lower temperatures. All three regions of the cholesterol molecule are important for its function. The equitorially orientated β-hydroxyl group defines the alignment of the molecule within the bilayer. The rigid, puckered planar ring then interacts with the adjacent fatty acyl chains, restricting their motion in the fluid state, yet preventing them packing in the all-*trans* configuration in the solid state. The hydrocarbon tail, whilst having only small effects on the motion of the acyl chain terminal methyl groups is essential in the expression of the condensing effect of cholesterol on fluid lipid bilayers by promoting van der Waals contact between adjacent acyl chains. It will thus have a very marked effect on mobility gradients within the bilayer, all of which will be co-ordinated by the precise integration of cholesterol within the bilayer achieved by virtue of its β-hydroxyl group.

In binary mixtures of cholesterol and phosphatidyl cholines it has been suggested that at less than equimolar levels of cholesterol, short-lived 1 : 1 complexes of cholesterol and phospholipid can form. Changes in slope occurring at about 22 and 30 mol per cent on the solidus curve have been interpreted as reflecting respectively the non-ideal mixing of these complexes with pure lipid and the appearance of a highly disordered lipid state at high cholesterol levels.

An alternative explanation for the association of cholesterol with bilayer phospholipid has been mooted to explain both the variety of changes in physical properties associated with the bilayer and the occurrence of phase boundaries at around 22 and 30 mol per cent cholesterol. Here it is proposed that at concentrations below 22 mol per cent cholesterol each molecule is surrounded by 7 unshared acyl chains, requiring an average of 3.5 phospholipid molecules. One can readily calculate that the maximum incorporation of cholesterol into a bilayer whilst maintaining this structure is 22 mol per cent, suggesting that a change in the organization occurs above this point. It is then suggested that cholesterol–cholesterol dimer formation ensues. These dimers are surrounded by 9 unshared acyl chains, where the cholesterol molecules may interact either through van der Waals interactions between the two β faces or through close contact between the planar α faces. In both these structures the β-OH groups of the cholesterol molecule would remain free to hydrogen bond with adjacent phospholipids. Such an organization could accommodate up to 31 mol per cent cholesterol in the bilayer, above which one might expect the dimers to begin sharing acyl chains until finally the system becomes unstable and large aggregates of pure cholesterol begin to separate out as a separate phase. This model is attractive because one can mathematically predict the occurrence of discontinuities at specific mole fractions of cholesterol, where they have indeed been found to occur. Effectively, the only significant difference between this and other models is the formation of cholesterol dimers, the occurrence of which may hopefully be resolved

76

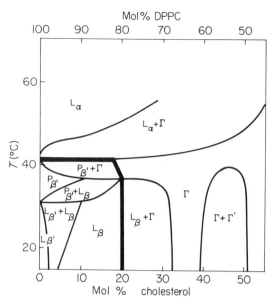

Figure 2.21 A phase diagram for cholesterol–DPPC mixtures. The addition of cholesterol to a phospholipid bilayer gives rise to a highly complicated phase diagram, even in a 'simple' mixture of cholesterol and a single phospholipid species. For explanation see text. (Reprinted with permission from Lentz, B. R., Barrow, D. A., and Hoechli, M. (1980) *Biochemistry*, 19, 1943–1954. Copyright (1980) American Chemical Society)

using ^{13}C nmr studies. These should enable a rotational relaxation time to be calculated for cholesterol, which should be sensitive to dimer formation.

A phase diagram summarizing the information obtained from a variety of different physical studies carried out on dipalmitoyl phosphatidyl choline (DPPC)–cholesterol mixtures is shown in Figure 2.21. There are three particularly notable features of this rather complex diagram. The first is that cholesterol (>5 mol per cent) can affect the angle of tilt of the fatty acid chains of solid ($L_{\beta'}$) phosphatidyl choline. Presumably the integration into the bilayer core of the rigid sterol ring which is aligned by virtue of the interaction of its β-hydroxyl group with the phospholipid overcomes the tendency of the acyl chains to tilt to the bilayer normal. This is defined as an $L_{\beta'} \rightarrow L_{\beta}$ transition where the fatty acid chains now align extended parallel to the bilayer normal. The second notable occurrence is the production of a new structure (Γ) occurring at high cholesterol content. The detailed molecular structure of Γ is not known, but it would seem that it is considerably disordered. It then has physical properties more akin to the L_{α} (fluid) than the L_{β} or $L_{\beta'}$ (solid) phase lipid. The third notable feature is the abolition of the pre-transition at concentrations >20 mol per cent cholesterol. Normally, over the temperature range at which the pre-transition occurs (see section 2.9) the bilayer becomes distorted into a monoclinic lattice with a periodic ripple ($P_{\beta'}$). In freeze–frac-

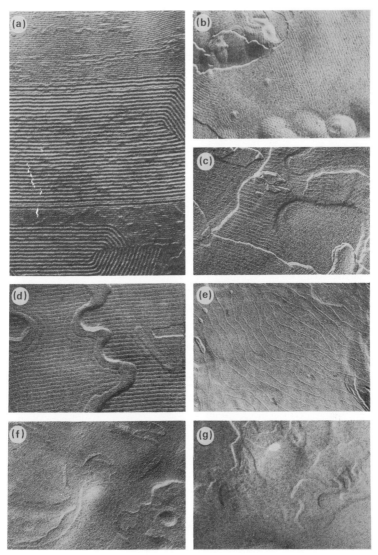

Figure 2.22   Freeze–fracture analysis of cholesterol–DMPC mixtures. All liposomes were quenched from a temperature (17 °C) intermediate between the pre-transition (13 °C) and the main transition (24 °C). (a) 0 mol per cent cholesterol. Rippled structure of parallel lines characteristic of $P_{\beta'}$ lipid. Over (b) 4 mol per cent, (c) 8 mol per cent and (d) 12 mol per cent cholesterol there is a progressive increase in the primary ripple repeat distance. At (e) 16 mol per cent cholesterol, there is a dramatic increase in the ripple repeat distance and the lines are wavy and distorted rather than being parallel. At 20 mol per cent (f) and above (g, 24 mol per cent cholesterol) there is no evidence of rippling and a smooth structure is evident. (Reprinted with permission from Copeland, B. R., and McConnell, H. M. (1980) Biochim. Biophys. Acta, 599, 95–109)

ture analysis the periodicity of this ripple repeat distance (Figure 2.22) can be shown to increase as the cholesterol concentration is raised up to 20 mol per cent, above which no rippling occurs at all. This point corresponds to a main transition from ordered to ordered plus disordered phases (Figure 2.21) and it correlates well with the dramatic increase in the lipid diffusion coefficient that occurs at this point. It has been suggested that for mixtures of phosphatidyl cholines containing <20 mol per cent cholesterol, then at temperatures just below that at which the main transition occurs, parallel strips of pure, rippled phosphatidyl choline become separated out from smooth strips of cholesterol-enriched lipid (Figure 2.23). These strips of cholesterol-enriched lipid are identifiable upon freeze–fracture analysis as smooth areas of disordered lipid (Γ) similar to the conformation that the entire lipid structure adopts above 20 mol per cent cholesterol. Thus, over this temperature range both $P_{\beta'}$ and r lipid are found (Figure 2.22). As the mole fraction of cholesterol within the liposome increases, then the amount of pure solid lipid ($P_{\beta'}$) decreases and the primary ripple repeat distance increases correspondingly (Figure 2.23). This is because the primary ripple repeat distance is that between the strips of pure, rippled lipid and the strips of smooth lipid which lie parallel to each other (Figure 2.23). Secondary ripple repeats, that is the distance between ripples *within* the pure lipid ($P_{\beta'}$) strips, only being observed

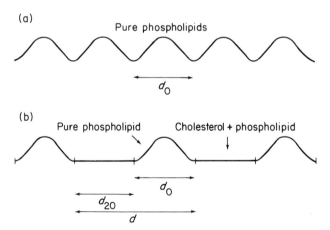

Figure 2.23 Schematic cross-sectional view of the bilayer ripple pattern. (a) Pure $P_{\beta'}$ phospholipid exhibits a periodic ripple, the ripple repeat distance of which is d_0. It has been suggested that at temperatures intermediate between the main and pre-transition and when cholesterol concentrations are < 20 mol per cent, the ripple lines consist of pure, rippled ($P_{\beta'}$) phospholipid, and the strips of smooth lipid found in between them are 'r' lipid. Thus, a hypothetical 'ordered, microscopic phase separation' is shown in (b). Rippled strips of pure PC of width d_0 are separated by strips of 'r' cholesterol–phospholipid mixture of width d_{20} and the total distance between the ripple lines is d. (Reprinted with permission from Copeland, B. R. and McConnell, H. M. (1980) *Biochem. Biophys. Acta*, **599**, 95–109)

when the cholesterol concentration is below 5 mol per cent. However, when cholesterol concentrations are raised above 15 mol per cent the primary ripple repeat distance is of such a magnitude that the strips no longer form parallel straight lines but instead, wavy lines of $P_{\beta'}$ lipid occur until at 20 mol per cent all of the lipid attains the smooth, disordered Γ form (Figure 2.23). This segregation into cholesterol-depleted and cholesterol-rich regions may be viewed as a microscopic phase separation occurring within what appears to be the main solidus region of the phase diagram (Figure 2.21). The reason for this is that even though disordered, Γ, lipid is present, it is packed between arrays of pure, solid $(P_{\beta'})$ lipid acting as a barrier to lipid diffusion. It is only then, when either the temperature is raised or the cholesterol content is increased above 20 mol per cent that a dramatic increase in lipid diffusion occurs due to the removal of these barrier strips of pure, solid phosphatidyl choline. Cholesterol can then have dramatic effects on solid-phase lipid even to the point of inducing microheterogeneity within it (non-ideal mixing).

In the fluid state the maximum concentration of cholesterol that bilayers can incorporate before a pure cholesterol phase separates out depends very much on the nature of the headgroups of the phospholipids present and their associated acyl chains. Usually this maximum level occurs at about 50 mol per cent cholesterol, but in the case of sphingomyelin or with highly unsaturated phosphatidyl cholines this can rise to 67 mol per cent, that is a 2 : 1 molar ratio of cholesterol to phospholipid. Such observations suggest that cholesterol may prefer to interact with certain species of phospholipids, and indeed this has been shown to be the case. In mixtures of phospholipids, cholesterol will preferentially react with lipids in the following sequence, sphingomyelin $\gg$ phosphatidyl serine, phosphatidyl glycerol $>$ phosphatidyl choline $\gg$ phosphatidyl ethanolamine. When mixtures of phospholipids with identical headgroups but different acyl chains are presented, then cholesterol will always associate preferentially with the species exhibiting the lower transition temperature. This effect can influence cholesterol's discrimination for particular headgroup types in lipid mixtures. However, its affinity for sphingomyelin in particular is so great that this interaction is favoured even when other types of lower-melting-point lipids are present. The molecular basis for such specific interactions is unclear at present, although it has been suggested that the overall geometry of the cholesterol and phospholipids are important as well as interaction through the β-hydroxyl group. Clearly, an appreciation of this phenomenon is important for our understanding of the structure of biological membranes as it may form the basis of lateral lipid and protein segregation into cholesterol-rich and cholesterol-poor domains (see section 3.3.4). Thus, for biological membranes containing cholesterol together with a heterogeneous mixture of phospholipids and proteins we can expect to see a considerably more complex phase diagram than that shown in Figure 2.21 for cholesterol–DPPC.

The condensing effect of cholesterol on the apolar core of the bilayer has important consequences for the passive diffusion of water and small

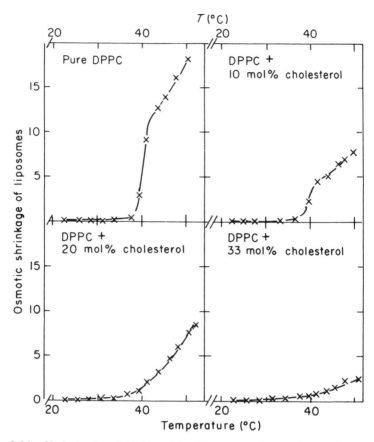

Figure 2.24 Cholesterol seals lipid vesicles. Liposomes of pure phospholipids, as well as being leaky to ions and other small molecules, are permeable to H_2O. This is conveniently demonstrated by following their osmotic shrinkage in glucose solutions. DPPC liposomes in the gel state are relatively impermeable to water. However, there is a dramatic increase in permeability over the lipid phase transition. This is due to the formation of interfacial lipid (see Figure 2.11). Above this temperature permeability continues to increase with rising temperature, but at a slower rate. The continued addition of cholesterol leads to a progressive reduction in the permeability of the liposomes to water. This is due to the ability of cholesterol to abolish the phase transition, preventing the formation of interfacial lipid and also by virtue of its condensing effect on the acyl chain region of the bilayer, inhibiting the formation of vacancies through which H_2O can pass. (Reprinted with permission from Blok, M. C., Van Deenen, L. L. M., and De Gier, J. (1977) *Biochim. Biophys. Acta*, **464**, 509–518)

molecules through the bilayer. Presumably the basis for such permeability is due to the presence of mobile defects or vacancies in the lipid structure. These reach a maximum at the lipid phase transition due to the formation of inter-facial lipid (see section 2.7). However, not only does cholesterol reduce the number of vacancies by virtue of its condensing effect but it also abolishes the lipid phase transition, preventing the formation of interfacial lipid. Thus, cholesterol dramatically reduces the passive permeability of lipid bilayer vesicles to water (Figure 2.24) and small molecules, e.g. K^+, Na^+, Ca^{2+}, glucose, glycerol and indoles. One very important function of cholesterol in biological membranes is then to seal them against the non-specific, passive leakage of small molecules. The concentration of cholesterol in the membranes is very important because at high levels, when a pure cholesterol phase separates out, then a considerable leak is induced due to the formation of interfacial lipid and unstable cholesterol aggregates. The level of cholesterol incorporated into a biological membrane may well be related to the phospholipid composition and function of the membrane (see Chapter 3). Clearly, the subtleties of cholesterol's interaction with phospholipids and its precise role in biological membranes have yet to be elucidated.

Summary

Cholesterol increases the rigidity of fluid-state lipid bilayers yet increases the fluidity of solid-state lipid bilayers.

Cholesterol smears out the lipid phase transition, destroying the co-operativity of the event and reducing the enthalpy change to zero.

The ordering effect of cholesterol on the fatty acyl chains is essentially limited to those interacting directly with the rigid steroid nucleus.

Cholesterol has a condensing, ordering effect on fluid phase lipid that decreases passive permeability to H_2O and small molecules.

2.12 MODES OF MOBILITY OF INTEGRAL MEMBRANE PROTEINS

Integral membrane proteins exhibit two distinct types of mobility within the bilayer. These are a fast lateral diffusion within the plane of the bilayer and a fast rotational diffusion about the long axis of the protein lying perpendicular to the plane of the bilayer (Figure 2.25). Rotation through the plane of the bilayer is a prohibited event.

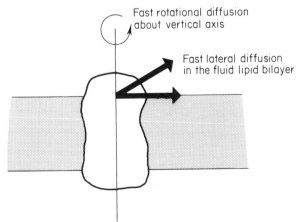

Figure 2.25 Allowed modes of mobility of proteins

2.12.1 Rotational diffusion

The pioneering studies were carried out on the visual pigment protein, rhodopsin (M_r = 28,000) which constitutes some 50 per cent by weight of the disc membrane. The chromophore, 11-*cis*-retinal, is covalently attached to a globular domain on the protein which is associated with the membrane through a pedicle region inserted into the lipid bilayer (see 7.4.3). In the native (unbleached) state the retinal rods are strongly dichroic when viewed from the side, but not at all when viewed end-on. This implies that rhodopsin is uniquely oriented parallel to the plane of the membrane and there is a random distribution of molecules. Partial bleaching of rhodopsin in the membranes, however, fails to induce dichroism. This implies that these molecules can undergo a fast axial rotation rather than being held immobile in the plane of the membrane. This is supported by the observation that if rotation is blocked, by cross-linking the molecules together using glutaraldehyde, then subsequent bleaching can induce dichroism.

The rotational motion can be measured directly by exposing the rods to extremely short (5 ms) flashes of polarized light and measuring the rate of decay of dichroism. From this a rotational relaxation time of 20 μs can be determined, indicating that the molecule is rotating in a lipid environment which has a viscosity like a light machine oil (approx. 2 p). Cytochrome a_3 and cytochrome P_{450}, both of which possess endogenous chromophores, also undergo fast axial rotations with relaxation times of the order of 100–300 μs (Table 2.3).

As the majority of proteins do not have chromophores, then a suitable chromophore must be attached in order to study their mobility. This has been carried out using a reactive isothiocyanate derivative of eosin which can couple covalently to proteins. The rotation of band 3, the major globular integral protein from erythrocytes (see Chapter 4), has been studied in detail with this probe. The erythrocyte membrane appears to be rather more viscous than the disc membrane with relaxation times of about 500 μs being obtained for band 3.

Axial rotation can also be studied using saturation transfer esr spectros-

Table 2.3   Rotational diffusion times of integral membrane proteins

Protein	Membrane	Rotational relaxation times (μs)
Rhodopsin	Rod outer segment	20
Cytochrome a_3	Mitochondrion (IM)	100–300
Cytochrome P_{450}	Mitochondrion (IM)	100–300
Band 3	Erythrocyte	500
Ca^{2+}-ATPase	Sarcoplasmic reticulum	60–100
Acetylcholine receptor	*Torpedo* electroplax	>0.7
Bacteriorhodopsin	*Halobacterium*	>20,000

copy. This extends the range of motions that can be detected by esr from those normally followed, with relaxation times of the order of 1 μs, to those occurring with relaxation times of 1 ms. Using a maleimide spin label to tag the sarcoplasmic reticulum Ca^{2+}-ATPase, a rotational relaxation time of 60 μs was obtained which agrees well with the value obtained from fluorescent studies using an eosin label (100 μs). The relaxation time obtained was dependent upon temperature and affected by cross-linking the proteins. Therefore, relaxation times may be taken as a measure of the axial rotation of the proteins (Table 2.3).

Rotation of proteins about an axis through the plane of the membrane does not occur at all. There are obvious thermodynamic difficulties that would have to be overcome in transferring bulky hydrophilic constituents through the apolar core of the bilayer. This is shown by the absolute asymmetry of membrane proteins (see section 4.4). At one time it was considered that transport proteins could function by rotating in such a fashion (see 7.5.1). However, observations that transport proteins could still function fully after cross-linking or reaction with bulky polar antibodies (M_r = 160,000) has made this proposal untenable.

2.12.2 Lateral diffusion of proteins

This was first demonstrated in a rather elegant experiment carried out with surface antigens on mouse and human cells which had been labelled with appropriate antibodies against cell surface glycoproteins. In this experiment mouse H-2 cells were treated with fluorescein (green)-tagged antibodies, whereas human cells were treated with tetramethylrhodamine (red)-tagged antibodies. These cells were then fused together using inactivated Sendai virus (see section 3.5) to form a single heterokaryon. Heterokaryons (mouse–human cell hybrids) were initially formed having both a large green and a large red area of fluorescence; however, after about 40 min complete mixing had occurred. This highly temperature-dependent intermixing process was unaffected by the addition of agents that blocked either respiration or protein synthesis, consistent with it reflecting the lateral diffusion of the membrane proteins within the plane of the bilayer. The mobility of the integral glycoproteins (M_r = 40,000) that were labelled in this experiment yielded a diffusion coefficient of 10^{-9} cm^2 s^{-1}. Modifications of this technique have been applied to muscle cells with similar results. It is important to note that univalent (Fab') fragments of antibody must be used to avoid cross-linking of surface antigen and its active removal through cap formation (see section 6.7.4).

The lateral diffusion of rhodopsin has been evaluated by bleaching one section of the membrane and following the rate of appearance of unbleached rhodopsin in this bleached area of the membrane. This is an exponential process yielding a value of 5×10^{-9} cm^2 s^{-1} for the diffusion coefficient. Modifications of this method have been used to evaluate the diffusion coefficients of proteins tagged with fluorescent probes. Indeed, this so-called photobleach recovery method (FPR) has been applied to a defined system consisting of a small (M_r = 5260), hydrophobic, phage (m13) coat protein inserted into dimyristoyl phosphatidyl choline vesicles by a detergent dilution procedure (see section 5.2). At temperatures above the phase transition (see sections 2.6 and 2.7), where the bilayer is in the fluid state, then diffusion coefficients of the order of 10^{-8}–10^{-9} cm^2 s^{-1} were observed. However, at temperatures below the phase transition, the value of the diffusion coefficient fell dramatically to only 10^{-10}–10^{-12} cm^2 s^{-1}. Thus, proteins are able to undergo fast lateral diffusion in fluid bilayers, but they are completely immobilized when the lipids crystallize out into a solid phase. Indeed at temperatures below the phase transition, integral proteins cluster out as a separate phase.

This technique has been applied to biological membranes with varying degrees of success. However, in some cases the values obtained for the diffusion coefficients were very small and the recovery of fluorescence into the bleached area was less than 100 per cent (see Table 2.4). These observations could be due to the immobilization of certain proteins (see below), although it

Table 2.4 Lateral diffusion coefficients (L_D) for some integral membrane proteins

Target proteins	Tissue	Recovery (%) (in FBR only)	T °C	D_L (cm^2 s^{-1})
Rhodopsin	Rod outer segment	100	20	2–6×10^{-9}
Acetyl choline receptors	Myotubes	75	35	10^{-12}
Thy-1 antigen	Lymphocytes	50–90	20	3×10^{-10}
Surface-Ig	Lymphocytes	50–90	20	3×10^{-10}
Insulin and EGF receptors	Fibroblasts	40–85	23	3–5×10^{-10}
M-13 coat protein	Liposomes	100	$\begin{cases} T > T_t \\ T < T_t \end{cases}$	7×10^{-9} 10^{-10}–10^{-12}
F_c receptors	Mast cells	50–80	25	2×10^{-10}
Anti-P388 receptors	Fibroblasts	50	25	3×10^{-10}
Surface antigens	Fertilized mouse egg	25–60	25	2×10^{-19}
	Unfertilized mouse egg	25–60	25	$<10^{-11}$
Integral surface proteins	Erythrocyte ghosts	—	20	$<3 \times 10^{-12}$
Cell surface major glycoprotein	Fibroblasts	<10	20	$<5 \times 10^{-12}$
Concanavalin A receptor	Myotubes	—	22	3×10^{-11}
	Myoblasts	70	23	3×10^{-11}
	Glia	—	22	10^{-11}
	Neurones	—	22	10^{-11}
	Fibroblasts	40	22	10^{-11}
Glycophorin A	Liposomes (DMPC)	100	$\begin{cases} T > T_t \\ T < T_t \end{cases}$	10^{-8} 10^{-11}

has been suggested that the intense burst of laser light used in the bleaching experiments could cause cross-linking of the proteins to adjacent species.

Summary
  Integral membrane proteins can undergo both a fast axial rotation and a fast lateral diffusion in the bilayer.
  Rotation of proteins through the plane of the bilayer does not occur.
  Mobility of membrane proteins is severely restricted in bilayers where the lipid is in the crystalline (solid) state.

2.13 RESTRAINTS UPON THE ROTATIONAL AND LATERAL DIFFUSION OF MEMBRANE PROTEINS

Investigations on the mobility of membrane proteins have demonstrated that whilst some species are extremely mobile, others exhibit little or no mobility. Limitations on the mobility of integral proteins can be imposed in a number of ways, such as: the preferential accumulation of protein species in laterally segregated lipid domains; the aggregation or association of integral proteins within the plane of the bilayer to form complexes; the interaction of integral proteins with peripheral membrane proteins and the interaction of integral proteins with components of the cell cytoskeleton (Figure 2.26).

2.13.1 Lipid-mediated effects

Integral membrane proteins are found preferentially in areas of the bilayer containing fluid lipid as they are excluded from solid lipid domains. In biological membranes where cholesterol may be segregated into 'cholesterol-rich' and 'cholesterol-poor' domains (see section 2.11 and 3.3) the integral membrane proteins are almost exclusively found in domains that are relatively cholesterol-poor. The occurrence of fluid–fluid immiscibility (see section 2.1) yielding areas of very different chemical composition may also affect protein distribution by preferential exclusion or incorporation into discrete phases (see Chapter 3).

2.13.2 Protein associations within the bilayer

Sustained interactions between proteins can produce stable complexes in the plane of the bilayer. In the extreme case of bacteriorhodopsin this leads to the formation of a paracrystalline array of the protein in the bilayer which has been used in its structure determination (see section 7.6.3). Whilst this example is quite unique, the plasma membrane of most cells contain structures where the interactions between integral membrane proteins overcomes the randomizing effect of lateral diffusion as embodied in the Fluid Mosaic model (see section 1.5). Thus junctional complexes (see sections 1.2.1 and 7.4.2) are common features of plasma membrane structure which aid the adhesion of

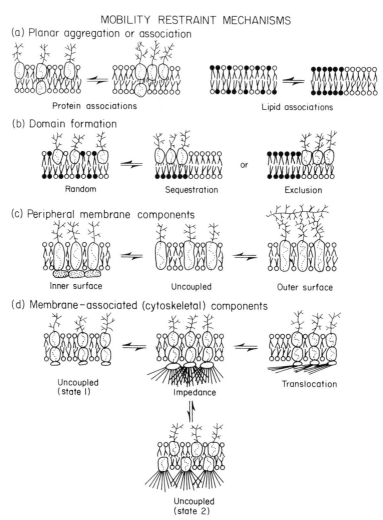

MOBILITY RESTRAINT MECHANISMS

(a) Planar aggregation or association

Protein associations

Lipid associations

(b) Domain formation

Random   Sequestration   or   Exclusion

(c) Peripheral membrane components

Inner surface   Uncoupled   Outer surface

(d) Membrane–associated (cytoskeletal) components

Uncoupled (state I)   Impedance   Translocation

Uncoupled (state 2)

Figure 2.26   Mobility restraint mechanisms. (a) Self-association of integral membrane components. (b) Restriction of proteins to domains of specific lipid compositions. This may be through a positive interaction or an exclusion based upon either the physical properties of the lipids or their physico-chemical properties. (c) Interaction of peripheral membrane components, e.g. protein or carbohydrate networks with integral components. (d) Interaction with cytoskeletal components. These may immobilize or direct the mobility of integral proteins. (Reprinted with permission from Nicolson, G. L. (1976) *Biochem. Biophys. Acta*, 457, 57–108)

adjacent cells or mediate the flow of ions and small molecules between them. Like the arrays of bacteriorhodopsin in *H. halobium*, junctional complexes contain a small number of characteristic integral membrane proteins to the exclusion of others and thereby show that the interactions between them exert a powerful constraint on their lateral mobility.

Synaptic junctions formed between axons of neuronal cells are another example. Specialized areas of these membranes contain arrays of a membrane protein which projects into the cytosol forming a grid-like structure. This protein array is believed to be involved in the release of transmitter into the cleft and in stabilizing the synapse. The pre-synaptic membrane is separated from the post-synaptic membrane by a cleft of some 20–25 nm, across which the released neurotransmitter diffuses and interacts with an array of receptor proteins exposed at the surface of the post-synaptic membrane.

The densely packed cluster of nicotinic, acetylcholine receptors is a remarkably stable structure with the constituent molecules showing no tendency for free lateral diffusion even when the muscle end plate is 'uncovered', to expose them. The inability of individual receptors at the synaptic junction to undergo free lateral diffusion contrasts strongly with receptors not involved at the synapse, (extra-junctional receptors) which have a density distribution some 100–1000 times lower than that seen at the synapse and are able to undergo free lateral diffusion at normal rates. The stability of the lattice may be related to the oligomeric nature of the asymmetrically disposed transmembrane receptor. This might allow the formation of such a cross-linked structure either by direct interaction or as a consequence of an interaction with cytoskeletal components (see section 2.13.4). Certainly there is evidence suggesting that extra junctional receptors can be recruited and organized into packed clusters during the formation of the neuromuscular junction *in vitro*. However, although the pharmacological characteristics of junctional and extra-junctional receptors are very similar, there are differences in the isoelectric points of the solubilized receptors. This represents a net difference of about 20 charges per receptor molecule and may be due in part to the phosphorylated nature of components of the junctional receptor. It would seem from this that both receptors have a common origin but undergo some modification during junction formation. This allows them to pack together to form a tight lattice, inhibiting their lateral diffusion and reducing their turnover rate from about 20 h to 5 d or more. Such an assembly of clustered receptors is clearly quite different from that achieved by divalent antibody which merely causes receptor aggregation by cross-linking molecules at the surface. The net result of this is a fourfold increase in receptor turnover presumably by internalization with subsequent degradation (see Chapter 6).

2.13.3 Interaction with peripheral proteins

A number of peripheral proteins which bind with high affinity and specificity to integral proteins have been identified. If such a peripheral protein were multivalent, then it could initiate the clustering of integral membrane proteins within the plane of the membrane. The clearest example of this is given by spectrin, the peripheral protein of the erythrocyte plasma membrane. This interacts through another peripheral protein, ankyrin, with the transmembrane protein band 3 (see section 4.4). However, less than 10 per cent of the total band 3 molecules are involved as attachment sites for the spectrin net and those exhibit negligible rotational and lateral diffusion. The remaining molecules of band 3 exhibit rotational diffusion coefficients consistent with free molecules, yet their lateral diffusion appears to be restricted. This constraint on their free lateral diffusion is lifted when the spectrin is removed from the membrane, suggesting that although these molecules do not interact directly with spectrin, their ability to migrate over the membrane surface is restricted by this underlying network (see Figures 2.27 and 4.12). The free lateral diffusion of glycophorin is also restricted indirectly by this network. Indeed the chemical cross-linking of spectrin causes glycophorin to aggregate into confined clusters as the spectrin network becomes more enmeshed.

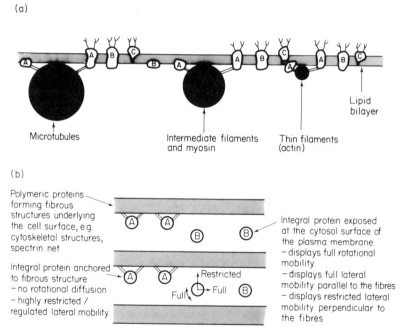

Figure 2.27 Possible modes of influence of cytoskeletal components on the rotational and lateral diffusion of integral membrane proteins. (a) A cross-sectional view of the plasma membrane. This demonstrates the relative dimensions of membrane components and cytoskeletal components. Protein A is anchored to the cytoskeleton and will not exhibit rotational diffusion. Its lateral diffusion may be negligible or actively directed by the cytoskeleton. Protein B will exhibit normal values for its rotational diffusion coefficient; however, its lateral diffusion in a direction perpendicular to the fibres of the cytoskeleton will appear to be restricted for steric reasons, i.e. if the membrane is stretched over the cytoskeleton then it will act as a barrier or 'fence' blocking migration. Protein C has little or none of its residues exposed at the cytosol surface and its lateral and rotational diffusion will be unaffected by the presence of underlying cytoskeletal components. (b) A view from below. This demonstrates that the lateral diffusion of protein B can be affected/directed by the cytoskeleton without any attachment between them

Spectrin has not been found in cells other than the erythrocyte. Ankyrin, on the other hand, has been identified in other cells, which implies that peripheral protein networks may not be unique to the erythrocyte.

2.13.4 The cell cytoskeleton as a controlling influence on membrane protein mobility

Examination of cells by transmission electron microscopy has identified bundles of filaments permeating the cell cytoplasm, some of which underlie and some of which interact with the cell plasma membrane. These polymeric structures, which form the cell cytoskeleton, have subsequently been shown to be composed of distinct protein subunits. These can be distinguished *in situ* by interaction with fluorescent-labelled antibodies. Thus, we can identify structures of varying diameter and organization, from microtubules (25 nm) and myosin filaments (15 nm), to actin filaments (6 nm), with a collective series of so-called 'intermediate' filaments (10 nm) lying in between these. The nature of these protein components and some of their functions is given in more detail in section 6.7.3.

Both the tubulin-containing microtubules and the actin-containing microfilaments form distinct structures within the cell which at one extremity may be anchored to the plasma membrane itself. Anchoring of the cytoskeleton appears to occur at particular loci, and in the case of the microfilaments is achieved using a protein called α-actinin. By virtue of the interaction of these components with the plasma membrane, their permeation of the cell cytosol and their clustering around the cell nucleus, they provide both structural support to the cell and a means of achieving movement and changes in cell shape.

By far the most convincing functional role for the cell cytoskeleton in controlling the mobility of cell surface proteins is related to cap formation. This is the active movement of aggregated protein species to one pole of the cell (see section 6.7.4). However, with perhaps one exception, there is no evidence to suggest a direct involvement of the cell cytoskeleton in controlling the mobility of individual proteins. Indeed, the size of cytoskeleton component structures compared with individual integral proteins makes this rather unlikely (Figure 2.27). It is more probable that the underlying cytoskeleton confines integral proteins to certain areas of the cell surface. It may achieve this by trapping them within boundaries, as suggested for spectrin in the erythrocyte membrane (see section 2.13.3 and Figure 2.27). This may be most pronounced at intercellular junctions where plasma membrane domains of different characters are apposed.

It has been suggested that cell surface glycoproteins on fibroblasts that can act as receptors for the lectin, concanavalin A, may be linked to components of the cell cytoskeleton. These glycoproteins do not appear to undergo rotational motion which is consistent with their being anchored in some way. Similarly, their lateral diffusion is severely restricted and indeed these glycoproteins have been shown to align themselves parallel to 'stress-fibres' that underlie the membrane surface. By assessing the degree of lateral mobility in two dimensions, it can be shown that these proteins can move along an axis parallel to the 'stress fibres' at a rate an order of magnitude faster than motion perpendicular to the fibres. This suggests that the motion of concanavalin A receptors is closely linked with the underlying cell cytoskeleton. However, the majority of membrane proteins appeared to be excluded from areas of membrane close to these stress fibres. Thus the association of integral proteins with the cytoskeleton may be the exception rather than the rule.

Summary

Integral membrane proteins can form associations with each other in the bilayer in ways which affect their diffusion.

Peripheral proteins can interact directly with integral proteins to form aggregated species.

Networks of peripheral proteins and cytoskeletal components can form barriers which inhibit free lateral diffusion but not rotational diffusion of integral proteins.

FURTHER READING

Chapman, D. (1975) Phase transitions and fluidity characteristics of lipids and membranes, *Quart. Rev. Biophys.*, **8**, 185–235.

Chen, C. A., Dale, R. E., Roth, S., and Brand, L. (1977) Nanosecond time-dependent fluorescent depolarisation of DPH in DMPC vesicles and the determination of 'microviscosity', *J. Biol. Chem.*, **252**, 2163–2169.

Cornell, B. A., Chapman, D., and Peel, W. E. (1979) Random close-packed arrays of membrane components. *Chem. Phys. Lipids*, **23**, 223–237.

Demel, R. A., Jansen, J. W. C. M., Van Dijck, P. W. M., and Van Deenen, C. C. M. (1977) The preferential interaction of cholesterol with different classes of phospholipids, *Biochim. Biophys. Acta*, **465**, 1–10.

Edidin, M. (1974) Rotational and translational diffusion in membranes, *Ann. Rev. Biophys. Bioeng.*, **3**, 179–201.

Fleischer, S. and Packer, L., eds. (1974) *Methods in Enzymology*, Vol. 32, Part B, Academic Press, New York. Section D Biophysical techniques (esr, nmr, XRD, ORD, fluorescence, IR, laser Raman and DSC).

Lee, A. G. (1975) Functional properties of biological membranes: a physical–chemical approach, *Prog. Biophys. Molec. Biol.*, **29**, 3–56.

Lee, A. G. (1977) Lipid phase transitions and phase diagrams: I Lipid phase transitions, *Biochim. Biophys. Acta*, **472**, 237–281.

Lee, A. G. (1977) Lipid phase transitions and phase diagrams: II Mixtures involving lipids, *Biochim. Biophys. Acta*, **472**, 285–344.

Lee, A. G., Birdsall, N. J. M., and Metcalfe, J. C. (1974) Nuclear magnetic relaxation and the biological membrane, in *Methods in Membrane Biology*, Vol. 2 Korn, E. D., ed., pp. 1–156.

March, D. (1975) Spectroscopic studies of membrane structure, *Essays in Biochemistry*, **11**, 139–180.

Martin, R. B., and Yeagle, P. L. (1978) Models for lipid organisation in cholesterol–phospholipid bilayers including cholesterol dimer formation, *Lipids*, **13**, 594–597.

Martonosi, A., ed. (1976) *The Enzymes of Biological Membranes*, Vol. I, John Wiley, New York. Physical and chemical techniques (includes XRD, nmr, esr and fluorescence studies).

Nicolson, G. L. (1976) Transmembrane control of the receptors on normal and tumour cells. I. cytoplasmic influence over cell surface components, *Biochim. Biophys. Acta*, **457**, 57–108.

Nicolson, G. L. (1976) Transmembrane control of the receptors on normal and tumour cells. II. surface changes associated with transformation and malignancy, *Biochim. Biophys. Acta*, **458**, 1–72.

Nicolson, G. L., Poste, G., and Tai, H. J. (1977) The dynamics of cell membrane organisation, in *Dynamic Aspects of Cell Surface Organisation* Poste, G., and Nicolson, G. L., eds, *Cell Surface Reviews*, Vol. 3 Elsevier/North Holland Biomedical Press, pp. 1–73.

Oldfield, E., and Chapman, D. (1972) Dynamics of Lipids in Membranes: heterogeneity and the role of cholesterol, *FEBS Lett.*, **23**, 285–297.

Podo, F., and Blasie, J. K. (1975) in *Biochemistry of Cell Walls and Membranes*, Vol. 2, Fox, C. F. ed., MTP International Review of Science, Biochemistry Series One, Butterworths, University Park Press, pp. 97–122.

Poznansky, M. J., and Lange, Y. (1978) Transbilayer movement of cholesterol in phospholipid vesicles under equilibrium and non-equilibrium conditions, *Biochim. Biophys. Acta*, **506**, 256–264.

Sauerheber, R. D., Gordon, L. M., Crosland, R. D., and Kuwahara, M. D. (1977) Spin label studies on rat liver and heart plasma membranes: Do probe–probe interactions interfere with the measurement of membrane properties? *J. Membrane Biol.*, **31**, 131–169.

Schreier, S., Polnaszek, C. F., and Smith, I. C. P. (1978) Spin labels in membranes: Problems in practice, *Biochim. Biophys. Acta*, **515**, 375–436.

Seelig, J. (1977) Deuterium magnetic resonance: theory and application to lipid membranes, *Quart. Rev. Biophys.*, **10**, 353–418.

Tanford, C. (1973) *The Hydrophobic Effect*, John Wiley, New York.

Chapter 3

Lipid–protein interactions

3.1   INTRODUCTION

The biological membrane is a mixture of lipids and proteins all able to undergo rapid lateral diffusion within two dimensions of the bilayer structure.

However, the various components of the biological membrane are not distributed homogeneously. Within the bilayer structure, clusters or pools of lipid of various compositions may be found existing in the bulk fluid lipid pool, and localized concentrations of specific proteins may be associated with either these lipid domains or tethered by structures outside the cell membrane. This ability to create organized areas of the cell membrane gives rise to phenomena such as cell polarity, junctional complexes and phase separations.

Integral proteins are found almost exclusively with fluid (liquid-crystalline) state lipid. If they are inserted in a bilayer and the temperature is reduced below the phase transition, then they are excluded from the solid-state lipid and form a separate phase with the clustered protein molecules surrounded by a shell of disordered lipid. This is presumably because proteins act as an 'impurity' in the system, depressing the 'freezing point' and disrupting packing of the lipid bilayer. So they are accommodated in the more flexible structure of fluid lipid, which is perhaps fortunate as the activity of most integral enzymes or transport proteins ceases as the lipid solidifies about them. From this we can see that the physical state or fluidity of the lipid surrounding integral proteins can not only affect their conformation, but it can also regulate their activity. Thus changes in the fluidity of a biological membrane, caused either by the cell itself or by external factors, can lead to dramatic effects on the activity of the integral proteins and hence on cellular metabolism and responses. Indeed, because biological membranes are asymmetric then the modulation of one or other half of the bilayer has selective effects on the integral proteins dependent on whether they actually penetrate into that particular half.

A fundamental problem is that of integrating proteins into and across lipid bilayers whilst still retaining the functioning of the protein and not creating a 'leaky' membrane. A simple way of achieving a sealed bilayer is to have pure lipid in the crystalline state. However, by inserting proteins into a solid bilayer, not only would integral proteins fail to function but the structure would be mechanically unstable and high rates of leakage would occur at the lipid–protein interface. Biological membranes are therefore flexible, consisting of mixtures of fluid lipids which allow proteins to function whilst leakage is kept to a minimum (see section 3.3).

Proteins embedded in the bilayer pose a problem themselves as they are a potential focus of disruption. Indeed they do perturb the bilayer and this is most markedly seen in the restricted mobility of the immediate ring of lipids around the integral proteins. This layer of lipids, which is intimately involved in sealing an integral protein into the bilayer and allowing it to function, has been termed the lipid annulus. The properties of both the annular and bulk lipid pools can regulate the functioning of integral membrane proteins. This is most acute for the globular integral proteins where changes in the physical state of the lipid environment are likely to influence their conformation. However, for those proteins which are anchored merely by a hydrophobic tail to the bilayer, changes in bilayer properties are unlikely to have dramatic consequences on protein function.

3.2 PERIPHERAL PROTEINS AND THEIR ASSOCIATION WITH BIOLOGICAL MEMBRANES

Peripheral proteins bind to biological membranes through predominantly electrostatic interactions (see section 1.4.2) and so their release does not require detergents. Merely changing either the ionic strength or elevating pH or adding chelating agents (EGTA/EDTA) is usually sufficient to remove them from the membrane. However, they possess the ability of being able to rebind tightly to those membranes from which they were released. This distinguishes them from adsorbed proteins which can be found associated with membranes and which are readily eluted by washing with isotonic buffer at neutral pH.

Classes of peripheral proteins may be expected to bind either to the lipid or protein or to both components of the membrane. However, we would expect that those exclusively associated with the headgroups of phospholipids in the membrane would be found distributed throughout the various intracellular membranes. This is because the different phospholipid species which provide the attachment sites are found in varying proportions in all eukaryotic cell membranes. The only notable exception of course is cardiolipin which is found exclusively in the mitochondrial inner membrane and so this particular phospholipid could provide a unique anchor. The way to achieve a definite location, not only to a particular membrane but to a particular side and lateral situation on the membrane, is for the peripheral protein to bind specifically to a site on an integral protein. As integral proteins, by virtue of their biosynthetic mechanism, exhibit an absolute asymmetry and are associated with specific cell membranes, then these properties are immediately conveyed upon any peripheral protein that attaches itself to it (see section 6.2 and 4.4).

Two well-defined examples of peripheral enzyme proteins found associated with specific integral proteins are the glyceraldehyde-3-phosphate dehydrogenase (GAPDH) from human erythrocyte plasma membranes and the insulin-stimulated cyclic AMP phosphodiesterase (PDE) from rat liver plasma membranes. GAPDH (M_r = 37,000) binds to a single class of high affinity ($1 \times 10^7 \, M^{-1}$) protein sites found exclusively on the cytosol side of the erythrocyte plasma membrane. Approximately 2×10^6 molecules of GAPDH bind reversibly to these sites which are now known to be the globular transmembrane protein, band 3. However, the properties of the enzyme do not appear to be altered upon binding and so the physiological significance of this interaction remains unclear. On the other hand the liver plasma membrane PDE is associated with the membrane so as to allow insulin, after binding to its receptor, to trigger the cyclic AMP-dependent phosphorylation and activation of this enzyme. This PDE is a monomeric protein (M_r = 52,000) found attached, with high affinity ($2 \times 10^9 \, M^{-1}$), to a single class of sites on an as yet unidentified integral plasma membrane protein. It is found exclusively on the cytosol side of the hepatocyte plasma membrane to which approximately 10^4 molecules of PDE can bind. However, as with GAPDH, the kinetic properties of the enzyme are relatively unaffected by the specific attachment to this binding site. Two other peripheral proteins, spectrin and ankyrin, bind either directly or indirectly to a transmembrane protein (band 3) in the red cell and this is discussed in detail in 4.4

Examples of proteins that bind to membranes by strong electrostatic interactions with bilayer phospholipids are myelin basic protein and another basic protein, cytochrome c. At neutral pH, myelin basic protein exerts such an attraction for acidic phospholipids that its association with bilayers can lead to the phase separation of the acidic phospholipids in the membrane. If there are sufficient acidic phospholipids available then each molecule of myelin basic protein can cluster and bind about 27–34 molecules of them. Cytochrome c is found exclusively at the cytosol side of the mitochondrial inner membrane where it is responsible for transferring electrons to cytochrome oxidase. It binds to both cytochrome oxidase and to mito-chondrial phospholipids where it is able to undergo fast lateral diffusion. In broken mitochondria it binds avidly to the matrix side of the inner membrane, where the acidic phospholipids are exposed (see section 4.7.2, 4.9.5). The basic nature of this protein is also reflected in its preferential binding to liposomes that exhibit the greatest negative surface charge. Thus, its localization on the cytosol side of the mitochondrial inner membrane reflects its site of synthesis on cytosol ribosomes, its inability to penetrate the inner membrane and its interaction with certain specific integral proteins exposed at the external surface of the inner membrane. Indeed the ability of cytochrome c to interact (physically) with proteins involved in the mitochondrial electron transport chain is essential for its functional role. Cytochrome c is found associated with various cell membranes due to its ability to interact with acidic phospholipids. However, it predominates at the C-side of the mitochondrial inner membrane because of the integral proteins exposed there that can bind it.

The interaction of peripheral proteins with the lipid and protein components of cell membranes is a relatively neglected field. There is a general feeling, based on little or no evidence, that peripheral proteins are anchored to membranes exclusively by binding to the headgroups of phospholipids. Such a mode of interaction would not yield a specific localization of a peripheral protein. Indeed in cases which have been rigorously examined such as GAPDH, PDE, spectrin and ankyrin (see section 4.4) it is clear that a specific localization for functional and structural purposes is achieved by their interaction with distinct integral proteins. The only clear example of a peripheral protein exhibiting functional high-affinity binding sites for bilayer lipid is the enzyme pyruvate oxidase from *Escherichia coli* whose activity is increased some 25-fold by interaction with membrane lipids or some suitable amphiphile. However, even this protein interacts, like other peripheral proteins, with integral proteins. In this case they are integral proteins involved in the electron transport chain that are exposed at the inner surface of the cytoplasmic membrane. Thus, there are no alternative membrane sites for this enzyme inside the bacterium.

Besides the specific association of peripheral enzymes with the polar surfaces of biological membranes, there is also the absorption to the membrane of certain cytosol proteins that exist at high concentrations within the cell. Of course, during subcellular fractionation which involves a considerable dilution, these will be desorbed and hence they will be found in the 'soluble' fractions obtained in such preparative experiments. However, there is good evidence in model systems that soluble enzymes such as lactate dehydrogenase, malate dehydrogenase, glutamate dehydrogenase and fructose 1,6 biphosphatase can all adsorb, to varying extents, to lipid bilayers at high protein concentrations. This can result in changes in their thermal stability, sensitivity to proteases, and kinetic properties. It remains to be shown whether this happens *in vivo*, and if so its physiological relevance needs to be elucidated.

Summary

Peripheral proteins can bind to both lipid and protein sites on the membrane surface.
In order to achieve a unique localization the peripheral protein must bind to a unique anchor.
Such an anchorage point can only be supplied by an integral protein whose location and asymmetry in the membrane is then immediately conferred on the peripheral protein.
Certain soluble proteins may be non-specifically adsorbed on to cell membranes.

3.3 THE MODULATION OF INTEGRAL PROTEIN FUNCTION BY LIPIDS OF THE BILAYER

Integral membrane proteins are inserted into the lipid bilayer where they exhibit extensive hydrophobic and electrostatic interactions with the surrounding lipids, and in some cases with each other. These proteins act, of course, as impurities within the lipid bilayer and as such we might well expect them to perturb its structure. For the majority of biological membranes the molar ratio of lipid : protein is rather high, about 200 : 1, and any 'classical depression of the freezing point', i.e. the lipid phase transition/separation temperature, would be correspondingly small ($< 0.5\ °C$). However, in those membranes with high molar concentrations of proteins, such as occurs in for example various junctional complexes, then these effects become more marked. A general increase in membrane viscosity can occur and the lipid phase separation is depressed and broadened, even to the extent of being smeared out. This is because van der Waals interactions between the protein and lipid components predominate to such an extent that the co-operativity of the lipid phase transition/separation, which depends so much on lipid–lipid interaction, is lost.

Integral proteins will of course be sensitive to the nature of the lipid environment in which they are embedded. This may be reflected by their interaction with particular headgroups of phospholipids and also by their reaction to the physical state of the bilayer. Strangely enough there is only one clear-cut case of a specific lipid being absolutely essential for the catalytic activity of an integral protein. This is the enzyme, β-hydroxybutyrate dehydrogenase, which is integrated into the mitochondrial inner membrane. Its active site is exposed to the matrix and is responsible for converting β-OH butyrate to oxaloacetate in an NAD^+-dependent reaction. This enzyme requires the headgroup of phosphatidyl choline to bind and utilize NAD^+. The only species that can be substituted for phosphatidyl choline are the chemically related N,N,N-trimethyl phosphatidyl butanolamine and the detergent stearoyl phosphatidyl choline.

The viscosity of the bilayer, which may be likened to a light machine oil, can affect the dynamic properties of integral proteins. In particular, it will affect their lateral mobility and their conformational flexibility, namely the ability to undergo large-amplitude internal motions required for enzyme or transport functions. These changes in the physical characteristics of the bilayer will therefore modulate the activity of enzymes and transport proteins embedded within it. The particular structure of individual integral protein as well as their disposition in the membrane decides their sensitivity to the lipid environment. Thus peripheral proteins (Figure 1.6a) will not be affected by changes in bilayer fluidity, while integral membrane proteins with their globular domain only partially embedded in the bilayer (Figure 1.6d and e) will be sensitive to changes in one half of the bilayer only. Cytochrome b_5 reductase is an example of an integral membrane protein which is relatively insensitive to changes in the physical state of the bilayer (see section 7.8). In this case the functional globular domain of the enzyme is decoupled from changes occurring in the folded structure of its hydrophobic pedicle. In fact, this enzyme can readily be released from the bilayer in a fully soluble form after mild proteolysis, which merely clips off the globular region containing the active centre from its hydrophobic anchor, whilst retaining full function. A very different picture is seen with integral proteins like the Ca^{2+}-ATPase, Na^+/K^+-ATPase and cytochrome oxidase whose globular, functional regions are fully integrated into the bilayer. For in their case a hydrophobic environment supplied by a phospholipid or suitable non-ionic detergent is essential to maintain functional integrity of the proteins. The activity of these integral proteins, like many others, is modified by the physical state of the bilayer. Indeed, the requirement for sealing integral proteins into the bilayer to prevent non-specific leakage around the proteins, means that their conformational flexibility is usually constrained by bilayer lipids. Thus, many enzymes are in fact activated upon detergent solubilization, which immediately releases them from physical constraints imposed upon them by the bilayer.

As well as constraining the conformational flexibility of integral proteins, the physical state of bilayer lipids can also regulate the degree and rate of lateral diffusion that they can undergo within the bilayer. This has a marked effect on the activity of systems that require to undergo free lateral diffusion in order to function. Examples of this are the interaction between specific hormone receptors and adenylate cyclase (see section 7.7) and that between cytochrome b_5 and cytochrome b_5 reductase (see section 7.8).

3.3.1 Annular lipid

Integral membrane proteins are solvated by the lipids of the bilayer. This is done in such a fashion as to allow the functioning of the inserted protein and yet to minimize the leakage of material across the membrane by reducing any mismatch at the lipid–protein boundary. This allows membrane potentials and solute gradients to be built up across, for example, the plasma membrane and mitochondrial inner membrane by functioning transmembrane proteins embedded in the lipid bilayer.

The immediate lipid ring surrounding a penetrant protein provides the interface between that protein and the bulk lipid pool. This lipid ring has been called the lipid annulus, boundary layer lipid and halo lipid. We prefer to call this immediate ring of lipid the lipid annulus, a word which is derived from the Latin, *anulus*, meaning a ring (Figure 3.1).

Integral proteins have a number of different effects upon the lipids of the membranes in which they are embedded. They increase the viscosity of a fluid lipid bilayer; they modify the degree of order of the acyl chains of annular

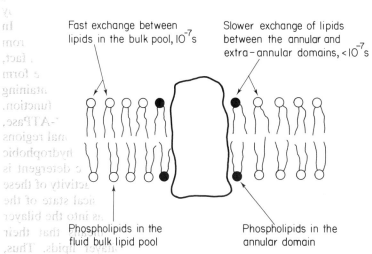

Fast exchange between lipids in the bulk pool, 10^{-7} s

Slower exchange of lipids between the annular and extra-annular domains, $< 10^{-7}$ s

Phospholipids in the fluid bulk lipid pool

Phospholipids in the annular domain

Figure 3.1 Annular and bulk lipid domains. The immediate ring of lipids that provide the interface between the protein and the bulk lipid pool has been termed the annular lipid domain. By virtue of the interaction of annular lipid with the protein, the rate of exchange of annular lipid with lipid in the bulk lipid pool is rather slower than the exchange between adjacent lipids within the bulk lipid pool

phospholipids; they depress or abolish the phase 'transition' of annular lipid and they can depress the phase transition of the bulk lipid phase. Perhaps one of the most interesting effects is related to the rate of exchange (V_{ex}) between neighbouring lipids, the so-called hopping frequency. In the bulk lipid pool this occurs extremely rapidly with $V_{ex} = 10^{-7}$ s. However, annular lipid, by virtue of its interaction with the integral protein, appears to exchange with its neighbours in the bulk lipid pool at rates at least an order of magnitude slower. The limits of the exchange between the annular and extra-annular lipid pools has been put at $10^{-4} > V_{ex} > 10^{-6}$ s for most of the integral proteins that have been studied so far. This is rather interesting, for the turnover numbers of enzymes are in most instances much slower than this exchange rate. Annular lipid is not then a static layer as lipids in the annular domain will change several times during one turnover of the enzyme. Therefore, if annular lipid is going to affect the activity of an integral enzyme it must do this by virtue of its average fluidity.

The majority of integral proteins are able to undergo rotational diffusion in the membrane with relaxation times of the order of 10^{-7}–10^{-4} s. This may be affected by the viscosity of the membrane environment. However, the rate of rotation is extremely fast compared with the molecular turnover numbers of enzymes. Thus, changes in rotation are unlikely to be able to regulate integral protein function. Indeed, the ATP-dependent Ca^{2+}-pump from sarcoplasmic reticulum can function perfectly well in large cross-linked complexes. This implies that rotation about either a vertical or horizontal axis in the membrane is unnecessary for it to function.

The amount of lipid that constitutes an annulus is peculiar to any particular protein species. By definition it clearly needs to be the amount necessary to form a single ring of bilayer around the hydrophobic domain of the protein. In the case of the Ca^{2+}-pump from rabbit muscle sarcoplasmic reticulum, 30 lipids constitute the annulus, of which 15 can be shown to be present at each side of the bilayer. These then form a symmetrical structure around this transmembrane cylindrical protein (Figure 3.2). The functional integrity of the Ca^{2+}-pump and of cytochrome oxidase is totally dependent upon a hydrophobic environment supplied by bilayer lipids or a suitable non-ionic detergent. The progressive removal of lipids from bilayers containing these proteins has little effect on the activity of the enzymes until a critical point is reached. This is where there are insufficient lipids to provide a protective annulus. Indeed, the progressive stripping away of annular lipid leads to a rapid and irreversible loss of enzyme activity (Figure 3.3), unless during the process it is immediately replaced by a suitable non-ionic detergent which binds to, and masks, the hydrophobic sites.

The ability to prepare vesicles with different lipid to protein ratios (see section 5.4) can be used to identify features of their interaction. Integral proteins tend to increase the ordering of annular phospholipids at temperatures above the lipid phase transition, yet decrease their order at temperatures below the phase transition. As one might expect, they also

100

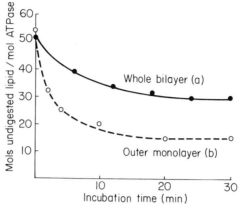

Figure 3.2   The symmetry of the phospholipid annulus of the Ca^{2+}-ATPase from sarcoplasmic reticulum. Treatment, with phospholipase D, of leaky, radio-labelled phosphatidyl choline vesicles (a) containing Ca^{2+}-ATPase resulted in the degradation of all of the phospholipid to phosphatidic acid except for that immediately surrounding the protein. The 30 molecules of annular phospholipid associated with each molecule of Ca^{2+}-ATPase remained sterically inaccessible to the action of phospholipase D. When intact, sealed vesicles (b) containing radiolabelled phosphatidyl choline only in the external half were treated with externally added phospholipase D then all of the phospholipids in the external half of the bilayer could be degraded except those interacting with the Ca^{2+}-ATPase. In this instance 15 molecules of phospholipid per Ca^{2+}-ATPase were unavailable for reaction. This suggests that the Ca^{2+}-ATPase spans the membrane as a symmetrical cylinder with 15 molecules of phospholipid in each half of the bilayer providing the annular domain. (Reprinted by permission from Bennett, J. P., McGill, K. A., and Warren, G. B. (1978) *Nature (Lond.)*, 274, 823–825, copyright © 1978 Macmillan Journals Limited)

increase the viscosity of fluid lipid bilayers. Thus, at high protein concentrations or at temperatures near or below the lipid phase transition there is an increasing tendency for the protein molecules to aggregate and form a protein-enriched phase. Under these conditions a significant fraction of lipid will become trapped between adjacent annular domains and some lipid will be directly trapped between adjacent proteins, thus becoming shared annular lipid. Lipids in these environments exhibit distinct physical properties and are readily identified by physical probes. These demonstrate that the most highly perturbed lipids are those occupying the annular domain, and next those lipids trapped between adjacent annular domains. Lipid distal to the annular domain which is not trapped between adjacent annuli is hardly perturbed at all (Figure 3.4). This means that the perturbation of the bulk lipid phase by integral proteins, at the lipid to protein ratios found in most biological membranes, is minimal. The effect of proteins on fluid biological membranes is then essentially restricted to an increase in the order of lipid occupying the annular domain.

   In general, then, lipids in the annular domain exchange rapidly with those

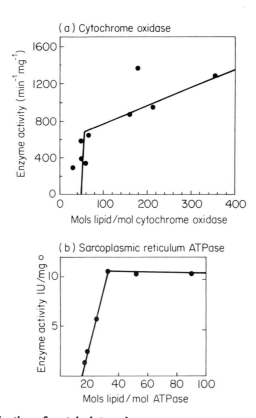

Figure 3.3 **Inactivation of certain integral enzymes can occur upon removal of lipid below that required to provide an annulus. Using the ionic detergent cholate to strip away phospholipid from complexes with either yeast cytochrome oxidase or sarcoplasmic reticulum Ca^{2+}-ATPase results in the loss of enzyme activity when < 55 moles or 30 moles respectively, of lipid remain associated with the protein. Their removal leads to the exposure of hydrophobic regions of the molecule to the aqueous environment, causing denaturation. (Reprinted with permission from Knowles, P. F., Watts, A., and Marsh, D.,** *Biochemistry*, **18, 4480–4487, copyright (1979) American Chemical Society; and Hesketh, T. R., Smith, G. A., Houslay, M. D., McGill, K. A., Birdsall, N. J. M., Metcalfe, J. C., and Warren, G. B. (1976)** *Biochemistry*, **15, 4145–4151, copyright (1976) American Chemical Society)**

in the bulk lipid pool, but at a slower rate than exchange occurs between neighbours within the bulk lipid pool itself. This lipid gives rise to a so-called 'immobilized component' upon esr analysis. However, this only appears immobilized compared with the rapid motions of bulk phase lipids detected upon esr analysis. For example, an event occurring with a relaxation time of 10^{-5} s would appear 'immobilized' on esr analysis but not on nmr analysis, and it is by observing both nmr and esr spectra that limits can be placed on the rate of exchange between annular and extra-annular lipids. In kinetic terms we can give rate constants to the entry and exit of lipid from the annular

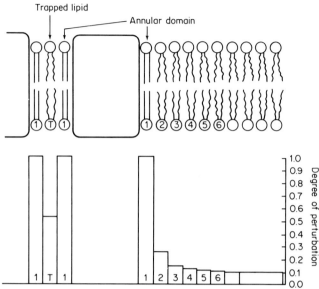

Figure 3.4 The degree of perturbation of lipid shells by an integral protein. By using physical probes it is possible to evaluate the effects of an integral protein on bilayer lipids. Rapid motions can be detected by esr analysis using phospholipids with nitroxide spin labels attached to their acyl chains. Three very distinct lipid environments show up in esr spectra, that is, lipid in the annular domain; lipid trapped between adjacent lipid annuli and bulk lipid. (Reprinted with permission from Knowles, P. F. *et al.* (1979) *Biochemistry*, 18, 4480–4487. Copyright (1979) American Chemical Society)

domain which will reflect the affinity of lipids for the protein. Thus a relative equilibrium constant (K_r) can be defined as, $K_r = k_{on}/k_{off}$, where k_{on} is the rate constant for entry into the annular domain and k_{off} is the rate constant for exit. In this exchange equilibrium process between annular and extra-annular (bulk fluid phase) lipids we can assume that k_{on} is diffusion controlled and will be identical for all lipids and proteins. Any changes in k_r will then be dependent upon the rate constant for exit from the annular domain (k_{off}). These will be influenced by electrostatic and van der Waals interaction between particular protein and lipid species. In such a situation the equilibrium composition of lipids in the annular domain need not reflect that of the bulk lipid pool (Figure 3.5). A protein may then exhibit segregation for or even against particular lipid species (Table 3.1).

β-Hydroxybutyrate dehydrogenase, which depends upon the headgroup of phosphatidyl choline in order to function, segregates this lipid out of the total lipid pool. Given a choice of mitochondrial lipids or the highly fluid dioleoyl phosphatidyl choline, it chooses the latter to provide its working environment. This is reflected in the extremely slow rate of exchange of annular phosphatidyl choline with the bulk lipid $(V_{ex} < 10^{-3}\,s)$. Other

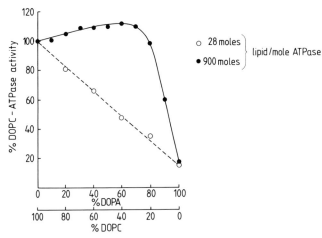

Figure 3.5 Ca^{2+}-ATPase can segregate lipids necessary for its function into its annular domain. The acidic phospholipid, dioleoyl phosphatidic acid (DOPA) is a powerful inhibitor of Ca^{2+}-ATPase. Complexes of this enzyme with this lipid exhibit less than 20 per cent of the activity of complexes with the neutral phospholipid, dioleoyl phosphatidyl choline (DOPC). In mixtures of DOPA and DOPC where there is only just enough phospholipid present to form a single bilayer ring around the protein, then Ca^{2+}-ATPase activity is linearly related to the fraction of DOPA/DOPC in the mixture (○). This is because every lipid must interact with the protein. However, when there is a vast excess of lipid available to the protein (900 mol phospholipid/mol Ca^{2+}-ATPase) then activity is not linearly related to the DOPA/DOPC content (●). Only at high DOPA content is the activity of the enzyme inhibited. This is because when a choice of lipids is available, the Ca^{2+}-ATPase can select appropriate lipids from the bulk lipid pool to constitute its lipid annulus. The composition of the lipid annulus will then depend on the rate constant for exit of particular lipid species from the annular domain and also on the molar amount of lipid in the pool. The smaller the rate constant for exit from the annular domain, the greater the chance of the lipid being found there. (Reprinted with permission from Bennett, J. P., McGill, K. A., and Warren, G. B. (1980) *Current Topics in Membranes and Transport*, Vol. 14, Academic Press)

evidence for the segregation of lipid species by integral proteins is given by the visual pigment rhodopsin and the ion-transport protein, Na$^+$/K$^+$-ATPase. Both of these proteins preferentially accumulate acidic phospholipids within their annular domain even though they do not appear to be essential for the functioning of the protein. This segregation, which can be demonstrated using both spin-label and cross-linking techniques, diminishes at high salt concentrations which confirms the role of electrostatic interactions in the process.

A detailed investigation of the segregation of particular headgroup phospholipid species into the annular domain has been carried out with yeast mitochondrial cytochrome oxidase. Using spin-labelled phospholipids this protein was shown to select either cardiolipin or phosphatidic acid in its annular domain rather than phosphatidyl choline. These lipids exhibited rate constants for exit (k_{off}) which were five times (cardiolipin) and twice

Table 3.1 Properties of the annular lipid domain of integral membrane proteins

Protein	Number of annular lipids	Indications of segregation[a]
β-Hydroxybutyrate dehydrogenase	30	Phosphatidyl choline (+ ve)
Ca^{2+}-ATPase (sarcoplasmic reticulum)	30	Phospholipids (+ ve) cholesterol (− ve)
Cytochrome oxidase	55	Cardiolipin (2+ ve) and acidic phospholipids (+ ve)
Glycophorin	30/dimer	Acidic phospholipids (+ ve)
Na^+/K^+-ATPase	?	Acidic phospholipids (+ ve)
Rhodopsin	24	Acidic phospholipids (+ ve)

[a]Here, +ve, indicates a segregation into the annular domain and, −ve, indicates an exclusion from the annular domain.

(phosphatidic acid) as slow as that for phosphatidyl choline. The rate of k_{off} was similar to phosphatidyl choline for phosphatidyl serine, phosphatidyl glycerol and phosphatidic acid. The specificity of the interaction with cardiolipin and phosphatidic acidic cannot be simply due to electrostatic interactions as not only was there no preference for the acidic phosphatidyl serine but the experiments were carried out at high ionic strength. This suggests that the protein is recognizing the shapes of the molecules. Indeed, cardiolipin does have a rather different molecular geometry from other phospholipids. This might well complement the structure of the protein, so forming the basis of the preferential interaction of the enzyme with this lipid (see section 3.3.4).

So far, cholesterol is the only membrane lipid which is preferentially excluded from the annular domain. This has been demonstrated for the Ca^{2+}-ATPase, band 3, β-hydroxybutyrate dehydrogenase, Mg^{2+}-ATPase and succinate dehydrogenase, indicating that it may well be a fairly general property of globular integral protein. Such a segregation may reflect both the poor complementation of the rigid, planar sterol ring of cholesterol with an unaccommodating protein surface, and also the strength of the association of cholesterol for phospholipids (see section 2.11). Thus, it would seem that both cholesterol and the integral proteins would prefer to interact with phospholipids rather than with each other. If, however, cholesterol is forced to interact with these proteins, by adding excess cholesterol in the presence of the detergent Na cholate (see section 5.3), a reversible inhibition of enzyme activity occurs. There are undoubtedly many reasons for this, amongst them being the rigid conformation that cholesterol adopts which would put constraints upon the conformational flexibility of the protein, and also the lack of a significant polar region akin to the phospholipid headgroup.

Thus, cholesterol, which is used in many membranes to reduce non-specific leakage through the lipid bilayer, does not significantly interact with penetrant proteins. This is perhaps fortunate as not only would it inhibit the functioning of enzymes but the inability of this rigid molecule to adapt to the protein surface would make it a poor agent for sealing the protein efficiently into the bilayer. Indeed, sealing integral proteins into lipid bilayers such that non-specific solute leakage at the lipid–protein interface is minimal is of considerable importance. Relatively little consideration has been given to this problem even though it is well known that the incorporation of integral proteins into phosphatidylcholine liposomes actually increases leakage, yet leakage across the protein-containing plasma and mitochondrial membranes is so small that membrane potentials and solute gradients can be maintained across them. Perhaps the variety of phospholipids present in biological membranes allows integral membrane proteins to 'select' those that complement their structure best, allowing them to be sealed into the bilayer. The extent of the hydrophobic domain on a transmembrane protein may require lipids of specific dimensions to complement these surfaces and seal the protein into the bilayer. Thus the thickness of the annular region could be either larger or smaller than that of the bulk lipid phase (see section 3.3.4). The importance of lipids in the annular domain in sealing an integral protein into the bilayer is known from studies on the efficiency of an ATP-driven Ca^{2+}-pump reconstituted in liposomes of various compositions (see section 5.5.2). The number of Ca^{2+} transported for each ATP molecule hydrolysed, was greatly increased if a small amount of phosphatidyl ethanolamine was incorporated in a defined reconstitution of the Ca^{2+}-ATPase plus dioleoyl phosphatidyl choline. This lipid, however, did not affect the activity of the enzyme or the Ca^{2+} leakage from liposomes consisting only of dioleoyl phosphatidyl choline. So it appears that phosphatidyl ethanolamine can achieve the microscopic sealing of the Ca^{2+}-pump into the bilayer, such that the leakage of solute at the protein–lipid boundary, i.e. the annulus, is at a minimum. Lipids within the annular domain should then allow the protein both to function and yet seal it efficiently into the bilayer.

3.3.2 Temperature-dependent changes in membrane fluidity and protein function

By attaching spin labels, nmr probes or fluorescent tags to integral proteins or by using circular dichroism or optical rotatory dispersion analysis it is possible to see changes in protein folded structure that are triggered by changes in the physical state of the bilayer. However, we need to know whether the physical properties of the lipid environment can regulate the functioning of an integral protein, whether there are means available to the cell to regulate protein function by modifying its membrane lipids, and whether external influences can affect protein function by interacting with lipids in the bilayer. Over the past decade considerable effort has been applied in trying to answer these questions.

One of the easiest ways of seeing whether a membrane function is influenced by the physical state of the bilayer is to alter the temperature and observe its effect as temperature modulates membrane fluidity. Membrane fluidity is a convenient, if rather imprecise term embracing all the changes that occur in bilayer phospholipids when either the temperature is altered or agents that increase order or disorder are added. Lipid fluidity encompasses such phenomena as changes in the lateral and rotational mobility of phospholipids; changes in the frequency of *trans–gauche* isomerization of C—C bonds in the phospholipid acyl chains; changes in the swinging motion of the phospholipid fatty acyl chains, and changes in phospholipid headgroup conformation. So fluidity increases with increasing temperature and this corresponds to an overall increase in molecular motion throughout the bilayer. Clearly, all or some of these particular changes could influence the activity of integral proteins. However, until further work has been carried out on defined, reconstituted systems we cannot define their relative contributions. It is therefore convenient to talk about changes in membrane fluidity and its effect on protein functioning. In general, increases in membrane fluidity lead to corresponding increases in the activity of functioning membrane proteins.

Arrhenius plots of enzyme and transport activity

Temperature is a convenient means of manipulating the fluidity of natural and model membranes. Defined synthetic phospholipids when formed into liposomes exhibit distinct, highly co-operative lipid phase transitions where all of the lipid passes from a fluid to a solid state at a particular temperature. This is known as the lipid phase transition temperature (see section 2.6). In contrast to this, biological membranes contain complex mixtures of lipids which do not go from a fluid to a solid state during a single, highly co-operative event. Instead a decrease in temperature can induce certain species of the bulk lipid pool to preferentially segregate out and form clusters or islands of solid lipid within the fluid, bulk lipid pool. This occurs at a particular point called the lipid phase separation temperature (see section 2.10) and can result in a change in the composition and physical properties of the fluid lipid pool. The activity of certain integral proteins which, as freeze–fracture electron microscopy shows, are excluded from solid lipid domains, can be affected to a marked extent by lipid phase separations and it is this that can be detected by Arrhenius plot analysis.

The Arrhenius equation is given by

$$\ln k = \ln \frac{KT}{h} - \frac{\Delta H^*}{RT} + \frac{\Delta S^*}{R} \tag{3.1}$$

where k is the rate constant for the reaction, K is Boltzmann's constant, h is Planck's constant, R is the gas constant, T is the absolute temperature, ΔS^* is the entropy of activation, and ΔH^* is the enthalpy of activation. However, for most practical purposes it is convenient to express this equation as

$$\ln k = \frac{-E_a}{RT} + \ln A \qquad (3.2)$$

where E_a is the energy of activation and A is a constant called the pre-exponential term. Thus a plot of the logarithm of the reaction rate against the reciprocal of the absolute temperature will give a straight line. The activation energy can then be calculated from the slope and the pre-exponential term from the y-intercept.

The Arrhenius plots of the activity of many membrane-bound enzymes exhibit abrupt changes in slope occurring at a particular temperature known as the break point. In many cases the temperature at which the break point occurs corresponds closely to the lipid phase separation temperature of the membrane in which the enzyme is integrated. Indeed, Arrhenius plots of some transmembrane proteins actually exhibit two distinct breaks corresponding to distinct lipid phase separations occurring in each half of the bilayer (see section 4.3.2).

By using this method it is relatively easy to obtain information on the sensitivity of functional proteins to lipid phase separations occurring within the bilayer. As Arrhenius plots are widely used (and abused) it is worth considering in a little detail the experimental precautions and supplementary information necessary to interpret them correctly.

Constructing an Arrhenius plot

The number of data points. Arrhenius plots are commonly constructed over a range of temperatures varying from about 1 °C to about 45 °C. It is imperative that as many data points as possible are collected over this range in order to assess linearity and to identify either break points or discontinuities in the plot. Preferably an initial rate of reaction should be measured every 1 °C, for failure to obtain sufficient data points can lead to erroneous conclusions (see Figure 3.6).

Linearity of assays. Using continuous spectrophotometric or titrimetric assays it is very easy to discern whether initial rates are being measured. However, many systems utilize 'stopped' assays, where substrate utilization or product production is assessed at a single time interval. It is then assumed that a linear rate is maintained over this period and that this reflects the initial rate of reaction. Because of the wide range of reaction velocities that are observed over the temperature range it is often convenient to alter the length of incubation, making it essential to ascertain that linear rates are indeed observed during the incubations. The effects of temperature on activity

108

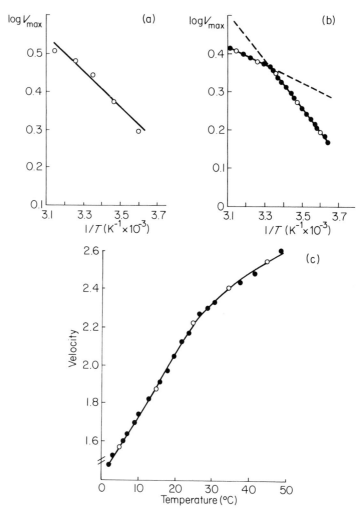

Figure 3.6 The temperature dependence of enzyme and transport functions. (a) If five observations of V_{max} are taken at equally spaced intervals over 5–45 °C it is impossible to determine the form of the Arrhenius plot. Given experimental error this type of data is often interpreted by drawing a straight line through it (correlation coefficient, 0.982). (b) By increasing the number of observations, these data which comprise the Arrhenius plot can be resolved into two intersecting lines. A data point taken every one or two degrees is necessary. Statistical analysis can be used to identify the break point at the intersection of the two lines which yields a minimum error (sum of residuals). These data are now fitted excellently by two intersecting straight lines of correlation coefficients 0.994 and 0.999 respectively (cf. a single line, 0.981). (c) The data of (b) when plotted as velocity versus temperature exhibit a smooth curve

should be shown to be reversible. Checks of this sort ensure that non-linearity caused for example by thermal denaturation, limiting substrate concentration or product inhibition can be spotted readily.

Substrate concentration. Ideally, we need to monitor the change in V_{max} of the reaction with temperature. However, because of the need to collect data points every 1 °C, it would be extremely difficult to carry out a full kinetic analysis at every temperature. Usually a compromise is achieved by using a single saturating substrate concentration with which to perform the assays. However, if there is a dramatic change in K_m over the temperature range then a substrate concentration that was saturating at one temperature might not be saturating at another. This can lead to artefactual shifts in break points or even the occurrence of breaks in what should be linear plots. Precautions should be taken to ensure that the substrate concentration used is saturating ($\times 10 K_m$) at either end of the temperature range. Another possible source of error is, of course, substrate inhibition where the K_i may be markedly temperature dependent. It is then often worth performing the experiment at a few substrate concentrations well above the K_m ($\times 5 - \times 20 K_m$).

pH. Many buffers change pH with temperature. Not only can this affect the rate of enzyme reactions but it can also affect lipid phase separations (see section 2.6).

Interpretation of Arrhenius plots

It is commonly assumed that breaks in Arrhenius plots of membrane-bound enzymes are due to lipid phase separations occurring within the bilayer. Breaks, however, can occur for a variety of other reasons, and because of this are even displayed by Arrhenius plots of the activities of some soluble enzymes, such as fumarase.

There are many possible reasons for not obtaining a linear Arrhenius plot, some of these being:

(i) a phase change in the solvent, for example a lipid phase separation;
(ii) two parallel reactions carried out by two different active centres, on one or more types of enzyme, with different energies of activation;
(iii) an overall process with two successive reactions with different activation energies;
(iv) two forms of the same enzyme which each exhibit different activation energies for the reaction and are in a temperature-dependent equilibrium;
(v) equilibrium of an enzyme between active and inactive forms.

Clearly, the observation of a break in an Arrhenius plot of the activity of a membrane-bound enzyme or transport function is not in itself sufficient to indicate either that the activity of the enzyme responds to changes in its lipid environment or that a lipid phase separation occurs at the break point. It is therefore essential that the suspected lipid phase separation is independently identified using physical chemical methods (e.g. esr and fluorescence polarization studies) or alternatively that the break in the Arrhenius plot may be generated using different methods to modulate the bilayer fluidity. In addition to the methods for manipulating the fluidity of membranes *in situ* which are described in section 3.4, it is possible to solubilize the protein of interest and reconstitute it in different lipid environments (see section 5.4.1). Care must be taken however since the detergents, when substituted for lipid molecules in the annular layer, can modulate the activity of an enzyme due to temperature dependent structural changes of the detergents themselves.

The use of a number of these procedures should demonstrate whether the activity of an integral protein is sensitive to its lipid environment and responds to any lipid phase separation occurring in the bilayer.

It is important to emphasize that, for biological membranes, a break in the Arrhenius plot of a membrane-bound enzyme which occurs due to a lipid phase separation does not mean that an enzyme has moved from a fluid to a solid environment, or vice versa. Proteins will preferentially occupy fluid-lipid domains (see section 3.3.3). The formation or disappearance of clusters or islands of solid lipid within the fluid–lipid domain will merely alter the composition and physicochemical properties of the fluid–lipid pool (see section 2.10) within which the proteins migrate. Such a change in the lipid environment of the proteins triggers a new phenomenon at temperatures on either side of that at which the lipid phase separation occurs. This is reflected in a change in the activation energy for the reaction and hence in the slope of the Arrhenius plot. Whilst this can appear to be very dramatic, the Arrhenius transform is conceptually misleading. For the same data can show a smooth increase in velocity against temperature when these two parameters are plotted directly against each other (see Figure 3.6).

Non-linear Arrhenius plots are exhibited by the activities of a number of membrane functions (see section 3.4). These show relatively sharp breaks occurring at one or more particular temperatures (T_s) with, in most cases, the activation energy for the reaction being less at temperatures above T_s. In some instances, however, inflections (i.e. actual jumps in the Arrhenius plots) can be seen at T_s, corresponding to a sharp increase in the velocity of the reaction at that particular temperature. It is worth while discussing how these differences might arise. If we consider equation (3.1) then, for example, if ΔH^* was decreased by 30 kJ mol^{-1} at a T_s equal to 20 °C we would expect the reaction rate, k, to increase by a factor of 10^5 at T_s. This would give rise to an enormous inflection in the Arrhenius plot at T_s. However, in the majority of instances inflections in Arrhenius plots of enzyme or transport activities have either not been observed or have been very small (less than twofold). From equation (3.1) it is clear that one way of explaining the apparent absence of a marked change in k at T_s is for there to be an exactly compensating effect in the entropy of activation (ΔS^*). For example, in a situation where there is no inflection, then at T_s the velocity of the reaction for the high-temperature state (suffix h) is identical to that of the low-temperature state (suffix l), thus

$$\ln k_h = \ln k_l \tag{3.3}$$

and so from equation (1) we can write:

$$\frac{\Delta S_h^*}{R} - \frac{\Delta H_h^*}{RT_s} = \frac{\Delta S_l^*}{R} - \frac{\Delta H_l^*}{RT_s} \tag{3.4}$$

which upon rearrangement gives,

$$\Delta H_l^* - \Delta H_h^* = T_s(\Delta S_l^* - \Delta S_h^*) \tag{3.5}$$

So in order to prevent a large increase in the velocity of the reaction (k) as the temperature passes through T_s, any change in the enthalpy of the reaction ($\Delta\Delta H^*$) must occur hand-in-hand with an exactly compensating change in the entropy of reaction ($\Delta\Delta S^*$).

That this model puts such constraints on the processes which give rise to the changes in enthalpy and entropy of the reactions at the break temperature (T_s) makes it rather unsatisfactory, although it may well be obeyed in certain cases.

An alternative, highly adaptable approach of explaining this phenomenon is by the use of an enzyme (protein)–lipid phase diagram. This for biological membranes is going to be a highly complex (see sections 2.10 and 3.3.3) system of a number of phases which will be influenced by changes in their immediate environment caused by temperature, pH, divalent cations and ionic strength. If we take a simple case of an enzyme and a lipid and make the assumption that they are only partially miscible in the fluid and solid lipid states (see 2.10 and 3.3.1, 3.3.4), then the membrane will contain two phases at temperatures above and below T_t, namely a solution of enzyme and lipid together with a pure lipid phase. At T_t three phases will co-exist, these are a solution of enzyme and fluid lipid together with pure solid lipid and pure fluid lipid. Over the change in temperature that spans the phase transition the lipid content of the enzyme–lipid solution will change (see section 3.3.3). If we assume that the maximal activity of the enzyme is proportional to the nth power of the lipid content of the enzyme–lipid solution then the apparent activation energy of the reaction, ΔH_{app}^* is given by

$$\Delta H_{app}^* = \Delta H^* + n\,\Delta H \tag{3.6}$$

where ΔH is the heat of solution of the lipid. However, although it can be demonstrated that the solubility of a solute (lipid) is a continuous function of temperature even through the phase transition of the pure substance, the rate of change of solubility will alter at the transition temperature. Thus, ΔH will change when the lipid undergoes a phase transition, being less above than below the transition temperature. On this basis we would expect that the apparent activation energy observed for membrane-bound enzymes will change at T_t, being greater at temperatures below T_t than above. In the majority of cases this is indeed what happens.

The conformation of enzyme and transport molecules undoubtedly changes during their reaction sequence. It is reasonable therefore to expect the conformational flexibility of the protein to be influenced by the physical properties of the lipid around it and the compressibility of the enzyme lipid solution. For pure lipids, it has been shown that at the transition temperature where two lipid phases (solid and fluid) co-exist, the bilayer is highly compressible, so the compressibility of an enzyme–lipid solution might then be approximately proportional to the lipid content of this solution.

If this latter model is correct, then it is clear that the activation energy of the rate-limiting step of the reaction cannot be deduced from the Arrhenius plot as an apparent activation energy (ΔH^*_{app}) is in fact being measured. The change in activation energy, above and below a lipid phase separation, given by most enzymes is roughly in the range of $20–120$ kJ mol^{-1}. If one compares this to the latent heat of the solid–fluid phase transition of dipalmitoyl choline, 41 kJ mol^{-1}, then from equation (3.6) we can see that n would have a value in the region of $\frac{1}{2}$–3. The phase change causing the alteration in the observed activation energy need not necessarily originate in the enzyme–lipid solution, for a change occurring in the pure lipid phase will induce changes in the enzyme–lipid solution. This has particular relevance to biological membranes which consist of several distinct phases that are not completely miscible even in the fluid state observed at physiological temperatures (see section 2.10). Thus, changes in the concentrations of cholesterol, integral proteins, divalent cations, pH and ionic strength can all affect the composition and states of various phases with concomitant effects on the activity of functioning integral proteins. These effects can readily be discerned by Arrhenius plot analysis as this model would indeed predict. However, in order to understand it all fully we need to know a great deal more about the precise physical interactions of lipids with proteins, their influence on protein structure and function, and the changes that occur in the protein–fluid-lipid solutions during lipid phase separations.

Arrhenius plots of enzyme and transport activities in model and biological membranes

The best-defined model system that has been examined so far is the ATP-dependent Ca^{2+}-pump from rabbit muscle sarcoplasmic reticulum. This pure protein has been inserted into vesicles consisting of a single, synthetic phospholipid species (see section 5.5.2). When the bilayer consisted of dipalmitoyl phosphatidyl choline and the protein was present at similar lipid to protein ratios to those seen in the native membranes, then two distinct breaks could be seen in Arrhenius plots of its ATPase activity. The upper break at about 42 °C reflects the main lipid phase transition of the bulk lipid pool and the lower break, at around 30 °C, a structural rearrangement of the annular lipid. Both of these phenomena could be identified by esr analysis using either a fatty acid spin probe or TEMPO, and the temperature of onset of these events could be depressed by benzyl alcohol.

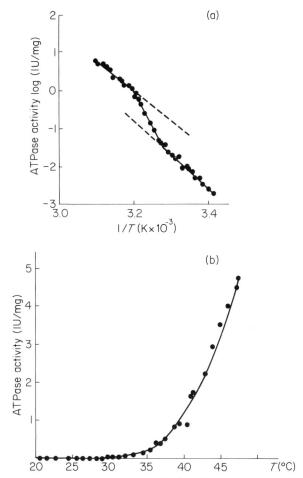

Figure 3.7 Arrhenius plot of the activity of Ca^{2+}-ATPase in a model lipid bilayer. In native sarcoplasmic reticulum the Arrhenius plot of the Ca^{2+}-ATPase exhibits a single break which corresponds to a lipid phase separation (cluster formation) in the bilayer. After lipid substitution with dipalmitoyl phosphatidyl choline each, Ca^{2+}-ATPase can be obtained in a complex with 1500 lipid molecules. Arrhenius plots (a) exhibit a well-defined break at around 40 °C which reflects its sensitivity to the lipid phase transition occurring in the bulk lipid pool of DPPC. A further break occurs at around 29 °C. This is due to a structural rearrangement of annular lipid which influences the activity of the protein. Activity versus temperature plots (b) exhibit a smooth change. (Reprinted with permission from Hesketh, T. R. *et al.* (1976) *Biochemistry*, 15, 4145–4151. Copyright (1976) American Chemical Society)

114

At temperatures below 42 °C, the lipid phase transition of dipalmitoyl phosphatidyl choline, the Ca^{2+}-ATPase molecules together with their annular lipids cluster into a protein-rich phase within the confines of the bilayer. The lipids occupying the annular domain are sufficiently perturbed, by virtue of their interactions with this integral protein, that they cannot undergo a highly co-operative transition. The increased disorder of their acyl chains allows the Ca^{2+}-ATPase to continue functioning well below the bulk lipid phase transition at 42 °C. However, at 30 °C these annular lipids undergo a structural rearrangement whereupon they become markedly more rigid. This dramatically reduces the Ca^{2+}-ATPase activity at all temperatures < 30 °C (see Figure 3.7), presumably by constraining the conformational flexibility of this integral protein.

The appearance of breaks in Arrhenius plots of the activity of a functioning integral protein that reflect changes in the conformation of lipids in the annular domains depends very much on the nature of the individual protein and the lipid environment in which it is placed. Certainly for proteins embedded in their native biological membranes there is no evidence, as yet, for their exhibiting breaks other than those reflecting lipid phase separations in the bulk lipid pool. Indeed, a theoretical study (see Figure 3.8) which admittedly only considers hydrophobic interactions, demonstrates that for

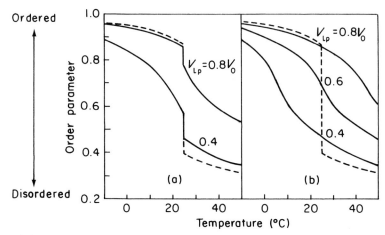

Figure 3.8 The temperature dependence of lipid order in the annular domain. This shows a theoretical consideration of the lipid ordering that occurs in the annular domain of an integral protein inserted in a fluid lipid bilayer at low protein concentrations (a) and at the higher protein concentrations more likely to be encountered in biological membranes (b). Calculations have been carried out assuming various strengths of lipid–protein interactions. The larger the value of V_{Lp} the greater the interaction. The transition of the bulk lipids is given by the dotted line. Through interactions with protein the annular lipid loses its highly co-operative change in order and the event is diminished and can be smeared out. (Reprinted with permission from Marcelja, S. (1976) *Biochem. Biophys. Acta*, 455, 1–7)

relatively weak interactions between the lipid and proteins, the region of rapid change in lipid order occurring in the annular domain would occur at a temperature below the transition of the bulk lipid pool. In systems with stronger interactions between the lipids and integral proteins, the temperature at which the region of rapid change occurs approaches that at which the transition occurs in the bulk lipid pool. However, further increases in interaction lead to such a broadening of this effect that any transition in the annular lipid pool is completely smeared out. So except for situations where the lipid–protein interaction was particularly weak, any annular transition would be experimentally indistinguishable from the bulk transition either because it occurred at a temperature very close to it or because it was smeared out. The Ca^{2+}-pump is normally efficiently sealed into the sarcoplasmic reticulum membrane, suggesting that it interacts tightly with the lipids in the annular domain. Indeed, Arrhenius plots of its activity in native bilayers only exhibit a single break, reflecting cluster formation within the bulk lipid pool, with no distinct break due to rearrangements of annular lipid. However, when the Ca^{2+}-pump is reconstituted in a bilayer consisting only of phosphatidyl choline then a considerable leak (section 3.3.1) occurs at the lipid–protein interface (annulus). Presumably in the model system the single lipid species only interacts weakly with the protein and this is reflected by the distinct, depressed annular lipid transition.

Many studies have been carried out on enzyme and transport processes associated with biological membranes. In a large number of cases breaks in the Arrhenius plots of these activities have been correlated with lipid phase separations occurring in the membranes. It is worth discussing here a few examples of membranes which have been investigated in some detail.

In *hamster liver plasma membranes* there are two quite distinct lipid phase separations which occur at 13 °C and 26 °C and which have been attributed to lipids in the outer and inner halves of the bilayer respectively (see section 4.9). These lipid phase separations can be detected with a variety of physical probes and may be manipulated by lipid fusion techniques or by the addition of benzyl alcohol. It is thought that this asymmetry arises out of the known asymmetry of the phospholipid composition of the membranes. From detailed Arrhenius plots of the activity of a number of integral membrane enzymes it was found that transmembrane enzymes were sensitive to both lipid phase separations. However, integral enzymes with their functional, globular portion predominantly associated with one or other half of the bilayer were only sensitive to the lipid phase separation that occurred in that particular half (see section 4.3.2). A similar situation has been noted in the plasma membrane from rat liver. These observations imply that the two halves of the liver plasma membrane express very different fluidities and that the proteins that penetrate them sense the lipid phase separations occurring within the bulk lipid pool. In no instance was there evidence for distinct transitions due to lipids in the annular domain.

It may at first sight seem rather strange that a number of biological

116

membranes containing high concentrations of cholesterol exhibit lipid phase separations. For cholesterol at the concentrations found in many plasma membranes, typically 0.6–0.8 moles of cholesterol per mole of phospholipid, smears out the lipid phase transition of a single defined phospholipid species (see section 2.6). However, as we have seen before (section 2.11), in bilayers consisting of mixtures of phospholipids, cholesterol exhibits a distinct preference for interaction with certain species of phospholipids, notably sphingomyelin and those with unsaturated acyl chains. The net effect of this is the lateral segregation of cholesterol within the bilayer, producing cholesterol-rich and cholesterol-poor domains. This lateral organization within the biological membrane allows lipid phase separations to occur either within the cholesterol-poor pool or between these two pools. From freeze–fracture studies (see section 3.3.4) we know that the integral proteins segregate into cholesterol-poor regions of the bilayer where they exclude cholesterol from their annular domains and sense lipid phase separations occurring in the biological membranes (Figure 3.9).

3.3.3   Lateral distribution of proteins in membranes

When glycophorin, Ca^{2+}-ATPase, band 3 and rhodopsin have each been reconstituted into vesicles of a single defined synthetic phospholipid, the proteins clearly segregate into distinct fluid-lipid domains at temperatures below that of the phase transition (Figure 3.10). Two distinct phases are formed, one of pure lipid and the other of an aggregated protein–lipid solution. The (annular) lipid surrounding the protein at these temperatures in these model systems is in a more disorganized, fluid state than the crystalline bulk lipid by virtue of its interaction with the protein (sections 3.3.1, 3.3.3).

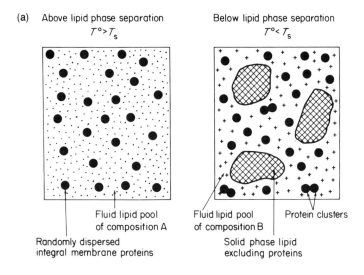

(a)   Above lipid phase separation
$T° > T_s$

Below lipid phase separation
$T° < T_s$

Fluid lipid pool
of composition A

Fluid lipid pool
of composition B

Protein clusters

Randomly dispersed
integral membrane proteins

Solid phase lipid
excluding proteins

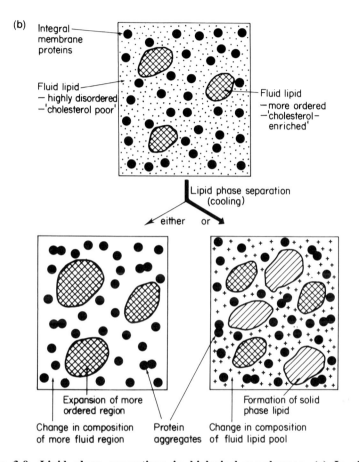

(b)

Integral membrane proteins

Fluid lipid
— highly disordered
—'cholesterol poor'

Fluid lipid
—more ordered
—'cholesterol-enriched'

Lipid phase separation
(cooling)

either   or

Expansion of more
ordered region

Formation of solid
phase lipid

Change in composition
of more fluid region

Protein
aggregates

Change in composition
of fluid lipid pool

Figure 3.9 Lipid phase separations in biological membranes. (a) In cholesterol (sterol)-free membranes, as the temperature is lowered below the phase transition temperature, clusters of solid phase lipid form within the bulk lipid pool. Over a finite temperature range both solid and fluid lipid co-exist and the composition and properties of the fluid lipid pool continually change as the high-melting-point (less fluid) lipids are selectively recruited into the growing solid clusters. Integral membrane proteins are excluded from regions of solid phase lipid and sense the changed composition and size of the environment. (b) In cholesterol-containing membranes which exhibit lipid phase separations sensed by integral membrane proteins, both lipids and proteins are inhomogeneously distributed within the bilayer. The proteins preferentially occupy cholesterol-poor regions. Lowering the temperature causes lipid phase separations to occur as in (a). These produce further areas of lipid from which protein is excluded and affect integral protein functioning both by changing the properties of its environment and by inducing aggregation due to the small fluid lipid pool in this hypothetical model.

This prevents it undergoing the highly co-operative fluid to solid phase transition.

The aggregation of proteins forming protein-enriched, lipid–protein solutions can happen at temperatures below that at which phase separations occur in biological membranes. This has been observed by freeze–fracture analysis of the plasma membranes of erythrocytes, *Acholeplasma laidlawii*, *Tetrahymena pyriformis*, *Bacillus stereothermophilus* and the *lac* carrier proteins in *Escherichia coli*.

A rather striking experiment that emphasizes the preference of globular integral proteins for fluid-phase lipid has been carried out with the sarcoplasmic reticulum Ca^{2+}-ATPase reconstituted in defined lipid environments. This enzyme exhibits a far greater activity when inserted into the highly fluid lipid, dioleoyl phosphatidyl choline (DOPC), than in dioleoyl phosphatidic acid (DOPA). The addition of high (20 mM) concentrations of a divalent cation (Mg^{2+}) inhibits the enzyme only when it is inserted in pure DOPA and not in pure DOPC. This is because divalent cations increase the rigidity of acidic phospholipids by promoting the formation of the gel-phase, whereas they have little effect on neutral phospholipids (see section 2.9). Reconstituted in a DOPA : DOPC mixture, the enzyme exhibits an activity intermediate between that seen by complexes with the pure lipid species. This reflects the mixing of DOPA and DOPC to form a bilayer of intermediate fluidity. However, the addition of high Mg^{2+} to the enzyme in the DOPA : DOPC mixture produces an increase in activity to a level close to that exhibited by the enzyme in the pure DOPC complexes. This is because the divalent cation interacts with the acidic phospholipid, causing it to separate out as a distinct phase of solid lipid. The net effect of this is to produce a fluid lipid phase enriched in DOPC. Freeze–fracture analysis shows that the Ca^{2+}-ATPase is excluded from gel-state lipid and is found to occupy the fluid-state lipid regions. It thus exhibits an increased activity due to the heightened fluidity of its DOPC-enriched environment.

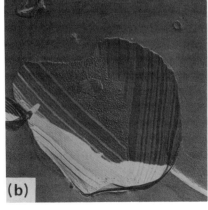

(a)

(b)

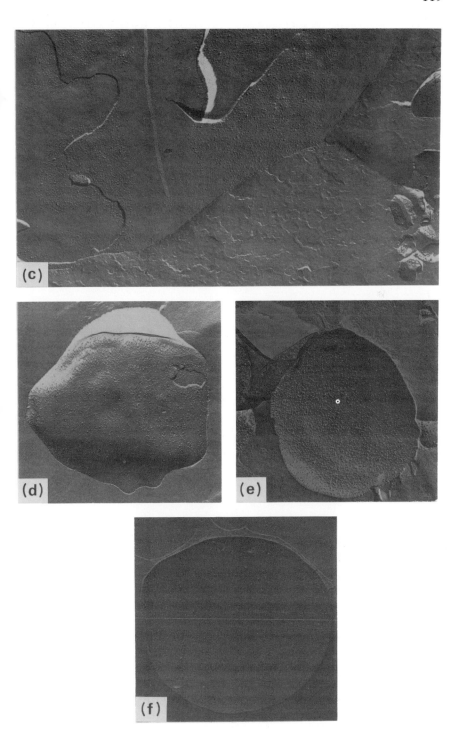

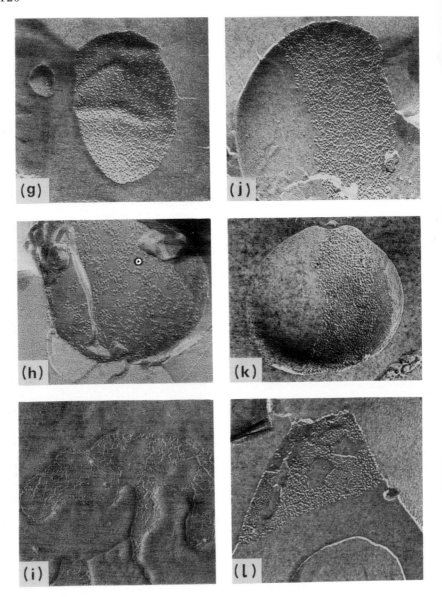

Figure 3.10 Protein clustering. In freeze–fracture micrographs of a two-component mixture of DMPC plus rabbit sarcoplasmic reticulum Ca^{2+}-ATPase, the protein particles are randomly distributed (a) in the fluid (jumbled) lipid when quenched from a temperature (30 °C) above the phase transition (23 °C). However, when quenched from below the phase transition (13 °C) the protein particles are excluded (b) from the banded regions of solid ($P_{\beta'}$) lipid. They form a separate phase of aggregated proteins surrounded by relatively disordered lipid. Similar results have been obtained for other globular proteins, e.g. bacteriorhodopsin and rhodopsin in model systems. In mixtures of the Ca^{2+}-ATPase plus DMPC, but containing 10 mol per cent cholesterol, two phases one 'protein-rich' and the other 'protein-poor' (c) were observed at temperatures (15 °C)

which were below the lipid phase transition of DMPC. A reversible decrease in the amount of 'protein-poor' phase occurred as the sample temperature was raised (d, 19.5 °C) until a random distribution was obtained at $T > T_t$ (e, 31 °C; f, 25 °C). If mixtures of the globular protein, bacteriorhodopsin and DMPC are quenched at temperatures (30 °C) above the lipid phase transition then the particles (of protein) are randomly dispersed (g). However, if cholesterol is incorporated into these bilayers a marked segregation takes place which is seen at 10 mol per cent (h) and 29 mol per cent (i, j) cholesterol. At very high cholesterol concentrations of 39 mol per cent and above (k, l) then the rhodopsin packs together in extremely dense clusters. These studies show that cholesterol can induce protein segregation into 'cholesterol-poor' phases both above and below the phase transition of the lipid. (a–f) Reprinted with permission from Kleeman, W., and McConnell, H. M. (1976) *Biochem. Biophys. Acta*, 419, 206–222; (g–l) reprinted with permission from Cherry, R. J., Muller, U., Holenstein, C., and Heyn, M. P. (1980) *Biochem. Biophys. Acta*, 596, 145–151

It is possible that certain membrane proteins could exhibit a non-random distribution within the fluid-lipid bilayers exhibited at physiological temperatures by most biological membranes. This might occur if the lipid composition gave rise to fluid–fluid immiscibility (see section 2.10) and the membrane protein exhibited a preference for interaction with a lipid species that was non-randomly distributed in the plane of the bilayer. At present, examinations have not been sufficiently detailed to assess whether this does occur to any significant extent in biological membranes. However, the use of bifunctional reagents to cross-link PE and PS to integral erythrocyte proteins, have demonstrated that the fatty acyl chain composition of this bound lipid differed from that found with free PS and PE. This is consistent with integral proteins being able either to segregate out specific lipids from the bulk lipid pool to constitute their annulus or to associate preferentially with specific fluid–fluid immiscible lipid phases. Indeed both processes may be occurring.

A further factor that influences the lateral distribution of integral proteins within the membrane is cholesterol. The incorporation of cholesterol into lipid bilayers at the concentrations found in many biological membranes (> 0.6 mol cholesterol/mol phospholipid) produces cholesterol-poor and cholesterol-rich domains. This can occur even in model systems containing but a single species of phospholipid, but is exacerbated in bilayers containing a heterogeneous mixture of phospholipid species, as cholesterol can preferentially interact with particular species (see section 2.11). Freeze–fracture analysis of model membranes containing defined lipid and protein species clearly demonstrates that integral proteins preferentially segregate in the cholesterol-poor or deficient domains (Figure 3.10). This would also appear to be true of integral proteins inserted into biological membranes. Evidence supporting this has been obtained using the polyene antibiotics, filipin and amphoptericin B. These avidly interact with cholesterol (see section 7.4.3) to form complexes which yield identifiable lesions upon freeze–fracture analysis. The distribution of these lesions over the fracture face is non-random and can be used, in biological membranes, to give an indication of areas of membrane that are particularly rich in cholesterol. The areas that are rich in lesions, and so presumably in cholesterol, are devoid of integral proteins, giving further support to the notion that integral proteins preferentially reside in cholesterol-poor domains.

Model systems using the Ca^{2+}-ATPase and bacteriorhodopsin (see section 5.5) have been used to investigate this. In vesicles of dimyristoyl phosphatidyl choline (DMPC) and cholesterol, under conditions where a phase of pure DMPC can be made to separate out from DMPC : cholesterol, these integral proteins are found exclusively in the pure DMPC phase. This segregation into cholesterol-free or cholesterol-poor regions of the bilayer is sensitive to both temperature and the cholesterol content of the vesicles. Indeed very high concentrations of cholesterol can lead to aggregation of the proteins, mimicking the effect of decreasing temperature.

The major erythrocyte integral protein, band 3, also appears to exhibit an aversion to the cholesterol-rich areas of the erythrocyte membrane. For indeed quite large variations in the cholesterol : phospholipid ratio appear to have only small effects on its rotational diffusion. This would be quite consistent with it occupying relatively cholesterol-poor regions of the bilayer, and is supported by freeze–fracture analysis which indicates its non-random distribution. However, its clustering is aggravated by lowering the temperature and this leads to a reduction in its rotational diffusion.

Thus, integral membrane proteins need not exhibit a uniform distribution over the membrane surface but can be laterally segregated. This is undoubtedly of functional significance and open to perturbation by agents or processes that alter membrane fluidity or the composition of the membrane.

Mechanisms, other than the organization of bilayer lipids, which can affect the lateral organization of membrane proteins are discussed fully in section 2.13.

3.3.4 Bilayer thickness

Integral proteins that span the bilayer have defined polar surfaces exposed at either end of the molecule. These interact with both the headgroup region of the bilayer and the aqueous solvent. Between the two polar regions is an apolar area on the protein that interacts with the phospholipid fatty acyl chains in the core of the bilayer. Sealed in its native biological membrane we would expect that these surfaces on the protein would be complemented exactly by lipids within the bilayer (Figure 3.11). However by altering the thickness of the bilayer this interaction would be disturbed (Figure 3.12), leading to an alteration in the functioning of the protein. Indeed inserting the Ca^{2+}-pump or Na^+/K^+-ATPase (see section 5) into bilayers of synthetic short-chain phosphatidyl cholines fails to reconstitute their function. However, by merely adding an apolar agent like decane, which partitions into the centre of the apolar core of the bilayer and increases its thickness, the activity of the Ca^{2+}-pump can be regained. This activating effect of 'bilayer thickening

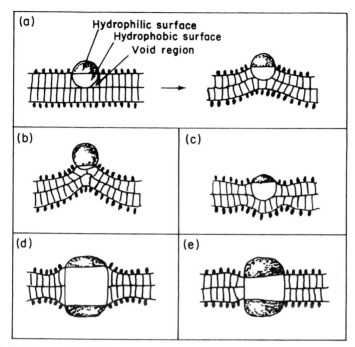

Figure 3.11   Interaction of integral proteins with the lipid bilayer. In order for integral membrane proteins to pack into bilayers without the occurrence of void regions the bilayer may become distorted (a–c) to accommodate them. With transmembrane proteins the size of the apolar domain may lead to increases or decreases in bilayer thickness at this interface (d, e). (Reprinted with permission from Israelachvili, J. N. (1977) *Biochim. Biophys. Acta*, 469, 221–225)

agents' is only seen in model systems using a single species of short-chain (< 12 C) phosphatidyl cholines. When the bilayer is composed of a single phospholipid species whose acyl chain length is similar to that found *in vivo* (C_{18}–C_{22}) then these agents actually inhibit enzyme activity. Presumably in this case the 'over-thickening' of the bilayer again leads to lack of complementation of polar and apolar surfaces supplied by the protein and the bilayer.

Although decane can have very marked effects on the functioning of the Ca^{2+}-pump in certain model systems, its effects are much less pronounced in the native membrane. It is likely that a biological membrane can buffer a protein against such changes. This can be achieved by the segregation of specific lipids to the annular domain, by the induction of a lipid phase separation of the phospholipids with shorter acyl chains into areas enriched in decane, and by the presence of cholesterol. Integral proteins other than transmembrane proteins should not be affected by agents that alter bilayer thickness.

Agents that alter bilayer thickness may also have effects on bilayer fluidity and so the relative contributions of these effects must be assessed. The converse is also true. However, small molecules like the neutral anaesthetic benzyl alcohol and the charged local anaesthetics (see sections 3.4.1, 3.4.9), which all increase bilayer fluidity, are located at or near the headgroup region of the bilayer. This results in a net lateral expansion of the bilayer and not an increase in its thickness.

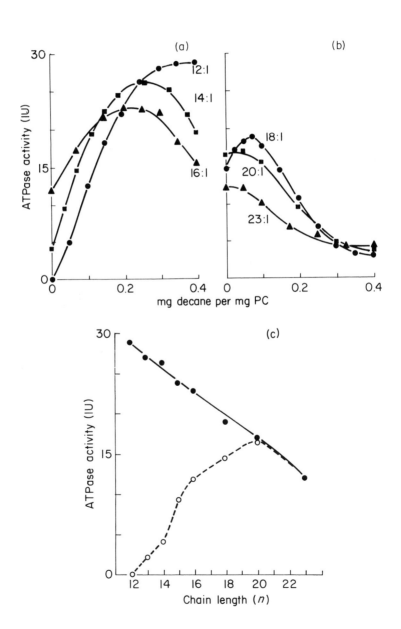

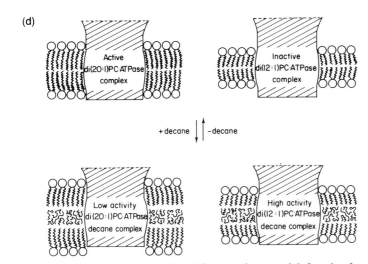

Figure 3.12 A bilayer of appropriate thickness is essential for the functioning of Ca^{2+}-ATPase. The Ca^{2+}-ATPase from sarcoplasmic reticulum when isolated by lipid substitution in defined complexes exhibits activities which are dependent upon the acyl chain length of the phosphatidyl cholines used (a, b). Treatment of short-chain phospholipids with decane, which can thicken bilayers, activates the complexes (a). However, treatment of long-chain phospholipid–Ca^{2+}-ATPase complexes with decane, inhibits activity (b). This is summarized in (c) which shows the activity for the complex alone (○) and with optimal decane (●). For the Ca^{2+}-ATPase to function efficiently it needs a bilayer with a thickness the equivalent of a phospholipid with at least a 20 carbon chain. If the bilayer gets any thicker, then this 'over-thickening' leads to inhibition (see d). The highest activities are seen with short-chain phospholipids thickened with decanol. This is because such bilayers are extremely disordered and highly fluid compared with those containing longer-chain phospholipids. (Reprinted with permission from Johannsson, A., Keightley, C. A., Smith, G. A., Richards, C. D., Hesketh, T. R., and Metcalfe, J. C. (1981) *J. Biol. Chem.*, 256, 1643–1650)

Summary

Integral membrane proteins have a lipid annulus which exchanges with lipid molecules in the bulk lipid pool more slowly than adjacent lipids in the bulk pool.

Specific lipids may be segregated into the annulus which help to seal the protein into the membrane.

The activity of integral membrane proteins with their functional domains embedded in the bilayer increases with elevations in bilayer fluidity.

When phase separations occur in the bulk lipid phase these may result in breaks in the Arrhenius plots of enzyme activity.

Annular lipids, however, do not undergo a normal phase transition.

3.4 METHODS OF MANIPULATING MEMBRANE FLUIDITY

Arrhenius plots of the activity of membrane functions can tell us whether or not the lipid bilayer can influence the activity of an integral protein, but it cannot describe the magnitude of the effect of changes in membrane fluidity on the activity. For although lipid fluidity does increase with increasing temperature, we would expect the activity of the membrane function to increase irrespective of this: until of course the protein was denatured by elevated temperature. This means that we require methods of manipulating bilayer fluidity at a constant temperature. A major problem in ascertaining the role of membrane fluidity in regulating a membrane function, however, is that many of the agents used to modify membrane fluidity exert other, more complex effects either directly on the protein itself or on the organization of the bilayer. It is important to be aware of these factors, not only because they can confuse the issue but also because they lend insight into the structure of the bilayer, the functioning of the proteins and the toxicity of the agents.

3.4.1 Anaesthetics

A property that all local anaesthetics have in common is the ability to perturb bilayer fluidity: the majority acting to increase it. The neutral, local anaesthetics are particularly useful as their neutrality precludes any electrostatic interactions with either the phospholipds or the proteins in the membranes. Examples of these types of compounds which act to increase bilayer fluidity are benzyl alcohol, short-chain ($< C 10$) n-alkanols and carbocaine. These molecules are all soluble to various extents in water and are thus readily added to test systems, where they partition into the membrane bilayer. Conversely, washing the treated membranes with buffer readily removes the anaesthetic, allowing the reversibility of any effects to be demonstrated.

Treatment of synthetic phospholipid vesicles with such compounds causes a characteristic depression of the lipid phase transition by virtue of the fact that they act as an impurity in the bilayer achieving a 'classical depression of the freezing point'. In biological membranes they depress the temperature at which lipid phase separations occur (see section 3.3.3) and increase the fluidity of the bilayer (see Figure 3.13). Benzyl alcohol (40 mM), for example, increases the fluidity of rat liver plasma membranes by an amount equivalent to a 6 °C rise in temperature. Such an increase in bilayer fluidity caused by benzyl alcohol has very different effects on various integral proteins associated with this membrane; some are much more sensitive to changes in bilayer fluidity than are others. For example, phosphodiesterase I and the Mg^{2+}-ATPase, both of which are ectoenzymes, are relatively insensitive to changes in bilayer fluidity yet Arrhenius plots of their activity clearly sense lipid phase separations occurring in the bilayer.

In any study of this nature it is important to assess whether the anaesthetic itself has a direct effect upon the protein. This is most conveniently carried out by solubilizing the membrane protein and then testing this preparation with the anaesthetic. Table 3.2 shows that benzyl alcohol had little effect on the majority of the solubilized enzymes from rat liver plasma membranes,

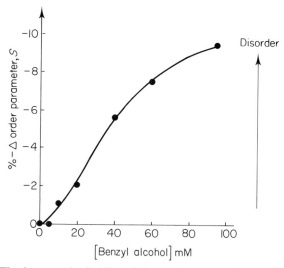

Figure 3.13 **The increase in fluidity of liver plasma membranes that is caused by benzyl alcohol. Benzyl alcohol increases the fluidity of liver plasma membranes as indicated by a decrease in the value of the order parameter, S, of a nitroxide-labelled fatty acid spin probe incorporated into the membrane. (Reprinted with permission from Gordon, L. M.** *et al.* **(1980)** *J. Biol. Chem.***, 255, 4519–4527)**

Table 3.2 Sensitivity of integral liver plasma membrane enzymes to benzyl alcohol-mediated increases in bilayer fluidity

Enzyme activity (optimal benzyl alcohol concentration)	Percentage of original activity in the presence of an optimal concentration of benzyl alcohol (membrane-bound state)	
	Membrane-bound	Detergent-solubilized
Basal adenylate cyclase (20 mM)	120	N.D.[a]
Fluoride-stimulated adenylate cyclase (40 mM)	140	95
Glucagon-stimulated adenylate cyclase (40 mM)	160	N.D.
5'-Nucleotidase (20 mM)	140	100
Cyclic AMP phosphodiesterase (50 mM, integral enzyme)	170	100
Na^+/K^+-ATPase (10 mM)	260	N.D.
Phosphodiesterase I (100 mM)	120	110
Mg^{2+}-ATPase (100 mM)	15	15

[a]N.D.—not determined.
From Gordon, L., M. Sauerheber, R. D., Esgate, J. A., Dipple, I., Marchmont, R. J., and Houslay, M. D. (1980) *J. Biol. Chem.*, **255**, 4519–4527.

some were even inhibited. Clearly then, those that were activated in the membrane-bound state by benzyl alcohol were responding to increases in membrane fluidity.

A further potential complication when using local anaesthetics is that at high concentrations they can lead to inhibition of membrane-bound enzymes. This is not due to either a direct effect upon the protein or to a 'too-fluid' bilayer but is thought to be caused by the anaesthetic competing for sites on the protein that are normally occupied by annular lipid. Occupancy of these sites by the anaesthetic leads to the loss of function of the protein (see section 3.5.8). The susceptibility of a protein to such inhibition clearly depends on the strength of the lipid–protein interaction and the affinity of the sites for the anaesthetic. Na^+/K^+-ATPase would seem to be particularly susceptible, for although it is dramatically activated by the small increases in bilayer fluidity achieved by low concentrations of benzyl alcohol, it is then strongly inhibited by increasing benzyl alcohol concentrations above this level (see Table 3.2).

One intriguing property of charged anaesthetics is that they possess the ability to perturb one or other halves of the bilayer selectively since, in the majority of cases, the acidic phospholipids predominate in the cytosol facing half (see section 4.5). The result of this is that the cationic anaesthetics will preferentially act on this half of the bilayer, whereas the anionic anaesthetics will preferentially act on the other half. This asymmetric perturbation of the bilayer can lead to changes both in the shape of cells and also in the direction in which cilia beat in the *Paramecium*. In the latter case it is the interaction of cationic anaesthetics with the inner surface of the plasma membrane that opens a voltage-sensitive Ca^{2+}-gate. This leads to Ca^{2+} influx and ciliary reversal. However, anionic anaesthetics are ineffective, demonstrating that the Ca^{2+}-gate is opened by an asymmetric perturbation of the bilayer. Consistent with these anaesthetic effects being modulated through the Ca^{2+}-gate protein is the observation that ciliary reversal cannot be effected in the *Paramecium Pawn* mutant. For this organism lacks a functional voltage-sensitive Ca^{2+}-gate oriented within its membrane. The ability to fluidize preferentially one or other half of the bilayer yields a tool for selectively modulating the activity of particular proteins. An interesting example of this is adenylate cyclase which is inserted asymmetrically in the plasma membrane with its functioning, globular portion localized at the cytosol side where the negatively charged phospholipids predominate. Under these conditions cationic anaesthetics activate the enzyme by increasing the fluidity, whereas anionic ones have no effect (Figure 3.14). However, in the presence of hormone which binds to a site on the receptor exposed at the opposite side of the bilayer, a transmembrane complex of receptor and adenylate cyclase is formed (see section 7.7). The activity of adenylate cyclase in this complex now becomes sensitive to anionic anaesthetics, which increase the fluidity of the external half of the bilayer and can activate the enzyme in this complex. Thus, anionic anaesthetics can selectively activate the hormone-stimulated enzyme.

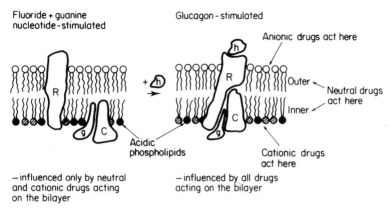

Fluoride + guanine
nucleotide-stimulated

Glucagon - stimulated

Anionic drugs act here

Outer

Neutral drugs
act here

Inner

Acidic
phospholipids

Cationic drugs
act here

– influenced only by neutral
and cationic drugs acting
on the bilayer

– influenced by all drugs
acting on the bilayer

Figure 3.14 **The selective effects of charged anaesthetics on adenylate cyclase activity. In liver plasma membranes the hormone, glucagon (h) stimulates the enzyme (c) adenylate cyclase. In the absence of hormone, the uncoupled catalytic unit only senses the lipid environment of the inner side, whereas in the presence of hormone the transmembrane receptor (R)–guanine nucleotide regulatory unit (g)–catalytic unit coupled complex that is formed senses the lipid environment of both halves of the bilayer. Thus neutral fluidizing agents activate both coupled and uncoupled activities (see Table 3.5) as do cationic drugs which selectively fluidize the inner half of the bilayer where the catalytic unit resides. However, anionic fluidizing agents will act selectively on the external leaflet, thus only having an effect on the coupled activity. (From Houslay, M. D., Dipple, I., and Gordon, L. M. (1981)** *Biochem J.***, 197, 675–681. Reprinted with permission of the Biochemical Society of London)**

Charged anaesthetics, especially cationic species, by virtue of their ability to interact specifically with charged phospholipid species and also to displace membrane-bound Ca^{2+}, will not only increase bilayer fluidity but can alter or induce lipid lateral segregation in biological membranes. The susceptibility of biological membranes to perturbation by charged anaesthetics will thus depend on their phospholipid composition and their cholesterol content.

3.4.2 Genetic methods

Two categories of mutants of *Escherichia coli* which are unable to synthesize unsaturated fatty acids have been identified. In the first of these the activity of β-ketoacyl ACP synthetase, the enzyme responsible for unsaturated fatty acid synthesis, is low. Those in the second category are deficient in β-hydroxydecanoyl thioester dehydrase, the enzyme which introduces *cis* ethylenic bonds at C_{10} of the nascent hydrocarbon chain. These unsaturated fatty acid auxotrophs usually succumb after one generation, but they survive and grow normally if given a monoenoic or other suitable fatty acid. Under normal circumstances unsaturated fatty acyl chains contribute approximately 20 per cent of the total in the membrane, but by coupling a mutation in the synthesis with one in the breakdown of fatty acids, then the fatty acid in the supplement will dominate the acyl chains of the membrane phospholipids. This type of manipulation, not surprisingly, has marked effects on the fluidity of the cell membranes, the temperature at which any lipid phase separation occurs and the activities of functioning membrane proteins. The changes in the temperature at which lipid phase separations occur is reflected in the Arrhenius plots of the activities of the transport processes and integral enzymes associated with the membranes. Such methods have been used in various other micro-organisms, in mycoplasma and in yeast with essentially similar results.

Fatty acid supplements can, however, lead to adaptive changes (see section 3.5.1) in sterol content and the ratio of different headgroup species of phospholipids in an attempt by the organism to maintain a constant bilayer fluidity. This can of course result in a change in bilayer asymmetry and lateral organization. For these reasons this method cannot be used to assess rigorously any changes in fluidity on membrane function. However, it does provide an extremely useful means of manipulating membrane lipid composition *in vivo* and testing for the sensitivity of membrane processes to changes in their lipid environment.

Headgroup composition changes have been effected using auxotrophic mutant strains of *Neurospora crassa* defective in either S-adenosyl-methionine : phosphatidyl ethanolamine methyl transferase or S-adenosyl-methionine : phosphatidyl monomethylethanolamine (dimethylethanolamine) methyl transferase and myoinositol-1-phosphate phosphatase. These mutants, grown in culture with supplements just adequate to support growth, had bizarre phospholipid compositions, but interestingly the total phospholipid content and the ratio of zwitterionic to anionic lipids remained constant. Again many parameters are being altered and this technique, whilst being of some value in ascertaining the requirement of a particular headgroup phospholipid for a membrane function, is again limited to denoting whether or not sensitivity to the lipid environment occurs.

3.4.3 Dietary methods

These can be applied either to cells in culture where the composition of the medium is controlled or to whole animals where the composition of the ingested food is controlled.

Mouse LM cells have been widely used in such studies because they can be grown in a serum-free, defined medium. By supplementing this medium with either choline, serine or ethanolamine, the headgroup composition of membrane phospholipids can easily be altered. However, whilst this does lead to a change in the ratio of the various phospholipid species present, it never leads to total domination by a single type of phospholipid. At the same time that changes in the phospholipid headgroup composition occur, there are concomitant changes in membrane sterol content and in the nature of the phospholipid acyl chains (see section 3.5). However, changes in the fluidity (viscosity) of the membranes do occur, with those from ethanolamine-supplemented cells appearing to be less viscous than those from choline-supplemented cells. These membranes can of course be manipulated further by giving fatty acid supplements.

Yeast and some other eukaryotic cells can be induced to take up and incorporate defined unsaturated fatty acids into their membrane phospholipids by growing them under anaerobic conditions on glucose. This is because these organisms normally produce unsaturated fatty acids by desaturating long acyl chains, provided usually by stearoyl CoA in an oxygen-dependent reaction. Again this produces complex changes in membrane composition.

Using whole animals the membrane lipid composition can be altered by manipulating the diet. The methods used have included the addition or elimination of either saturated or unsaturated fatty acids, the elimination of choline, the elimination or addition of cholesterol and the addition of cholesterol, together with unsaturated or saturated fatty acids. These methods have been shown to alter the lipid composition of various cellular membranes from a variety of organs, in, as one might expect, a rather complex fashion. Most investigations have centred on identifying changes in lipid phase separations and correlating them with breaks in Arrhenius plots of membrane functions rather than defining any changes in fluidity that may have occurred. Again, because of the complex changes that occur in such a situation, it is clearly impossible to relate changes in the activity of a membrane function to altered bilayer fluidity.

These dietary manipulations have, however, provided two important pieces of information, namely that changes in diet can affect membrane composition and also that many membrane functions are sensitive to their lipid environment and hence to their diet. Indeed we know that certain fatty acids, such as linoleate and linolenate, are essential nutrients for animals. This is because animals, unlike plants, are unable to desaturate fatty acids in the w^3 and w^6 positions of the hydrocarbon chain of monoenic fatty acids. The lack of such fatty acids leads to poor growth and the failure of the liver and nervous system to develop normally.

3.4.4 Enzyme inhibitors

By using drugs which act upon the enzymes concerned with the biosynthesis of lipids, the membrane lipid composition can be altered.

Phenethyl alcohol inhibits the synthesis of the various phospholipid classes in *E. coli* to very different extents (phosphatidyl ethanolamine > phosphatidyl glycerol ≫ cardiolipin). As the turnover rates for these species are also very different then this compound leads to remarkable changes in the lipid composition of the cell membrane. This not only affects the ratio of the various headgroup species but also the types of fatty acyl chains associated with them. It would appear that although the fluidity of the cell membranes is changed somewhat, the organism attempts to maintain a constant fluidity even though the headgroup composition is changing (see section 3.5.1).

The antibiotic, cerulenin, blocks the synthesis of both saturated and unsaturated fatty acids in *E. coli* by inhibiting the enzyme β-keto acyl-ACP synthetase. Thus, these cells will take up defined fatty acids from the medium. Another antibiotic, 3-decyonyl-*N*-acetyl cysteamine, specifically inhibits the synthesis of unsaturated fatty acids in *E. coli* by covalently inhibiting the enzyme β-hydroxydecanyl thioester dehydrase. Growth of *E. coli* and various other bacteria in the presence of this antibiotic, only occurs if an unsaturated fatty acid is added to the media. Eukaryotic cells are unaffected by this antibiotic because this mechanism of desaturation does not occur.

Rather complex changes occur in LM cell membranes if the choline analogue, *N*-isopropylethanolamine, is present in the culture medium. This reversibly blocks the uptake of choline into the cells and is itself incorporated into cell lipids as 1,2 diacyl-*sn*-glycero-3-phosphoisopropylethanolamine, leading to a reduction in the amount of phosphatidyl choline and phosphatidyl ethanolamine present. Furthermore, it apparently blocks the desaturation of stearic acid by specifically inhibiting the terminal oxidase activity of the desaturase enzyme complex. Thus, complex changes in both the headgroup ratio, acyl chains and bilayer fluidity occur.

The cholesterol content of membranes can be reduced by inhibiting desmosterol reductase. This leads to a rise in desmosterol, the precursor of cholesterol, which has a double bond in its alkyl side chain. This causes an increase in bilayer fluidity which has been demonstrated to activate the Na^+/K^+-ATPase. Such an effect would be consistent with the observations found using the local anaesthetic benzyl alcohol (see section 3.4.1).

3.4.5 Catalytic hydrogenation

Homogeneous catalysts containing complexes of the transition metal, rhodium I have been used to hydrogenate *cis*-double bonds in the fatty acyl chains of phospholipids in cell and artificial membranes (Figure 3.15). This leads to decreases in bilayer fluidity which have been demonstrated to inhibit the function of transport proteins, enzymes and agglutination reactions occurring in the membranes. The early studies were performed with a catalyst which had to be dissolved in dimethylsulphoxide (DMSO) or tetrahydrofuran in order to introduce them into the system. This suffered from the dual disadvantages that the solvents themselves perturbed the membranes and also that once the catalyst was incorporated into the bilayer it was extremely difficult to remove. More recently, however, a water-soluble homogeneous catalyst has been devised which works with great success. This method now shows great promise as a tool for manipulating membrane fluidity provided that suitable conditions are available both to make the membrane susceptible to penetration, and hence action of the catalyst, and for removal of the catalyst after hydrogenation. Unfortunately, cholesterol severely reduces the amount of catalyst able to partition into the membrane. This, together with the ability of cholesterol to complex unsaturated phospholipids (see section 2.6), reduces the degree of hydrogenation that can be achieved. It has the advantage, however, that it does not result in changes in the ratios of phospholipid species or in cholesterol content of the membranes and so is potentially an excellent method to study the fluidity effects on integral proteins in biological membranes.

3.4.6 Pressure

An increase in pressure favours the close packing of the phospholipid acyl chains, thus it decreases bilayer fluidity. Pressure changes offer a useful, non-perturbing method of manipulating bilayer fluidity and observing changes in protein functioning (see section 3.3.3). Unfortunately, these effects have not been rigorously examined, presumably because of the technical difficulties encountered in studying reactions in pressure vessels.

3.4.7 Calcium

This divalent cation interacts with acidic phospholipids, promoting the formation of solid phase (crystalline state) lipid and in so doing, elevates the lipid phase transition temperature (see section 2.9). In bilayers consisting of mixtures of neutral and acidic phospholipids, calcium clusters the acidic species. This will change the composition of the fluid bulk lipid pool in which the integral proteins are segregated (see section 3.3.3). Depending upon the physical nature of the acidic phospholipids we can expect the fluidity of the fluid lipid pool either to increase or decrease (see section 3.3.3). Millimolar Ca^{2+} has been demonstrated to increase reversibly both the rigidity and the temperature of the lipid phase separations occurring in plasma membranes from rat liver and heart. These effects are sensed by integral proteins whose Arrhenius plots show altered break temperatures. However, although this decrease in fluidity exerts a small inhibitory effect on proteins such as Na^+/K^+-ATPase and adenylate cyclase, by far the greatest proportion of the inhibition caused by Ca^{2+} is due either to its interaction with the substrate, ATP, or by binding to regulatory sites on the protein. Indeed Ca^{2+}-binding studies on biological membranes show two classes of sites of very different affinities. The high-affinity sites ($K_D = 10^3-10^6$ M^{-1}) have been attributed to proteins and the low-affinity sites ($K_D = 10^2-10^3$ M^{-1}) are due to binding to acidic phospholipids. So although Ca^{2+} can have rather interesting effects on bilayer lipid organization and on fluidity, it is not a particularly useful probe for fluidity as it is likely to have more potent effects on proteins for other reasons. To some extent the relative contributions of the various effects of Ca^{2+} can be assessed by studying membrane-bound and soluble sources of an integral enzyme protein.

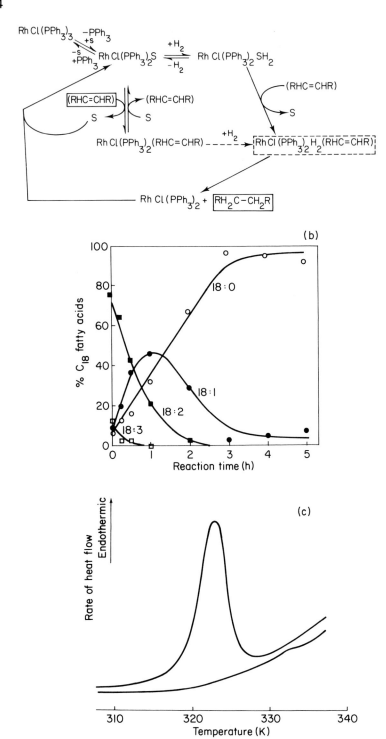

(b)

(c)

3.4.8 Cholesterol

Cholesterol is often used as a tool for manipulating membrane fluidity. Its level in biological membranes may be manipulated *in vitro* by incubation with multilamellar liposomes enriched in cholesterol. These liposomes can be prepared with cholesterol : phospholipid molar ratios of approximately 2, and over a period of hours this exogenous pool equilibrates with that of the membranes. After this treatment, membranes with cholesterol : phospholipid molar ratios in excess of 1 can be isolated. In most instances whole cells have been incubated for prolonged periods at physiological temperatures with the cholesterol-enriched liposomes, but more recently these methods have been adapted so that isolated membranes, incubated at low temperatures (4–10 °C), can be used. A major problem of these methods is fusion of the cholesterol-enriched liposomes with the membranes under study. For whilst this will of course increase the cholesterol content of the membrane it will also introduce exogenous phospholipid into the lipid pool. Fusion can, however, be kept to a minimum by raising the ionic strength of the medium and by adding chelating agents and serum albumin. It is important, however, to assess the extent to which fusion occurs and this can be done by using radio-labelled phospholipids. The effect, if any, of the fraction of fused exogenous lipid that is inevitably introduced into the membrane can be evaluated by using synthetic phospholipids of very different physical properties to form the bilayer or avoided by using liposomes having an idential phospholipid : cholesterol ratio to the original membranes. Treatment of membranes with cholesterol-free liposomes can in a similar fashion be used to deplete membranes of cholesterol.

As an alternative to liposomes, serum lipoproteins can be used to manipulate membrane cholesterol levels by an exchange process (Figure 3.16). Fusion can again be a problem as can the action of contaminating serum components.

Figure 3.15 Catalytic hydrogenation decreases the fluidity of lipid membranes. (a) This outlines the reaction scheme for a homogeneous catalyst based on a transition metal complex. This catalyst, chlorotris (triphenyl phosphine) rhodium I is introduced into membranes with a solvent vector, S, such as dimethyl sulphoxide or tetrahydrofuran. This allows hydrogenation of unsaturated (olefinic) bonds to proceed. (b) Hydrogenation of liposomes of phosphatidyl choline from soya beans. Note the increase in unsaturated 18 : 0 at the expense of C_{18} unsaturated lipids. (c) Unlike the native extract of phosphatidyl cholines from soya beans, which do not undergo a well-defined phase separation/transition, the hydrogenated liposomes display a dramatic endothermic event at a temperature consistent with the production of palmitoyl stearoyl phosphatidyl choline. This process has been successfully applied to biological membranes, although the presence of cholesterol dramatically reduces the effectiveness of hydrogenation. (From Chapman, D., Peel, W. E., and Quinn, P. J. (1978) *Annals. N.Y. Acad. Sci.*, **308**, 67–84. Permission granted by the New York Academy of Sciences)

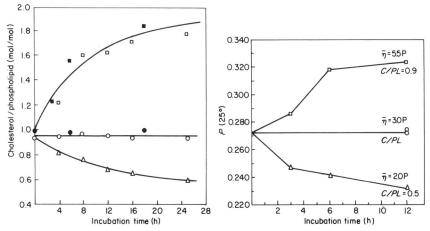

Figure 3.16 Manipulation of membrane cholesterol content. The cholesterol content of biological membranes can be manipulated by incubation with heat-inactivated serum. This takes advantage of exchange processes that occur physiologically between membranes and lipoproteins in the blood. The cholesterol content of the serum can be enriched by adding cholesterol in tetrahydrofuran. A cholesterol-depleting medium can be prepared by mixing cholesterol-free liposomes with the serum and using these as a pool to withdraw cholesterol from the test membranes. Incubation of erythrocytes with cholesterol-enriching (□), depleting (△) and control (○) media give rise to significant changes in cholesterol content of the membranes (a) with concomitant changes in fluidity (b) assessed by fluorescence polarization. (Reprinted with permission from Shinitzky, M. (1978) *FEBS Lett.*, 85, 317–320)

Cholesterol levels can also be modulated by dietary means (section 3.4.3) or by using inhibitors of its synthesis (section 3.4.4). However, these procedures also lead to changes in other lipid components of the bilayer and are unsatisfactory for defined studies. Similarly, the detergent-mediated lipid substitution or titration procedures (section 5.4) will affect the organization of the bilayer. This is because the detergent will 'scramble' the endogenous lipid, leading to the loss of asymmetry.

Polyene antibiotics such as filipin or amphoptericin B can also be used to manipulate cholesterol levels. These bind with high affinity to cholesterol to form membrane channel complexes within the membrane (see section 7.3). Indeed, polyene antibiotics bind so avidly to cholesterol that they can be used to investigate the effect of a reduction in cholesterol levels upon enzyme activities. However, as it is difficult to assess the interaction of both these antibiotics and their cholesterol complexes on the functioning proteins, this method has distinct limitations.

Manipulating cholesterol concentrations of biological membranes will cause complex changes in the lateral organization of bilayer lipids and of proteins. Often it is assumed that the only effect of changing the cholesterol concentration is to alter the fluidity of the membrane. However, at the same time that changes in cholesterol concentration are having effects on bilayer fluidity they also lead to changes in both the composition of the fluid lipid pool in which the proteins diffuse and in the size of this protein–lipid phase (see sections 2.11 and 3.3.2, 3.3.3). Both of these alterations can have equally dramatic effects on the functioning of integral proteins as can changes in bilayer fluidity.

Cholesterol is thus too complex a tool for studying the effects of bilayer fluidity on the functioning of integral proteins. The level of cholesterol in biological membranes changes in various disease, dietary and transformed states (section 3.5) making it worthwhile for study in its own right. As yet very few rigorous investigations have been carried out into its effects on the activity of membrane proteins. Of the proteins that have been studied in their native membranes it would seem that in most instances increasing the cholesterol content causes a progressive and reversible inhibition of their activity.

3.4.9 Fatty acids

The incorporation of fatty acids or fatty alcohols into the lipid bilayer can be used to modulate lipid fluidity. These agents have markedly different effects dependent upon their chain length and the degree of saturation, enabling them to be used both to fluidize and rigidize biological membranes. The main problem with the long-chain molecules is their insolubility in water, which means that to incorporate them in biological membranes they must either be added in an organic solvent or by partition from phospholipid liposomes. Often they are added dissolved in DMSO or ethanol, both of which are agents that can either perturb the bilayer or have direct effects on protein functioning. Once inserted into the bilayer, fatty acids or alcohols are not easily removed, which makes reversibility studies difficult to perform. They have, however, been shown to have dramatic effects on the activity of integral membrane enzymes, e.g. β-OH butyrate dehydrogenase and adenylate cyclase, and when applied with appropriate caution appear to be useful tools for modulating lipid fluidity. Problems can, however, arise with fatty acids which can exert effects directly on proteins, rather than through the manipulation of bilayer fluidity.

Summary

The sensitivity of an integral membrane protein to changes in lipid fluidity depend upon its nature and whether its functional globular region is inserted into the bilayer.

Care has to be taken with agents used to modulate bilayer fluidity that they do not have direct effects on proteins or exert complex changes in membrane composition.

3.5 THE RELATIONSHIP OF BILAYER FLUIDITY TO BIOLOGICAL PROCESSES

Changes in bilayer fluidity effected by either the cell itself or by some external influence can modulate normal cell function; a selection of these are described below.

3.5.1 Temperature acclimatization and homeoviscous adaptation

Fish, hibernators, mesophilic and thermophilic bacteria, and simple organisms like plankton, *chlorella* and tetrahymena, all possess mechanisms whereby their membranes can adapt to changes in their environmental temperature by altering their lipid composition. This adaptation is in fact an attempt, which is realized in *E. coli* and *B. stearothermophilus*, to maintain a constant lipid fluidity, called homeoviscous adaptation. Thus, any fall in the temperature initiates a response which causes a parallel increase in membrane fluidity and vice versa. The nature of this response predominantly takes the form of a change in the degree of saturation of the phospholipid acyl chains which maintains the plasticity of the membrane and effective sealing of integral membrane proteins. Bacteria can not only increase acyl chain unsaturation to decrease the lipid phase transition but also incorporate branched and cyclopropane-containing acyl chains into their phospholipids. The effectiveness of the homeoviscous adaptation mechanism is reflected in the ability of prokaryotic micro-organisms to grow under conditions of environmental temperature ranging from $-10\,^{\circ}C$ to almost $100\,^{\circ}C$. The changed composition of the membrane allows the bacterium to adapt to its environment and to grow normally. Defects in this mechanism give rise to mutants which cease to grow and often die when the environmental temperature is changed.

One of the best studied poikilotherms is the common goldfish whose membrane phospholipids not only exhibit changes in saturation in response to environmental temperature but also exhibit changes in their cholesterol : phospholipid ratio. Upon lowering the environmental temperature the degree of saturation of the phospholipids increases, whereas the cholesterol : phospholipid ratio decreases in an attempt to maintain a constant lipid fluidity.

Hibernating animals also exhibit a remarkable adaptation to changes in environmental temperature. Golden hamsters have a body temperature of about $37–38\,^{\circ}C$ under normal environmental temperatures $(20\,^{\circ}C)$, but their body temperature falls to only $4–5\,^{\circ}C$ when hibernating at similar or even lower environmental temperatures. Of course the metabolic rate of the animal drops dramatically, but it is kept tuned to a level sufficient to maintain cell integrity and to allow rapid arousal. Changes in membrane lipids do occur and again they display an increase in the unsaturated species in response to the lowered environmental temperature.

The rapidity with which adaptation can occur is shown in experiments with

T. pyriformis. Organisms grown at 28 °C exhibit a lipid phase separation in their microsomal membranes at 16 °C. If they are rapidly chilled to a temperature (10 °C) below which the lipid phase separation occurs then protein clustering, due to its exclusion from solid phase lipid (section 3.3.3), is clearly seen upon freeze–fracture analysis. However, if the organisms are maintained at 10 °C, then within 8 h the areas of solid-phase lipid disappear and the integral proteins once again become randomly distributed throughout the membrane. This adaptation is achieved by the desaturation of 50 per cent of the phospholipid acyl chains in the membrane, which results in the depression of the lipid phase separation to below the temperature of the environment (10 °C). A detailed analysis of a similar experiment is shown in Figure 3.17.

The mechanism or mechanisms by which these effects are controlled is unclear, but they are undoubtedly triggered by the temperature change itself. One obvious control point which would affect the degree of unsaturation of phospholipid acyl chain is the function of either or both the acyl desaturase enzymes and the acyl CoA transferases involved in the synthesis of phospholipids. In *Tetrahymena* and carp for example, this control is mediated by temperature-triggered change in the physical properties of the lipids surrounding the membrane-bound fatty acid desaturases which leads to the activation of these enzymes. In *E. coli* a different mechanism exists. Here the membrane-bound acyl transferases use an increasing proportion of unsaturated acyl CoA species as substrates in the synthesis of membrane phospholipids as the temperature is lowered. The mechanism for this selection is as yet unclear.

The cells of the majority of warm-blooded animals do not have to adapt to dramatic changes in environmental temperature, and experiments with cultured cells derived from them indicate that they have lost this ability to adapt. For any significant changes in the environmental temperature lead to the cessation of cell growth and often cell death.

Homeoviscous adaptation, however, need not only occur as a result of temperature manipulation but may also occur isothermally as a result of dietary challenge, although in many animals such effects are buffered by the digestive system and the liver. Cultured cells apparently possess a limited ability to respond in this way. For example, mycoplasma grown at sparing levels of cholesterol adapt by increasing the degree of phospholipid acyl chain saturation. Also, mouse LM fibroblasts grown in a defined synthetic medium with supplements of choline or either of its analogues, *N,N'*-dimethylethanolamine or *N*-monomethylethanolamine, were able to maintain the fluidity of the plasma membrane by altering the degree of saturation of acyl chains in all classes of phosphatides. Indeed, only an ethanolamine supplement caused a significant change (decrease) in the fluidity of the plasma membrane, and even in this instance changes in acyl chain saturation throughout the phosphatide classes had occurred. It is because of this ability of the cell to adapt to dietary changes that such methods are of little use in

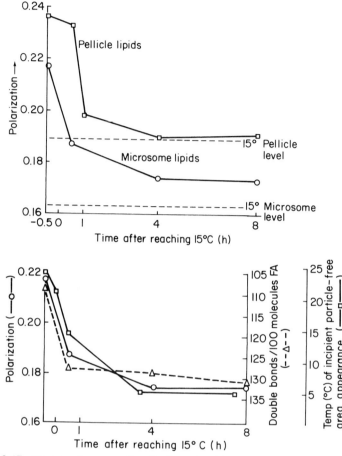

Figure 3.17 Temperature acclimatization in Tetrahymena. Cells grown and acclimatized at 39 °C were then shifted to 15 °C. The lipid fluidity of the lipids from the external membrane pellicle and also the microsomes was followed using the fluorescence-polarization of 1,6-diphenyl-1,3,5-hexatriene (DPH). The cells rapidly adapt to the decrease in temperature (a) by increasing the fluidity of their lipids. The fluidity of 15 °C adapted membranes is given by the broken line. In (b) the increase in fluidity of the microsomal membranes (○) is plotted with the increase in double bonds associated with the phospholipids (△) and the lowest temperature that can be achieved before membrane–protein aggregation occurs (□). (Reprinted with permission from Martin, C. E., and Thompson, G. A. (1978) *Biochemistry*, **17**, 3581–3586. Copyright (1978) American Chemical Society)

investigations of the effects of change in membrane fluidity (see section 3.4.2). However, although there may well be little significant effect on membrane fluidity at the growth temperature, there are effects on the lipid phase separation temperatures due to the changes in bilayer composition. It is usually these two changes that are subsequently monitored in studies on the functioning of integral proteins.

All of these studies indicate that cells attempt to maintain a particular membrane fluidity at the growth temperature.

3.5.2 Cell cycle, growth and differentiation

The change in microviscosity of the cell plasma membrane of a synchronized mouse neuroblastoma clone has been followed throughout the cell cycle using a fluorescence polarization technique. The microviscosity of the membrane varied by a factor of two, being maximal during mitosis and reaching a minimum during the stationary, S-phase. This confirms other observations that the cell membrane of mitotic cells is less fluid than interphase cells. As it is conceivable that the rate of lipid insertion is maximal during mitosis and the growth, G1 phase, it would seem that there is a distinct change in the nature of the lipids being inserted during this period. Indeed in the mesophilic yeast, *Candida utilis*, an increase in linoleic acid incorporation into phospholipids at the expense of oleic acid early in the growth cycle would support an initial rise in lipid fluidity occurring during the period of initiation of active DNA synthesis in G1.

As the proteins, adenylate cyclase, Na^+/K^+-ATPase and the Ca^{2+}-gates, are all inserted into the plasma membrane and are sensitive to changes in membrane fluidity, it is conceivable that ion fluxes and intracellular cyclic AMP levels could change during the growth cycle. Indeed, adenylate cyclase activity attains a maximum during the more fluid, S-phase.

These changes are pertinent to the action of mitogens which trigger lymphocyte proliferation. Upon binding to specific cell surface receptors these ligands can apparently increase bilayer fluidity by initiating the clustering of the more rigid sphingolipids and gangliosides within the plasma membrane. This is seen in the increase in the activity of adenylate cyclase, Na^+/K^+-ATPase and a number of other transport processes. This increase in the fluidity of the plasma membrane is a characteristic feature of proliferating cells, such as those found in tumours and regenerating liver.

Changes in membrane fluidity have also been observed in cells involved in a variety of other phenomena. For example, increases in fluidity are seen during the expansion of *Tetrahymena* nuclei in preparation for DNA synthesis; during the fusion of myoblasts to form multinucleate myotubes, and 10 min after fertilization of sea urchin eggs by a spermatozoon. In contrast to these increases in fluidity, neuroblastoma cells induced to form neurites by transfer from serum to phosphate-buffered saline show a marked decrease in fluidity occurring over the period of neurite formation. This is achieved by elevating the cholesterol : phospholipid ratio.

3.5.3 Disease

In general, transformed cells have more fluid plasma membranes than their normal counterparts. In hepatomas and in mouse 3T3 cells transformed with either SV40 or polyoma virus this effect is related to changes in both the cholesterol : phospholipid ratio of the membranes and in the degree of unsaturation of the phospholipid acyl chains. The relative changes in these parameters, however, differ depending upon the source of the cells, the nature of the transformation and the composition of the growth medium. This increase in membrane fluidity may well be related to the lack of growth control seen in these transformed cell lines. Indeed, by elevating the membrane cholesterol content of a lymphoma cell line to increase its rigidity, its ability to proliferate was markedly reduced. Similarly, raising the membrane cholesterol levels of lymphocytes reduced their susceptibility to the action of mitogens.

Atherosclerosis is a disease usually associated with high serum cholesterol levels which is reflected in the membrane composition of erythrocytes and other cell membranes as well as arterial walls. The increased rigidity and mechanical instability lead to alterations in cellular function and are believed to be important early steps in the aetiology of this disease. It is relatively easy to obtain animal models of this disease, merely by feeding a high-cholesterol (atherogenic) diet. This leads to rapid and dramatic increases in membrane cholesterol content: a doubling of the levels in rat liver subcellular membranes has been observed. The long-term effects of such a diet on membrane composition and function should prove to be very interesting.

There has of course been a considerable interest in developing drugs to depress plasma cholesterol levels, in particular because of its relationship with cardiovascular disease. One such clinically useful drug is clofibrate, which markedly reduces plasma cholesterol levels by methods which are unclear at present, but do not seem to involve inhibition of cholesterol synthesis. Unfortunately, this drug affects a number of other metabolic processes which may well limit its usefulness.

Increases in cell membrane fluidity have been noted in erythrocytes and skeletal muscle membranes from animals with congenital myotonia and myotonic muscular dystrophy. Such increases in membrane fluidity may well be responsible for some of the expressions of these syndromes, for if either animals or humans are fed 20,25-diazocholesterol, then cholesterol synthesis is blocked and desmosterol accumulates. This results in an increase in bilayer fluidity and the appearance of myotonia. In the plasma membranes of fibroblasts from human cystic fibrosis patients there is also an increase in bilayer fluidity and altered lipid phase separations. Thus, a study of fluidity changes may prove to be a rather profitable approach for an understanding of aspects of this disease.

Decreases in the sphingomyelin : phosphatidyl choline ratio of human pulmonary surfactant lipids occur during the final stages of gestation. This leads to an increase in the fluidity of the surfactant which is essential for the efficient inflation of the lungs at birth. However, failure to complete these changes can lead to respiratory distress, a condition which fortunately can be relieved simply by administering a dust of solid, synthetic dipalmitoyl phosphatidyl choline directly into the bronchial passage.

3.5.4 Development and ageing

During early development in mammals, dramatic alterations occur in the ratios of various phospholipid species, in the degree of saturation of their acyl chains, and in the cholesterol content of various biological membranes. These events correlate with the changes in diet that occur upon weaning. Surprisingly, little attempt has been made to correlate alterations in the activity of membrane-bound enzymes with the changes in the lipid environment of the membrane that occur over this period. The only well-defined example of an enzyme whose activity is regulated by changes in the physical state of the membrane lipid over this period is β-hydroxybutyrate dehydrogenase from brain whose activity progressively decreases after weaning due to a progressive increase in bilayer rigidity. This integral enzyme is of special importance to the brain, and also to the heart and kidneys, during weaning, since it allows ketone bodies, which are at high levels during this period, to be used both as an energy source and as a supply of acetyl CoA for biosynthetic purposes.

There is also evidence from changes in the form of Arrhenius plots of the activity of Na^+/K^+ ATPase, 5'-nucleotidase and adenylate cyclase from liver plasma membranes, that alterations in membrane lipid composition can alter the functioning of these enzymes over this weaning period.

The ageing process is also associated with changes in the lipid composition of biological membranes. Changes in lipid composition occur both qualitatively and quantitatively with age, taking the form of altered cholesterol phospholipid ratios, phosphatide ratios and the degree of saturation of phosphatide-associated acyl chains. No unique pattern of change is apparent, for differences occur dependent upon the tissue source and the particular membrane under study. In general, however, upon ageing there is a tendency for the occurrence of decreases in the more highly polyunsaturated fatty acids, increases in the monounsaturated fatty acids, and increases in the cholesterol : phospholipid ratio. Membranes in aged animals would therefore seem to be more rigid than their young counterparts.

3.5.5 Alcohol: tolerance and dependence

Ethanol is a drug with profound social implications. It can exert dramatic pharmacological effects on the sensory systems and metabolism of organisms. By virtue of its lipid solubility it can enter the apolar core of the bilayer, increasing its fluidity and like so many similar agents, effect local anaesthesia (see section 3.5.8). It is not unreasonable therefore, to expect that certain of its pharmacological effects may be related to its ability to perturb lipid bilayers. Indeed the chronic exposure of cells to ethanol may well induce an attempt to maintain a constant lipid fluidity (section 3.5.1). This has been suggested to lead, at least in part, to the initial functional tolerance and later physical dependence upon ethanol.

There is now considerable evidence that microsomal, mitochondrial and synaptosomal plasma membranes from brain and the erythrocyte plasma membrane, taken from ethanol-treated animals, are resistant to the fluidizing effect of ethanol. This is achieved by increasing the cholesterol : phospholipid ratio, and decreases in the amount of polyunsaturated acyl chains associated with phospholipids have been observed in membranes from ethanol-treated animals.

3.5.6 Hormone action

The anabolic hormone insulin binds to specific, cell-surface receptors on target cells. It has been suggested that insulin could exert its effects on such cells by increasing bilayer fluidity which could, for example, lead to activation of glucose transport. However recent studies have rigorously proved that insulin does not affect bilayer fluidity. This is not perhaps too surprising as such an action would lead to a non-specific perturbation of a variety of processes. Very recently in fact, insulin has been shown to increase glucose transport by actually recruiting new transport proteins to the plasma membrane. These molecules are supplied by specialized vesicles inside the cell. The identity of these vesicles and the mechanism of transfer of carrier molecules between them and the plasma membrane has yet to be determined.

It has been suggested that agonist occupancy of β-adrenergic receptors, benzodiazepine receptors and most cell IgE receptors, stimulates the cell plasma membrane to synthesize phosphatidyl choline from phosphatidyl ethanolamine. This reaction is purported to be carried out by endogenous S-adenosyl methionine-requiring methyl transferases. Although it has been claimed that this mechanism can lead to an increase in bilayer fluidity, which causes the activation of adenylate cyclase, this has been severely criticised on the grounds that only a very small amount of lipid is transformed. On this basis it would seem that both the magnitude of these effects and claims of their importance have been somewhat exaggerated.

3.5.7 Complement-mediated immune attack

The complement system is the method by which the immune system mounts a cytotoxic attack on invading cells. It comprises a number of soluble, plasma proteins that are activated, when antibody binds to foreign antigens on an invading cell, to form a complex which inserts into the membrane of the target cell and causes lysis (see section 7.4.4).

The serum of human beings who have contracted syphilis contains cardiolipin-specific (Wasserman) antibodies because, unlike mammalian cells, the invading bacteria have cardiolipin in their plasma membrane. The effect of fluidity on complement fixation has been investigated using this antiserum to interact with cardiolipin in phosphosphatidyl choline liposomes containing various concentrations of cholesterol. Only when the cholesterol : phospholipid ratio rose above 0.35 was there significant com-

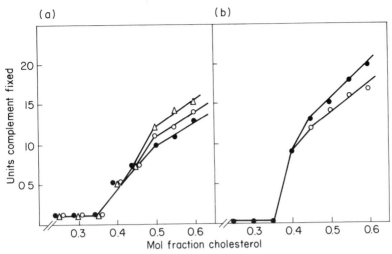

Figure 3.18 Complement-mediated immune attack requires a minimum cholesterol content. A monovalent lipid antigen, cardiolipin, was inserted into phosphatidyl choline liposomes containing various concentrations of cholesterol. These liposomes were tested for their ability to trigger (fix) complement activation in the presence of a specific antibody. Using various concentrations of liposomes (a) and various concentrations of specific antibody (b) it was found that a certain minimum concentration of cholesterol (35 mol per cent) was necessary for complement fixation to be triggered. Both of these experiments were carried out at temperatures above the phase transition temperature of the phospholipids. This suggests that a certain minimum cholesterol concentration is needed to produce cardiolipin clusters which can act as a multivalent ligand capable of interaction with antigen. This complex can then trigger complement fixation. (Reprinted with permission from Humphries, G. M. K., and McConnell, H. M. (1975) *Proc. Natl. Acad. Sci. U.S.A.*, **72**, 2483–2487)

plement fixation (Figure 3.18). At this level of cholesterol the mobility of the cardiolipin molecules became sufficiently restricted to enable it to interact with antibody and trigger complement fixation. As univalent antigen molecules do not normally trigger complement activation, it is likely that the formation of cardiolipin-enriched domains, containing clustered cardiolipin, acts to increase its apparent valency, facilitating the activation of complement. That only multivalent antigen triggers the complement sequence may be related to a requirement to distort the immunoglobulin molecule, principally IgM, by binding to critically spaced antigenic determinants. Thus the decrease in bilayer fluidity and the segregation into distinct lipid phases caused by cholesterol (see section 2.11) restricts the motion of the univalent cardiolipin sufficiently to allow it to trigger the complement sequence. Cholesterol also enhances the ability of complement to act by increasing the exposure of the cardiolipin. It presumably achieves this by expanding the headgroup region of the bilayer (see section 2.11) which not only increases accessibility to the antigenic determinant but also reduces steric hindrance to antibody binding by adjacent bilayer lipids.

3.5.8 Anaesthesia

Compounds which reversibly block the formation of an action potential, but have little effect on the resting potential of the target cell, induce anaesthesia. A wide variety of chemically diverse compounds can act as anaesthetics. These include many drugs which are used as tranquillisers, anti-arrythmics, narcotics, sedatives, anti-convulsants, antihistamines and β-blockers. Indeed it is this extraordinary range of chemically very different structures which suggests that anaesthesia is induced not through the specific reaction of these compounds with the target protein, but because they initiate a common structural perturbation which is sensed by the target protein.

Theories of the molecular mechanism of anaesthetic action have been increasing since, at the turn of the century, it was shown that the potency of an anaesthetic was directly related to its olive oil/H_2O partition coefficient. For this relationship, together with the observation that anaesthetics protect erythrocytes from haemolysis, suggested that anaesthetics might act by perturbing the cell membrane. Indeed the target for anaesthetic action is now believed to be the cell plasma membrane where they reduce conductance through the Na^+ channel. Hypotheses which have attempted to explain this fall into two main categories: (i) those where the anaesthetic alters either the lipid structure or the way in which lipids interact with the protein—this leads indirectly to changes in the structure and function of the target protein; and (ii) those where the anaesthetic interacts directly with the target protein causing a change in its function.

A great deal of attention has been focused on suggestions that changes in membrane lipid organization can lead to anaesthesia. The critical volume hypothesis suggested that anaesthesia (narcosis) ensues whenever a critical fraction of anaesthetic is present within the bilayer. This fraction will be influenced by the lipid composition of the membrane, but should be independent of the anaesthetic used. The incorporation of the anaesthetic into the membrane was presumed to cause bilayer expansion, thus leading indirectly to a perturbation of the target protein's structure and function. Anaesthetics do indeed cause a small expansion of lipid bilayers that is apparently due to their ability to increase disorder and hence the fluidity of the bilayer. Whilst changes in fluidity can alter the function of integral proteins, it is not now believed to be the mechanism through which anaesthetics act. The reasons for this are:

(i) many drugs initiate either general or local anaesthesia at concentrations at which they have no effect on bilayer fluidity;
(ii) n-alkanols with $C < 10$ increase fluidity and such effects are antagonized by n-alkanols with $C > 10$ which decrease bilayer fluidity; however, n-alkanols up to $C = 12$ can induce anaesthesia and this effect is not antagonized by n-alkanols with $C > 12$;
(iii) the increase in bilayer fluidity caused by neutral anaesthetics can be mimicked by a small rise in temperature; however, small increases in

temperature do not mimic anaesthesia and small decreases in temperature do not reverse anaesthesia;
(iv) agents which antagonize the anti-haemolytic action of short-chain (< 10 C) n-alkanols do not antagonize anaesthesia.

The very fact that stereoselectivity may be observed with certain anaesthetics such as the barbiturates, aromatic amines and in particular the steroid alphaxalone, whose action is antagonized specifically by its Δ^{16}-derivative, implies that an interaction between the anaesthetic and the target protein is of importance in the anaesthetic mechanism. Yet the wide variety of chemically unrelated compounds that can cause anaesthesia shows that the process must be highly degenerate at the molecular level. From such observations, when taken together with the correlation between anaesthetic potency and their olive oil/H_2O-partition coefficient, one might construe that anaesthetics interact with a site(s) on the target protein that is buried within the apolar core of the bilayer. Thus, increased pressure, which can reverse anaesthesia, could achieve its effect by reducing the concentration of anaesthetic within the bilayer membrane.

The loss of function of the Na^+-channel would result as a direct consequence of the anaesthetic interacting with a site on the protein previously

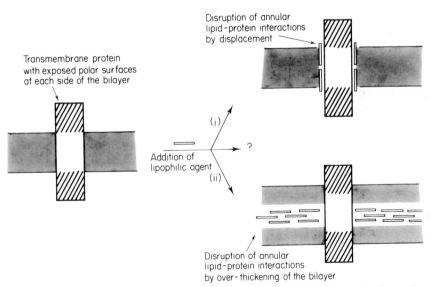

Figure 3.19 An annular disruption mechanism for local anaesthetic action. Anaesthetics achieve their effect by reducing Na^+ conductance, presumably by interfering with the functioning of the Na^+ channel. Their chemically diverse structures precludes their action at a site of high specificity, but suggests they act through a 'degenerate perturbation' of the protein structure. As anaesthetics are lipophilic agents that readily partition into biological membranes it is suggested that they interact with sites on the protein that are normally buried in the bilayer

148

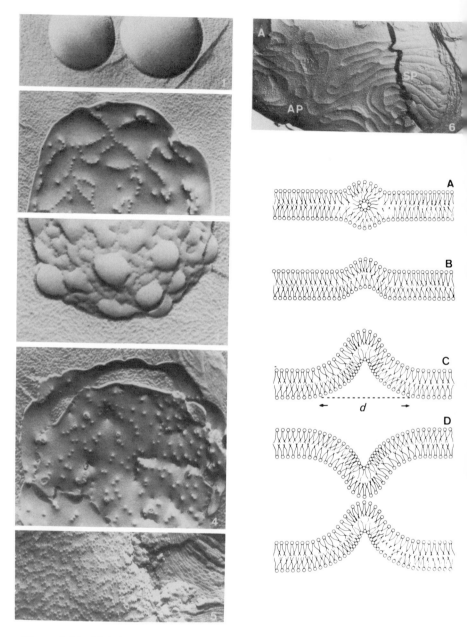

Figure 3.20   Lipidic particles. Integral membrane proteins give rise to characteristic particles in freeze–fracture. However, it is possible to obtain freeze–fracture patterns displaying similar 'particles' using liposomes which contain no proteins. This requires the addition of high Ca^{2+} concentrations usually to liposomes containing acidic phospholipids. Typical micrographs are shown in 1–6. Several hypothetical models have been proposed to account for the occurrence of this particle (A–D) including inverted micelles in the bilayer core (A). It has been suggested that the presence of

occupied by annular lipid. This has been termed the annular lipid disruption model (Figure 3.19). Presumably the reduced Na^+-conductance through the channel is a consequence of a conformational change in the protein caused either by its interaction with the anaesthetic or because of the displacement of phospholipid from these sites. This theory might explain why many membrane-bound enzymes can also be inhibited by certain concentrations of anaesthetics (see section 3.4.1). This sensitivity to inhibition depends upon the protein under study and presumably reflects a complex mixture of the strength of the annular lipid–protein interactions, the dependence upon specific annular lipid for function, and whether the anaesthetic can be inhibitory upon occupancy of the apolar regions of the protein. A direct test for this theory has yet to be devised. However, it has been demonstrated that at anaesthetically relevant concentrations benzyl alcohol can prevent the quenching of tryptophan fluorescence of integral membrane proteins caused by interaction with nitroxide-labelled fatty acids inserted in the membrane. This is consistent with the anaesthetic occupying annular lipid domains, preventing the interaction of the nitroxide-labelled fatty acid with the protein.

A change in bilayer thickness that can be caused by agents inserting themselves into the bilayer (see section 3.3.4) has also been suggested as a means of achieving anaesthesia. Indeed, this can be taken as a form of the lipid displacement model as it involves a perturbation of the normal lipid–protein interaction in the membrane, leading to changes in the functioning of integral proteins. However, as elaborated above (section 3.3.4), biological membranes undoubtedly possess a considerable capacity for buffering such changes.

3.5.9 Cell fusion

The paucity of multinucleated cells in organisms indicates that the fusion of adjacent cells is a restricted event. Only certain types of cells are found to fuse together, examples of which are myoblasts which yield multinucleated skeletal muscle fibres, macrophages which yield 'giant-multinucleated cells',

inverted micelles, which have subsequently been detected by nmr techniques, could act as an intermediate in the fusion process. (1) Cardiolipin : PC (1 : 1) in the absence of Ca^{2+} yields smooth surfaces ($\times$ 64,000); (2) as in (1), but plus 2 mM Ca^{2+} which induces particles in outer leaflet ($\times$ 48,000); (3) as in (2), but showing pits in inner leaflet ($\times$ 64,000); (4) pits and particles in CL : PC + 1 mM Ca^{2+}; (5) as in (1), but plus 10 mM Ca^{2+} causes precipitation of liposomes. Bilayer phases containing particles are seen with separate lamellae phase ($\times$ 48,000); unfixed and non-cryoprotected rat sciatic nerve showing axoplasm (A), axolemmal P-face (AP) and Schwann cell P-face (SP) at $\times$ 20,000 where ice formation has created quilt-like arrays of particles. Reprinted by permission from Miller, R. G. (1980) *Nature (Lond.)*, **287**, 166–167, copyright © 1980 Macmillan Journals Limited. (Reprinted with permission from De Kruijff, B., Verkley, A. J., Van Echteld, C. J. A., Gerritsen, W. A., Mombers, C., Noordham, P. C., and De Gier, *J. Biochim. Biophys. Acta*, **555**, 200–209) and gametes during fertilization. Clearly, there are barriers preventing inappropriate fusion and triggers initiating highly specific fusion events.

In these instances of specific membrane fusion, changes in the membrane which allow or trigger fusion can be identified. For example, the fluidity of the sperm plasma membrane increases upon maturation in the epididymis and this, together with an 'uncoating' process effected by residing in the female tract, are prerequisites for fusion to occur. The egg plasma membrane is then modified to prevent further fusion events.

Myoblasts proliferate in culture until they reach a certain cell density where they become susceptible to fusion. This fusion is dependent upon material released by the cells into the medium and can be inhibited by constantly placing them in fresh medium. The fusion event itself is initiated in the G1-phase and is dependent upon the extracellular Ca^{2+} concentration being $> 10^{-5}$ M.

Virus induced fusion. Enveloped viruses are the only types that can induce cell-fusion and as such they provide us with an extremely useful tool both for examining the fusion event itself and for fusing different membranes together. Two types of fusion have been observed. Fusion from without is initiated by extremely large doses of virus, it occurs within minutes and doesn't require the production of any new proteins. All that is necessary is that Ca^{2+} be present and the bilayer is in a fluid state. In contrast fusion from within occurs after infection of the cells and is dependent upon the synthesis of new viral proteins and their incorporation into the plasma membrane. The mechanisms of both these processes remains to be elucidated.

Chemically induced fusion. Polyethylene glycol has been shown to be an extremely useful agent (fusogen) for fusing animal cells and plant protoplasts together in such a way that the heterokaryon remains viable. Many other agents have since been found but they are usually less efficient and more cytotoxic. As in the cases of virus-induced fusion, the presence of Ca^{2+} is essential for the fusion process to proceed.

The mechanism of the fusion process remains obscure but there is some evidence to suggest that 'lipidic particles' form at the site of fusion. These consist of lipid in a state intermediate between a bilayer and hexagonal (H_{II}) phases which forms an inverted micelle sandwiched between the leaflets of the membrane (Figure 3.20).

Summary

The cell membranes of fish, hibernators, bacteria and simple organisms can be adapted, by the cell, to changes in their environmental temperature so as to keep membrane fluidity relatively constant.

This is called homeoviscous adaptation and it is achieved by altering the membrane lipid composition.

Changes in membrane fluidity can occur through the cell cycle; in transformation; in

151

certain disease states; during ageing and maturation; in obesity and during tolerance to ethanol.

Anaesthesia may result from a perturbation at the interface between the target protein and its annular lipids.

Many membrane fusion events depend upon a fluid lipid bilayer and the presence of Ca^{2+}.

Further Reading

Bennett, J. P., McGill, K. A., and Warren, G. B. (1980) The role of lipids in the functioning of a membrane protein: the sarcoplasmic reticulum calcium pump, in *Current Topics in Membranes and Transport*, Vol. 14, Academic Press, New York, pp. 127–164.

Brotherus, J. K., Jost, P. C., Griffith, O. H., Keana, J. F. W., and Hokin, L. E. (1980) Charge selectivity at the lipid–protein interface of membraneous Na^+/K^+-ATPase, *Proc. Natl. Acad. Sci. USA*, **77**, 272–276.

Caffrey, M., and Feigenson, G. W. (1981) Fluorescence quenching in model membranes: Relationship between Ca^{2+}-ATPase activity and the affinity of the protein for phosphatidylcholines of different acyl chain characteristics, *Biochemistry*, **20**, 1949–1961.

Chapman, D., Gomez-Fernandez, J. C., and Gori, F. M. (1979) Intrinsic protein–lipid interactions: physical and biochemical evidence, *FEBS Lett.*, **98**, 211–223.

De Pierre, J. W., and Ernster, L. (1977) Enzyme topology of intracellular membranes, *Annu. Rev. Biochem.*, **46**, 201–262.

Gennis, R. B., and Jonas, A. (1977) Protein–lipid interactions, *Ann. Rev. Biophys. Bioeng.*, **6**, 195–238.

Gordon, L. M. and Sauerheber, R. D. (1982) Calcium and membrane stability, in *Calcium in Normal and Pathological Biological Systems*, Anghileri, L. J., ed., CRC Press.

Jahnig, F. (1979) Structural order of lipids and proteins in membranes, *Proc. Natl. Acad. Sci., USA*, **76**, 6361–6365.

Kimelberg, H. K. (1977) The influence of membrane fluidity on the activity of membrane-bound enzymes in *Dynamic Aspects of Cell Surface Organisation*, Poste, G. and Nicolson, G. L., eds., *Cell Surface Reviews*, Vol. 3, Elsevier/North Holland Biomedical Press, pp. 205–293.

Knowles, P. F., Watts, A., and Marsh, D. (1981) Spin-labelled studies of head-group specificity in the interaction of phospholipids with yeast cytochrome oxidase. *Biochemistry*, **20**, 5888–5894.

Marsh, D., Watts, A., Maschke, W., and Knowles, P. F. (1978) Protein-immobilised lipid in DMPC-substituted cytochrome oxidase: evidence for both boundary and trapped-bilayer lipid, *Biochem. Biophys. Res. Commun.*, **81**, 397–402.

Melchior, D. L., and Stein, J. M. (1976) Thermotropic transitions in biomembranes, *Ann. Rev. Biophys. Bioeng.*, **5**, 205–238.

Sandermann, H. (1978) Regulation of membrane enzymes by lipids, *Biochim. Biophys. Acta*, **515**, 209–237.

Tanford, C., and Reynolds, J. A. (1976) Characterisation of membrane proteins in detergent solutions, *Biochim. Biophys. Acta*, **457**, 133–170.

Vaz, W. L. C., Kapitza, H. G., Stumpel, J., Sackmann, E., and Jovin, T. M. (1981) Translational mobility of Glycophorin in bilayer membranes of dimyristoyl phosphatidylcholine, *Biochemistry*, **20**, 1392–1396.

Chapter 4

Asymmetry in biological membranes

4.1   INTRODUCTION

The Fluid Mosaic model of membrane structure (see section 1.5) proposes the absolute asymmetry of the protein components of the biological membrane, that is, every copy of that polypeptide chain has the same orientation with respect to the lipid bilayer. This both achieves the localization of specific functions to a particular side of a biological membrane, and is the basis behind the vectorial transfer of solutes and receptor-mediated 'messages' across the lipid bilayer (see Chapter 7). Such an absolute protein asymmetry is created during biosynthesis and preserved throughout the lifetime of the membrane (see section 6.2).

Asymmetry is also observed for phospholipids. However, unlike the protein components of the membrane, the asymmetry is not absolute, instead individual phospholipids exhibit a preference for one or other side of the bilayer. In general the less reactive choline-containing phospholipids, sphingomyelin and phosphatidyl choline prodominate at the extracytoplasmic side of the bilayer, whilst the more reactive phosphatidyl serine, phosphatidyl ethanolamine and phosphatidyl inositol species are found at the cytoplasmic side.

In the plasma membrane many of the proteins and some of the lipids are glycosylated, and this carbohydrate appears exclusively at the extracytoplasmic surface of the membrane. Like the asymmetry of proteins, that of the carbohydrate appears to be absolute.

In this chapter we shall examine the evidence for asymmetry in biological membranes, discuss its origins and maintenance, and contemplate some possible functions. As asymmetry is difficult to measure we shall place our emphasis on methodology and a critical assessment of results together with appropriate illustrations.

4.2 MAINTENANCE OF MEMBRANE ASYMMETRY

Phospholipids found in nature will in an aqueous environment spontaneously form bilayers: thin flexible sheets whose physical properties are determined by their chemical composition. Biological membranes are composed of lipid bilayers studded with proteins which are solvated by their annular lipids (see Chapters 1 and 3). These membranes are thermodynamically stable structures and so we must ask why the concentration gradient imposed upon them by asymmetry is not dispelled by translocation of membrane components across the bilayer. Seen from the opposite perspective the presence of asymmetry itself in biological membranes is evidence that the translocation of molecules through the apolar core of the bilayer is a very rare event. Indeed all the available evidence supports this statement (see Chapter 3). For whilst a phospholipid molecule may take only 1 μs to diffuse 4 nm within the plane of the bilayer, translocation a similar distance across the bilayer (flip-flop) can take up to 10^{10} times longer. Furthermore, the absolute asymmetry of proteins indicates that protein flip-flop occurs at infinitesimal rates. Slow rates of translocation are undoubtedly due to the charge distribution of these amphiphilic molecules, where the required solvation of such charged groups in the apolar environment of the bilayer core during the process of translocation is clearly energetically unfavourable.

Since the thermodynamic basis for maintaining membrane asymmetry is the same for proteins, phospholipids and their glycosylated derivatives, the variation in their asymmetry must reflect a different specificity of insertion during biosynthesis of the membrane (see section 6.2, 6.4). Most integral membrane proteins are inserted directly into the membrane during biosynthesis where the biosynthetic mechanism and their primary structure determine orientation within the bilayer. Alternatively, in some instances membrane proteins are formed as pre-proteins which are synthesized free in the cytosol, but upon solvation in the bilayer are immediately modified, usually by proteolysis, whereupon their orientation is fixed (see Chapter 6).

Phospholipids are synthesized and inserted into the cytoplasmic side of the membrane only. Here enzymes called 'translocases, or 'flippases' catalyse the transfer of newly synthesized lipid from one side of the membrane to the other. Asymmetry arises out of the specificity exhibited for certain phospholipid species by such enzymes. However, it is apparent that such a mechanism cannot give rise to an absolute asymmetry, hence the phospholipids are distributed on both sides of the bilayer.

Glycosylation of proteins and carbohydrates is achieved by asymmetrically localized enzymes. In eukaryotic cells these line the lumen of the endoplasmic reticulum. The attachment of bulky hydrophilic moieties to lipids or proteins will further reduce the chances of flip-flop occurring, and the fact that absolute asymmetry is observed for carbohydrate attached not only to protein but also to phospholipid in the plasma membrane, suggests an absence of 'flippases' able to handle these species.

Underlying this discussion of the maintenance of membrane asymmetry is the assumption that membrane asymmetry is not spontaneously achieved. Reconstitution of membranes from detergent solubilized components in the majority of instances leads to a randomized orientation of components (see Chapter 5). However, certain proteins appear to be exceptions and it is also possible to create phospholipid asymmetry spontaneously in very small liposomes which possess a high radius of curvature.

Summary

Protein and carbohydrate asymmetry is absolute; lipid asymmetry is relative. This reflects the mechanism of membrane assembly.

4.3 ANALYSIS OF MEMBRANE ASYMMETRY

4.3.1 Principles and criteria

All the direct techniques used to investigate membrane asymmetry exploit the specific labelling or modification of one-half of the bilayer in sealed membrane vesicles. The components labelled under such conditions are then compared with the patterns obtained in preparations of broken membranes and if possible in inverted, sealed membranes.

The mammalian erythrocyte, often called the red cell, has proved a useful tool for many elegant studies, possessing the particular advantage that the only membrane present is in fact the plasma membrane. Red cells are readily available (about 10^{13} per investigator) and they can be simply and rapidly prepared from fresh blood using a bench centrifuge, yielding a cell preparation with a sealed plasma membrane. Pure plasma membranes can then be produced from these cells by simple hypotonic shock, and subsequent washing to remove the released haemoglobin yields a white fraction of pure plasma membranes, termed erythrocyte 'ghosts'. Using the procedures outlined in Figure 4.1 it is possible to obtain resealed ghosts in either an inverted or right-side-out configuration. This provides a perfect system for attacking the problem of membrane asymmetry, as each side of the cell plasma membrane can now be probed separately with non-penetrating reagents.

The distribution of phospholipid and carbohydrate between the two halves of the bilayer can be assessed and the disposition of proteins within the membrane can be determined (Figure 4.2). Transmembrane proteins, those that span the bilayer membrane, will react with labelling reagent placed at either side of the membrane, whereas proteins exposed only at one side will not react with labelling reagent placed at the opposite side of a sealed membrane. All of these species will of course be fully labelled if broken membranes are used, the only exception being a species of protein that is totally internalized within the membrane. As yet there are no clear-cut examples of such a species, certainly not of individual proteins, although it is possible that certain peptides forming part of the large protein complexes of components of the mitochondrial electron transport chain may be so buried within the bilayer.

Before any interpretations are made about the topography of any membrane protein it is vital to satisfy a number of criteria:

(i) There is good evidence that each cell type and each organelle within a particular cell contain membranes of different protein and lipid composition. *It is imperative, therefore, that studies on membrane asymmetry are only performed on membranes of the highest purity.*

(ii) The labelling or modifying reagent must only be allowed to react with one side of the membrane and this requires that *the vesicles be impermeable to the reagent and that all the vesicles have a unique sidedness.*

(iii) *The labelling reagent must not be able to enter the sealed vesicle by partition into the lipid component of the membrane.* Hydrophobic reagents which could cross lipid bilayers by partition must be modified to prevent their penetration by the addition of charged groups. The slow penetration of reagents during an experiment is a difficult problem to evaluate, although their interaction with soluble proteins inside the cell or vesicle can be a useful marker.

(iv) *The labelling reagent must not change the membrane permeability.* The leakage of soluble proteins or small molecules out of the cell or membrane vesicle is a useful indicator of membrane breakdown.

(v) *The labelling reagent must be able to react with the exposed membrane*

156

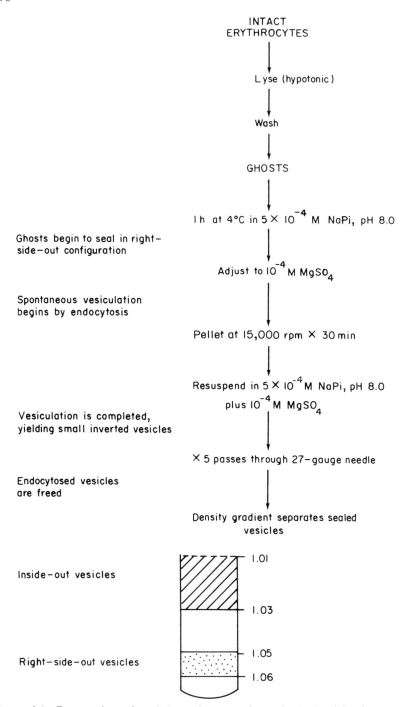

Figure 4.1 **Preparation of sealed erythrocyte ghosts in both right-side-out and inside-out configurations**

	Sealed right-side-out ghosts	Leaky ghost	Intact cell	Leaky ghost	Sealed inside-out ghosts
Transmembane	✓	✓	✓	✓	✓
Extracytoplasmic	✓	✓	✓	✓	No
Cytoplasmic	No	✓	No	✓	✓
Intramembrane	No	No	No	No	No

Figure 4.2 **The identification of asymmetrically disposed proteins in membrane vesicles of defined orientation. The topography of membrane proteins may be established from the pattern of labelling when a non-penetrating reagent is applied to right-side-out and inside-out preparations**

components. A major drawback of many protein labelling reagents is that they only react with certain amino acids, which are often minor or rare constituents of proteins.

The labelling reagents used should ideally be able to react with the target groups under physiological conditions of pH and osmolarity so that the state of the cell or vesicle can be maintained. Failure to observe these conditions can often result in the difficulties described in points (iii) and (iv).

Problems also arise in assessing lipid asymmetry where the reactivity of certain lipids to labelling or attack by phospholipases may be dependent upon the packing and organization of the bilayer.

To avoid criticisms concerning reactivity it is often best to use a number of different types of labelling reagents. If possible a consideration of labelling patterns in native and resealed vesicles of both orientations should be considered together in order firmly to establish the disposition of components. Unfortunately, such approaches are not always technically feasible.

(vi) *The methods of disrupting the sealed membranes must not redistribute the label or activate previously inaccessible sites*. If antibodies are used to label external antigens special care must be taken not to allow the redistribution of these antibodies during disruption of vesicles by homogenization or sonication.

4.3.2   Methods of labelling membrane components

There are a large number of chemical reagents which may be used in order to label the proteins, lipids and carbohydrate residues of membranes. Many of these suffer the disadvantage of being able to permeate through membranes or of causing lysis or disruption of the cell and vesicle membranes. This section illustrates some of the most common reagents used to label membranes.

Often the molar quantities of proteins labelled by these reagents are small, as is the molar incorporation of label into the target species. This necessitates a sensitive assay procedure for detection of the reagent, coupled with a means of separating the various lipid and protein species which provide possible targets for the reagent. To achieve the desired assay sensitivity, the reagent is usually radio-labelled or equipped with a fluorescent tag. This allows both a rapid identification of the labelled lipid or protein species and a quantification of labelling. The separation of lipid species is then readily achieved by using two-dimensional thin-layer chromatography (see section 1.7); and that of proteins by using SDS-polyarcylamide gel electrophoresis (see section 1.6).

Iodination of membrane proteins

One of the major reagents used for labelling membrane proteins is ^{125}I. Because of its high specific activity as a gamma emitter it is possible to detect labelling of even minor membrane proteins. Iodine (I_2) itself is permeable to membranes and so this rules out methods generating the free molecule. To obviate this difficulty the enzyme lactoperoxidase is used to catalyse the radio-iodination of exposed tyrosine residues on membrane proteins (Figure 4.3). Lactoperoxidase is an 80,000-dalton haemoprotein, stable in low ionic strength solutions over a wide pH range (3–10) and at temperatures up to 60 °C. It catalyses the oxidation of a wide variety of electron donors in an H_2O_2-dependent fashion. Its usefulness is that it can catalyse the direct iodination of tyrosine residues (Figure 4.3), independent of I_2 production, and as

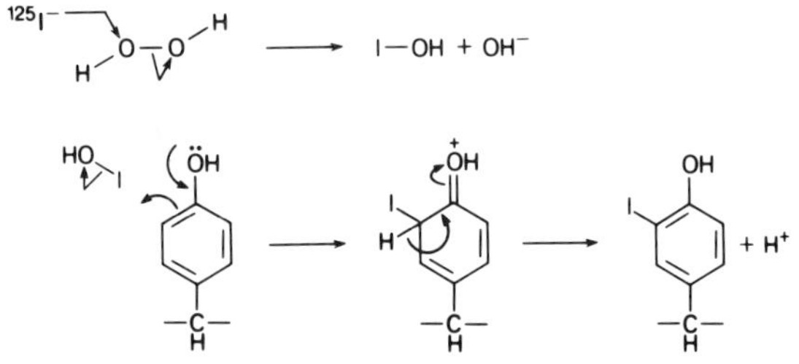

Figure 4.3   Iodination of tyrosine by lactoperoxidase. Lactoperoxidase catalyses the H_2O_2-dependent iodination of exposed tyrosine residues on proteins without forming free I_2. This localizes its action to accessible surfaces because it is a large protein and cannot permeate through membranes. It is used to iodinate selectively the proteins at the external surface of membrane vesicles

it is a large protein it cannot penetrate across membranes. Indeed, one can even use it immobilized on cross-linked dextran beads to ensure this and to aid removal of the enzyme after the labelling procedure is complete. This is essential as lactoperoxidase is self-iodinated during the procedure and must therefore be separated from the membrane vesicles prior to the examination of membrane proteins by SDS–PAGE. This procedure is exceptionally useful for the asymmetric labelling of protein species. Its main disadvantage, however, is that it only labels tyrosine residues. Preferably, these need to be presented by an exposed region of the target protein as there is evidence that lactoperoxidase itself needs to interact with the tyrosine residue in the protein to achieve iodination. This clearly limits the procedure, especially as tyrosine is not a commonly occurring amino acid in proteins. Furthermore, if there is insufficient acceptor substrate to saturate the enzyme it can catalyse the production of $^{125}I_2$ which will not only diffuse into the membrane and label internal protein residues, but will also label lipid fatty acyl chains.

To a certain extent lactoperoxidase will radio-iodinate phospholipids and steroids, probably due to the production of I_2, but because of the inefficiency of the labelling and uncertainties that it is restricted to the exposed side of the bilayer, this method is not used to assess lipid asymmetry.

Small molecules used to label lipids and proteins

The full names, formulae, targets and methods of identification of some of the reagents used to assess the disposition of membrane constituents are set out in Figure 4.4.

Labelling should ideally be carried out on separate preparations of sealed vesicles of both inside-out and right-side-out orientations, using more than one labelling technique. Usually vesicles are kept at low temperatures to minimize the permeation of these reagents into the sealed vesicles.

FMMP, often called 'Bretscher's Reagent', was first employed to demonstrate the asymmetry of aminophospholipids in erythrocytes. It reacts with amino groups (Figure 4.5) and can be used to label the amino phospholipids, phosphatidyl ethanolamine and phosphatidyl serine as well as proteins.

A commonly used reagent is TNBS which labels primary amino groups (Figure 4.5) on phospholipids and proteins. However, one of the problems encountered with this reagent is its inability to label all the surface aminophospholipids when they are present at a high density. This is because it attaches a bulky electronegative group to the lipid which poses spatial problems for further labelling.

A particularly useful pair of imidoesters for amidating proteins and amino phospholipids are IAI and EAI (Figure 4.5). IAI cannot penetrate the bilayer and so it only labels the external half of sealed vesicles, whereas EAI penetrates the bilayer and thus labels both sides. Treatment with this pair of reagents then readily yields information on the asymmetric disposition of membrane components.

Reagent	Target

FMMP
(formylmethionylsulphone methyl phosphate)

Primary amine on proteins and phospholipids. (radio-labelled, e.g. ^{35}S)

TNBS
(trinitrobenzene sulphonate)

Primary amine on proteins and phospholipid (radio-labelled, e.g. ^{3}H)

IAI
(isethionyl acetimidate)

Primary amine on proteins and phospholipid.
(non-penetrating, radio-labelled, e.g. ^{3}H)

EAI
(ethyl acetimidate)

Primary amine on proteins and phospholipid. (penetrating, radio-labelled, e.g. ^{3}H)

DABS
(diazobenzene sulphonate)

Amino, tyrosyl, histidyl or sulphydryl groups
(radio-labelled, e.g. ^{3}H)

SITS
(4-acetimido-4′-isothiocyano-2,2′-stilbene disulphonate)

Amino, tyrosyl, histidyl or sulphydryl groups
(radio-labelled, e.g. ^{3}H, ^{35}S or fluorescence)

DIDS
(4,4′-diisothiocyano-2,2′-stilbene disulphonate)

Amino, tyrosyl, histidyl or sulphonyl groups
(radio-labelled, e.g. ^{3}H ^{35}S or fluorescence)

NAP—Taurine
(N-(4-azido-2-nitrophenyl)-2-aminoethylsulphonate)

Non-specific, reacts with UV to form nitrene (radio-labelled, e.g. ^{3}H, ^{35}S)

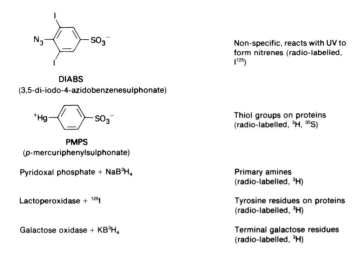

Figure 4.4 provides structures and descriptions:

DIABS
(3,5-di-iodo-4-azidobenzenesulphonate)

Non-specific, reacts with UV to form nitrenes (radio-labelled, I^{125})

PMPS
(p-mercuriphenylsulphonate)

Thiol groups on proteins (radio-labelled, ^3H, ^{35}S)

Pyridoxal phosphate + NaB^3H_4

Primary amines (radio-labelled, ^3H)

Lactoperoxidase + ^{125}I

Tyrosine residues on proteins (radio-labelled, ^3H)

Galactose oxidase + KB^3H_4

Terminal galactose residues (radio-labelled, ^3H)

Figure 4.4   Reagents commonly used for labelling membrane components

The reagents described so far have limitations in labelling proteins in that their specificity is restricted to amino groups. Diazonium derivatives such as DABS and the isothiocyanates, SITS and DIDS, extend this labelling to tyrosyl, histidyl and sulphydryl groupings (Figures 4.4 and 4.5) and PMPS labels sulphydryl groups exclusively.

Potentially the most useful reagents are the photoactive labels: these are azido derivatives which upon ultraviolet irradiation form nitrenes. The nitrene will then attack any C—H bond to form a secondary amine and so is independent of the type of amino acid exposed. NAP-taurine is a typical example of such a probe where ^{35}S is used as a radio-label for identification, although iodinated derivatives of NAP-azide are often used. One reagent of this type is DIABS (Figure 4.4). Photolysis is usually carried out for about 1 h on ice and exposed protein residues become labelled in a totally non-specific fashion to high specific radioactivity (Figure 4.5). This hydrophilic photolabel only reacts with surface groups on proteins and to a certain extent with the polar headgroup of phospholipids. Interestingly, it is taken up by kidney microvillar membrane vesicles in a Na^+-dependent fashion. This enables vesicles to be loaded up with photolabel in the dark, washed to remove the external label and then photolysed to modify proteins exposed at the cytoplasmic surface. Evidence for the asymmetric orientation of proteins can be made by resolution on SDS–PAGE and autoradiography of the gels. Of course, one disadvantage of photolabels is that all steps prior to photolysis must be carried out in the dark!

The reagents described above yield information about the localization of hydrophilic portions of membrane proteins. However, the depth of penetration of proteins into the bilayer varies, and it is possible that in the case of

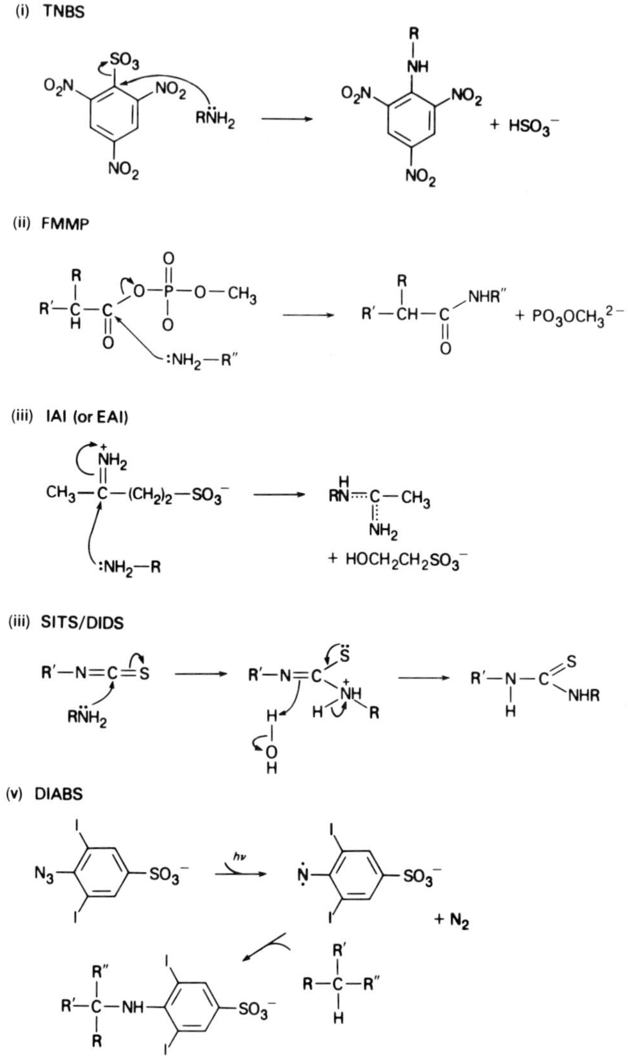

162

Figure 4.5 Mechanisms by which some small non-penetrant reagents chemically modify membrane components. Note that phosphatidyl serine, when present at high density is often incompletely labelled by TNBS due to charge repulsion between these two interacting species. TNBS can also react with H_2O to yield picric acid which may contaminate the preparation and lower the effectiveness of the reagent

some large multisubunit complexes, e.g. cytochrome oxidase, certain of these subunits are localized entirely within the bilayer. By synthesizing radioactive phospholipids bearing photoactive azido groups at different points down the fatty acyl chain, it is possible to obtain information on the disposition of proteins within the bilayer from their susceptibility to cross-linking with these agents.

Phospholipid exchange proteins (PLEP)

These are extremely useful non-perturbing tools for assessing bilayer lipid asymmetry in natural and artificial membranes. They are a family of proteins occurring in the cytosol of most if not all eukaryotic cells. The best-studied phospholipid exchange protein is a species from beef liver which specifically catalyses the transfer of phosphatidyl choline residues from the external leaflet of one vesicle to the external leaflet of another. It is monomeric (M_r = 22,000), has an acidic isoelectric point, and approximately 40 per cent of its amino acids have non-polar side chains. The isolated protein has a single phosphatidyl choline residue bound to it which is buried inside the molecule where it is resistant to phospholipase attack yet can be exchanged with other phosphatidyl choline molecules present in the external surface of membrane vesicles. By incubating sealed membrane vesicles with sealed liposomes containing synthetic, radio-labelled phosphatidyl choline in the presence of externally applied PLEP, it is possible to assess the size of the PLEP exchangeable lipid pool (Figure 4.6). This represents the fraction of phosphatidyl choline occurring in the external half of the membrane. As this transfer proceeds on a one-for-one exchange basis and merely involves replacing endogenous phosphatidyl choline for radio-labelled phosphatidyl choline, then minimal perturbation of the bilayer can be expected. The size of the protein precludes its leakage across the membrane, thus restricting access of the PLEP to the external surfaces.

Care must be taken when using such proteins to ensure that they are pure. This is because it is essential to exclude PLEP species of unknown specificity, proteins able to perturb the membrane, or proteins that can non-specifically bind lipid vesicles or membranes. One particular problem when carrying out such experiments is that some of the synthetic liposomes may fuse with the membranes. This would increase the lipid pool of the membranes and also incorporate radio-labelled lipid into both sides of the bilayer. Usually a non-exchangeable labelled lipid is also present in the liposomes so that incorporation of this into the membranes gives an indication of the amount of fusion occurring. This is usually minimized by manipulating experimental conditions such as pH, charge on the liposomes, liposome composition and ionic strength. The residual fusion may be corrected mathematically. Fusion is often encountered in such studies because it is usual to use an excess of labelled liposomes over membranes. This ensures that over 99 per cent of the phosphatidyl choline in the external half of the membrane is radio-labelled after exchange.

In some instances the reverse procedure is carried out. Here the target membrane contains the radio-labelled lipids which have been incorporated through additions of precursors to diet, culture media or by injection. The membranes are then incubated with unlabelled liposomes in the presence of exchange protein which serves to deplete the labelled lipids of the external side of the membrane.

164

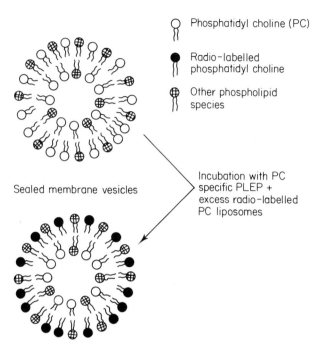

Phosphatidyl choline (PC)

Radio-labelled
phosphatidyl choline

Other phospholipid
species

Sealed membrane vesicles

Incubation with PC
specific PLEP +
excess radio-labelled
PC liposomes

Figure 4.6 **Determination of lipid asymmetry using phospholipid exchange proteins.**
Phospholipid exchange proteins (PLEP) catalyse the one-for-one exchange of
phospholipids, existing in the external side of the bilayer, between lipid vesicles.
Incubation of sealed membrane vesicles with excess liposomes containing radio-labelled
phosphatidyl choline will, in the presence of a phosphatidyl choline-specific PLEP, lead
to the replacement of all of the external, unlabelled phosphatidyl choline in the vesicles
with the labelled species. The vesicles are then separated and the lipids analysed. An
inspection of the specific radioactivity of the phosphatidyl choline in the vesicles will
allow a formulation of the asymmetry of distribution. This is because the specific
activity of the radio-labelled lipid used will be decreased by dilution with the unlabelled
phosphatidyl choline present at the internal side of the bilayer

Phospholipases as probes of bilayer asymmetry

Phospholipases A_2 and C from various sources, phospholipase D from cab-
bage, and sphingomyelinase from *Staphylococcus aureus* are usually employed to assess bilayer asymmetry. The reactions they catalyse are discussed
in Section 6.6. Phospholipases A_2 and C, depending upon the source of the
enzyme, have rather different substrate specificities and also vary in their
ability to attack biological membranes, depending upon the packing density
of the constituent lipids. They are all proteins of a sufficient size
($M_r = 15,000-30,000$) to prevent them penetrating across the bilayer membrane. This restricts their attack to those lipids occurring at the external side
of the bilayer and so evidence of degradation of specific phospholipids by
phospholipases can be taken as an index of the localization of the phospholipid.

Phospholipase treatment in effect removes the lipids forming the external monolayer of the membrane as well as producing reaction products which can potentially perturb bilayers, e.g. lysolecithins which can have surfactant properties. This makes it essential to monitor the permeability of vesicles over the usually protracted incubations necessary to carry out these experiments. Lysis would clearly lead to a degradation of lipids exposed at both sides of the membrane.

The inability of a phospholipase to degrade a particular species of phospholipid in sealed vesicles need not necessarily imply the absence of the phospholipid from the extracytoplasmic side and its exclusive presence at the internal face. In erythrocytes a number of phospholipase A and C enzymes can act on their substrates exposed at the extracytoplasmic surface, where these enzymes isolated from other species cannot act. This is related to the density of packing of the lipids in the membrane: some phospholipases can act on phospholipids that are far more densely packed than others. It would appear that for a phospholipase to act on the lipids at the extracytoplasmic side of the intact erythrocyte membrane they must be able to hydrolyse a monolayer of these lipids at a surface pressure of at least 30–32 dyne cm^{-1}. Those enzymes requiring more loosely packed lipids, as in monolayers with lower surface pressures, will be ineffective when acting on intact erythrocytes. The difference in packing density between the two surfaces of the erythrocyte membrane is well demonstrated by the phospholipase A$_2$ from pig pancreas which cannot hydrolise lipids in a monolayer with a surface pressure greater than 17 dyne cm^{-1}. This enzyme cannot degrade phosphatidyl choline at the extracytoplasmic side of the intact erythrocyte membrane. However, if the enzyme is sealed inside right-side-out ghosts it will hydrolyse the phosphatidylcholine exposed at the cytoplasmic side after addition of its co-factor, Ca^{2+}.

The susceptibility of lipid bilayers to attack by phospholipases is highly temperature-dependent as it would appear that certain phospholipases cannot act on gel phase lipid. This requires that studies be carried out at different temperatures as when using chemical probes. A further complication that may arise has been observed upon treatment of *Bacillus megaterium* protoplasts with phospholipase C. In this instance the loss of headgroups and the production of diglycerides perturbed the bilayer sufficiently to induce a trans-bilayer movement (flip-flop) of lipids from the cytoplasmic side to the extracytoplasmic side. It is not known at present whether or not this phenomenon is unique to *B. megaterium*, but it certainly demands that caution should be taken when carrying out such experiments.

NMR methods of assessing phospholipid asymmetry

The distribution of appropriately labelled synthetic phospholipids in either side of the vesicle bilayer can be determined by virtue of the fact that paramagnetic ions when added to these sealed vesicles only react with the polar headgroups at the exposed surface. This leads to a shift or broadening of the resonance observed, depending upon the ion used. Typically Eu^{3+}, Nd^{3+}. Pr^{3+} and $K_3Fe(CN)_6$ are used as shift reagents in ^{31}P and ^1H nmr studies, and Dy^{3+} is extremely effective in ^{13}C nmr studies. The proportion of each lipid species in either side of the bilayer can be calculated from the size of the separated peaks in the resonance spectrum.

Arrhenius plots of the activity of integral enzymes to detect bilayer asymmetry

This method has been used successfully to probe the lipid asymmetry of the liver plasma membrane from rat and hamster.

In the absence of stimulating hormone the catalytic unit of adenylate cyclase is sensitive to the lipid environment of the cytoplasmic side of the plasma membrane bilayer. However, in the presence of the stimulating hormone, glucagon, the glucagon receptor physically couples with the catalytic unit, forming a transmembrane complex, under conditions where it obeys a mobile receptor model, (see section 7.7) sensitive to the lipid enrivonement of both halves of the bilayer (Figure 4.7). By carrying out Arrhenius plots (see Sec-

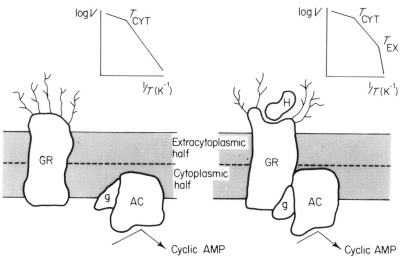

Figure 4.7 Hormone-stimulated adenylate cyclase detects bilayer asymmetry. In the absence of hormone, the catalytic unit is an independent entity sensitive to the lipid phase separation occurring in the cytoplasmic side of the bilayer. This may be detected in Arrhenius plots of its activity. In the presence of hormone, the receptor can couple to the catalytic unit of adenylate cyclase forming a transmembrane complex sensitive to the lipid phase separation occurring in both halves of the bilayer. In such a situation Arrhenius plots of the hormone-stimulated activity will exhibit two breaks: one representing the lipid phase separation occurring in the cytoplasmic side of the bilayer (T_{CYT}) and the other in the extracytoplasmic side (T_{EX}). H = glucagon' GR = glucagon receptor; AC = catalytic unit of adenylate cyclase; g = guanine nucleotide regulating protein which can associate/dissociate from adenylate cyclase (see section 7.7)

Table 4.1   The proposed disposition of some liver plasma membrane enzymes

Function	Location of active site	Location of protein within the bilayer
Adenylate cyclase	Cytoplasmic side	Cytoplasmic side only
Glucagon receptor	Extracytoplasmic side	Spans bilayer?
Cyclic AMP phosphodiesterase	Cytoplasmic side	Cytoplasmic side only
5'-Nucleotidase	Extracytoplasmic side	Transmembrane
Na$^+$/K$^+$-ATPase	Cytoplasmic side (ouabain binding side at the extracytoplasmic side)	Transmembrane
Mg^{2+}-ATPase	Extracytoplasmic side	Penetrates through to cytoplasmic side. Transmembrane?
Phosphodiesterase I	Extracytoplasmic side	Extracytoplasmic side only

Taken from Houslay, M. D. (1980) *Biochem. Soc. Trans.*, **7**, 843–846.

tion 3.3.2) of the activity of the enzyme, both in the presence and absence of hormone, it is possible to detect lipid phase separations occurring in both halves of the bilayer and to assign them appropriately. A number of other enzymes which appear to be asymmetrically disposed in the bilayer yield Arrhenius plots that support this procedure (see Table 4.1).

The limitations of this method are that it can only probe lipid phase separations occurring over a relatively narrow range of temperature (0–45 °C) and that it is highly likely that these enzymes monitor a restricted lipid environment. Independent tests must be made to confirm that the breaks in the Arrhenius plots are due to lipid phase separations (see section 3.3.2).

Proteases

The proteins of sealed, membrane vesicles of known orientation may be examined for their susceptibility to attack by externally added proteases. Modification may be monitored by a change in migration of individual proteins upon SDS–PAGE or by a loss of enzyme activity.

Proteases are large-molecular-weight proteins that cannot pass through membranes and thus are useful probes for assessing membrane asymmetry. They can also be immobilized on cross-linked dextran beads; this not only ensures that they do not penetrate into the vesicles but also provides a means of removing the protease after treatment. Trypsin is a particularly useful reagent because its action can be terminated by the addition of a specific inhibitor of the enzyme. This ensures that no further proteolysis of newly exposed residues occurs during analysis.

Owing to their different substrate specificities, various types of proteases may be employed to ensure that any protein having exposed residues at the vesicle surface has a chance of being modified. Clearly, a fibrous tail of a protein that may extend out on one side of the membrane by just a few residues is unlikely to be modified. On the other hand, long fibrous tails or globular portions of proteins should make suitable targets.

Proteases acting on identifiable target proteins can be extremely useful tools in providing information on the peptides exposed at either side of the membrane and those protected from attack, presumably because of their intramembranous nature. They often yield information on the asymmetric location of protein-bound carbohydrate as this can be liberated from the membrane in the form of glycopeptides.

Latency

The location of the active site of a membrane-bound enzyme may be assessed by virtue of its accessibility to non-penetrating substrates in sealed vesicles of a defined orientation. Enzymes with active sites exposed at the external surface should be fully active in the presence of externally added substrate, whereas those with active sites exposed at the internal surface should only exhibit activity in leaky vesicles. To make such judgements on the disposition of the active site it is imperative to demonstrate that substrate cannot pass across the membrane. Latency, the exposing of active sites at the internal side of the bilayer with subsequent unmasking of enzyme activity, can be achieved by making the impermeable vesicles leaky. This disruption of the membrane is carried out by either osmotic shock, freeze–thawing or treatment with low concentrations of detergent or organic solvent. Care must be taken using these last two agents as they may affect the activity of the enzymes under study. Latency and changes in activity may be distinguished by comparing the results obtained using different methods of membrane disruption.

This can be a very useful technique to determine the localization of the active site of membrane-bound enzymes. It is usual to confirm such findings by treating sealed vesicles with proteases or specific antibodies that can destroy the enzyme activity. This method, of course, cannot yield information on how deeply the protein penetrates the bilayer or whether the protein is spanning.

Antibodies

Antibodies can be used as reagents to label specific membrane components. They are too large ($M_r \geq 150,000$) to penetrate sealed membranes and are particularly versatile agents to use since antibodies of high specificity and affinity can be raised to virtually any protein. Binding of antibody to a target (antigen) on the cell surface may be assessed by neutralization of an enzyme active antigen or by labelling the antibody with [125]I. It is most convenient to do this by first obtaining immunoglobulin G (IgG) antibodies against the target antigen and then treating the sealed membrane vesicles with this specific IgG antibody. The antibody then binds to the exposed antigen molecules. After this the vesicles are washed before treating with an ^{125}I-labelled antibody that binds specifically to the IgG molecule (see Figure 4.8). This procedure obviates any need to consider changes in specificity of the antibody raised against the cell surface target antigen that may be caused by the iodination reaction.

The recent development of monoclonal antibody production provides the means of producing molecularly homogeneous antisera against defined membrane molecules or specific regions of these molecules. This is achieved by fusing a cell producing a single species of antibody molecule with a myeloma cell. This produces a proliferating clone of hybrid cells producing only the one type of antibody molecule. The precise location and orientation of specific antigens can then be determined.

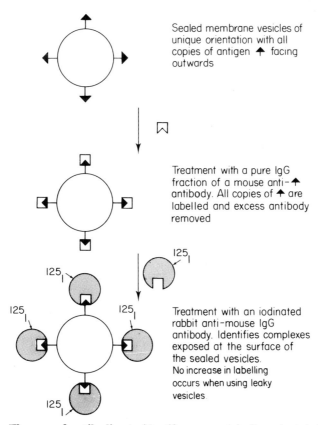

Sealed membrane vesicles of unique orientation with all copies of antigen ✦ facing outwards

Treatment with a pure IgG fraction of a mouse anti-✦ antibody. All copies of ✦ are labelled and excess antibody removed

Treatment with an iodinated rabbit anti-mouse IgG antibody. Identifies complexes exposed at the surface of the sealed vesicles. No increase in labelling occurs when using leaky vesicles

Figure 4.8   The use of antibodies to identify asymmetrically orientated membrane components

Table 4.2 Some commonly used lectins and their sugar specificity

Lectin	Mol. wt. (structure)	CHO binding specificity
Concanavalin A	110,000 (tetramer)	Me-α-D-glucosyl
		Me-α-D-mannosyl
Wheat germ lectin	36,000 (dimer)	NAc glucosamine
Lentil lectin	52,000 (2 non-identical dimers)	Me-α-D-mannosyl Me-α-D-glucosyl
Phytohaemagglutinin	100,000 (4–8 subunits— various types	Me-α-D-GalNAc
Helix pomatia lectin	79,000 (6 subunits)	Me-α-D-GalNAc

Lectins

Carbohydrate binding proteins, collectively called lectins, have been isolated from a wide variety of plants and invertebrates and have recently been detected in bacterial and animal cells. They were first isolated from plants and detected by their ability to agglutinate erythrocytes. Because each lectin molecule is multivalent, having a number of sites able to bind carbohydrate, they achieve this agglutination by cross-linking the surface carbohydrate residues of adjacent cells. Different lectins exhibit a specificity for certain types of sugar residues (see Table 4.2) to which they bind extremely tightly. Because of this they make suitable probes to detect and identify carbohydrate attached to both protein and lipid exposed at the vesicle surface. Visualization of lectin binding can be readily achieved in the electron microscope if the lectin is first conjugated to the iron-containing protein, ferritin: the bound lectin will be identifiable as dark, electron-dense blobs. Alternatively, the lectin can be radio-iodinated and visualized by autoradiography. Such radio-labelled lectins can be incubated with sealed membrane vesicles of either orientation to assess binding by the incorporation of radioactivity. Exposed carbohydrate residues can thus be detected and asymmetry ascertained.

Lectins can also be immobilized on inert supports of cross-linked dextran and used for preparing vesicles of both orientations by virtue of the absolute asymmetry of membrane carbohydrate (see section 4.4).

Lectins immobilized on agarose beads are also extremely useful agents for purifying membrane glycoproteins, which can subsequently be resolved by SDS–PAGE.

Labelling of glycolipids and glycoproteins with KB^3H_4

Galactose oxidase is a large protein and as such does not penetrate lipid bilayers. It oxidizes terminal galactose and galactosamine residues in carbohydrate chains attached to glycoproteins and glycolipids. They can subsequently be reduced by application of ^3H-labelled potassium boron hydride. This incorporates ^3H into the sugar molecules, allowing their localization to be determined.

Summary

Labelling of sealed membrane vesicles in order to determine membrane asymmetry must fulfil stringent criteria before interpretations are made.

Methods of labelling biological membranes rely on the impermeability of these membranes to large or highly charged molecules.

Ideally, sealed populations of membrane vesicles of both inside-out and right-side-out orientations should be used.

An approach to investigate the asymmetry of membrane components should employ a number of different methods to assesss each component under study.

4.4   PROTEIN ASYMMETRY IN THE MAMMALIAN ERYTHROCYTE MEMBRANE

The plasma membrane of the erythrocyte has been extensively studied because it comes close to fulfilling the criteria outlined in section 4.3.1; it is readily available and easily manipulated to form sealed vesicles of either orientation.

We should, however, take care in extrapolating from conclusions about membranes obtained by using the mammalian erythrocyte. It is an 'end' cell which cannot divide; it has no nucleus; it has restricted metabolic functions; and it has no intracellular organelles. It is specialized for its existence as a free living cell in the circulation by the presence of a rigid network of a polymeric, peripheral protein, called spectrin, underlying the membrane. This network helps to maintain the characteristic shape of the cell. The mammalian erythrocyte differs markedly from amphibian and avian erythrocytes which have a nucleus and intracellular organelles.

Having said this we think it will become clear that a lot of useful information can be obtained from a study of the mammalian erythrocyte, but the reservations expressed above must be carefully considered before extrapolation or generalization.

SDS–polyacrylamide gel electrophoresis (SDS–PAGE) of the red cell membrane can be used to separate protein species which can then be visualized by staining for protein or by their reaction with periodate–Schiff's (PAS) stain for carbohydrate. These bands are then classified according to their degree of migration along the gels. This yields four PAS staining sialoglycoproteins and seven other proteins that do not contain carbohydrate (Figure 4.9). In addition to these well-defined bands, about 20 faint bands can be discerned after staining for protein and these constitute approximately 1 per cent of the total protein. However, acetylcholinesterase and Na^+/K^+-ATPase, enzymes with important functional roles, account for <0.1 per cent of the total protein and are not detectable by SDS–PAGE.

All of the eight major polypeptides of the erythrocyte plasma membrane are available for modification or proteolytic digestion in leaky 'ghosts'. From this we may deduce that all of these proteins are exposed at one or other of the surfaces i.e. none are located totally within the hydrophobic core of the membrane. In intact erythrocytes only two of these proteins, called PAS 1

SDS-polyacrylamide gel of the coomassie-blue staining proteins of the human erythrocyte membrane

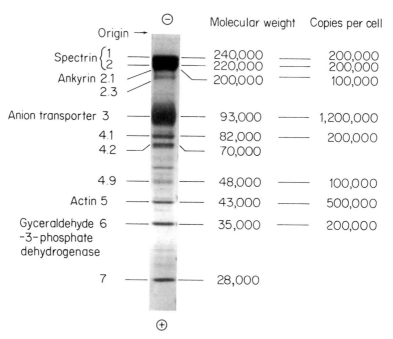

	Molecular weight	Copies per cell
Origin →		
Spectrin {1	240,000	200,000
{2	220,000	200,000
Ankyrin 2.1	200,000	100,000
2.3		
Anion transporter 3	93,000	1,200,000
4.1	82,000	200,000
4.2	70,000	
4.9	48,000	100,000
Actin 5	43,000	500,000
Gyceraldehyde 6 -3-phosphate dehydrogenase	35,000	200,000
7	28,000	

SDS-polyacrylamide gel of the PAS - staining proteins of the human erythocyte membrane

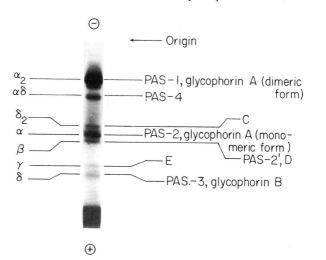

α_2 ——— PAS-I, glycophorin A (dimeric form)
$\alpha\delta$ ——— PAS-4
δ_2 ——— C
α ——— PAS-2, glycophorin A (mono-
β — meric form)
γ ——— E —— PAS-2', D
δ ——— PAS.-3, glycophorin B

and band 3, are labelled, but in unsealed membranes different regions of the two proteins become available for labelling, showing that these two proteins span the bilayer. Of the remaining six major polypeptides, five are located on the cytoplasmic face of the membrane and are peripheral proteins interacting with the membrane through predominantly charged interactions. These may be readily displaced from the membrane by treatment with either high ionic strength solutions or the chelating agent EDTA or by increased pH. Little is known about the other major polypeptide, band 7, save that it appears to be an integral protein, firmly bound to the membrane.

4.4.1 Properties of the integral proteins

Band 3 is a globular protein with a molecular weight of 90,000 daltons. It is a characteristic integral membrane protein, retaining considerable amounts of detergent bound to it after solubilization. Reconstitution of this extract into artificial phospholipid vesicles or black lipid films confers upon them the anion transport properties of the original membrane (see Chapter 5) and also the characteristic appearance of the membrane in freeze–fracture (see 1.5).

Band 3 protein is available for labelling or digestion by proteolytic enzymes in membrane preparations of both orientations. The regions which are attacked, however, are characteristic of whether inside-out or right-side-out sealed preparations are used. This demonstrates that the protein spans the membrane asymmetrically. The majority of the polypeptide chain is located at the cytoplasmic half of the bilayer since freeze–fracture electron micrographs show the particles on the cytoplasmic face of the bilayer and digestion of intact cells with pronase only reduces the molecular weight by one-third. Pronase digestion of the intact cell also liberates all of the carbohydrate residues attached to band 3, indicating the exclusively extracytoplasmic localization of this moiety. Indeed, labelling of the carbohydrate in intact cells leads to the modification of all copies of band 3, showing that every band 3 polypeptide chain is identically oriented in an asymmetric fashion in the membrane. These copies appear to be organized within the plane of the membrane as dimers or oligomers as they can readily be cross-linked together using bifunctional alkylating reagents (i.e. a molecule with two alkylating groups on

---

Figure 4.9 **Proteins of the human erythrocyte membrane. Here are shown typical SDS–PAGE gels stained with either Coomassie blue to identify proteins (a) or with PAS reagents to identify carbohydrates residues on glycoproteins (b). The Greek letters in (b) identify each glycoprotein bands in terms of their known composition, consisting of monomers, dimers or hybrids of α, β, γ and δ polypeptide chains (see Anstee, D. J., Mawby, W. J., and Tanner, M. J. A. (1979) *Biochem J.*, 183, 193–203). Pictures of the 6–20 per cent gradient gel (a) and the 10 per cent gel (b) were very kindly donated by Dr. M. J. A. Tanner, Department of Biochemistry, University of Bristol**

174

it separated by a molecular 'spacer') and this is supported by the size of the freeze–fracture particles.

A schematic representation of band 3 is shown in Figure 4.10.

PAS 1–4 are polypeptides, separated by SDS–PAGE, containing a high proportion of carbohydrate, so they stain strongly with PAS reagent.

PAS 1 and PAS 2 are interconvertible forms of the same sialoglycopeptide known as *glycophorin A*. Glycophorin A is a fibrous, integral protein that spans the bilayer. It consists of 131 amino-acid residues with 16 attached oligosaccharide chains which make up some 60 per cent of the total mass of the molecule. PAS 1 is a dimeric form of PAS 2 that occurs through associations between the hydrophobic segments of the chain, and this form is believed to exist in the membrane although cross-linking studies have so far

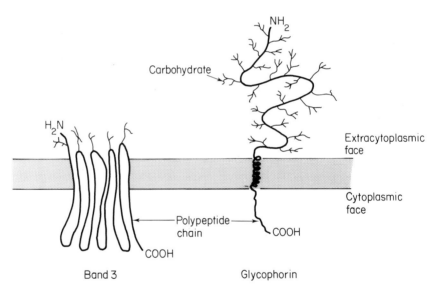

Band 3   Glycophorin

Figure 4.10 A schematic diagram of band 3 and glycophorin A. These are both integral, transmembrane proteins that are labelled in all orientations of erythrocyte ghosts. They are both asymmetrically disposed in the bilayer with the majority of the mass of band 3 at the cytoplasmic surface, and that of glycophorin A at the extracytoplasmic surface. All of the carbohydrate attached to both species is presented at the extracytoplasmic surface. All copies of each protein face the same way. These two proteins are representative of the two classes of integral membrane proteins: the globular proteins like band 3 and the fibrous ones like glycophorin A. The intramembrane portion of glycophorin A consists of about 36 amino acids which are all hydrophobic except for those which interact with the headgroup region of the bilayer. Interestingly, those charged amino acids at the cytoplasmic side of the bilayer are extremely basic and would be expected to interact with the negatively charged phospholipids found exclusively at this surface. It is likely that these intramembrane residues form eight or nine twists of σ-helix. This is sufficient to span the bilayer membrane

failed to yield any supporting evidence. This is likely to be explained in part by the small number of amino groups on the molecule that may be available for cross-linking rather than lack of close contact of molecules.

Like band 3, all copies of glycophorin A are oriented in the same direction with their carbohydrate-containing component attached to residues exposed at the extracytoplasmic surface. Since this region of the polypeptide containing the carbohydrate represents 55 per cent of the amino-acid residues, the bulk of the mass of the molecule is extracytoplasmic. This mass is connected by an α-helical region which spans the bilayer to the remaining 30 per cent of the polypeptide on the cytoplasmic side. It will be immediately appreciated that this structure, shown schematically in Figure 4.10, is quite distinct from that of band 3 and that predicted by the Fluid Mosaic model. It is, however, similar to the structure of viral coat proteins and suggests that it might be synthesized in a similar manner (see section 6.2). The thin thread of the α-helix region spanning the bilayer is presumably broken by freeze–fracture and since no intramembranous particles are associated with glycophorin A, the two halves of the molecule cannot penetrate into the bilayer to any great extent. This immediately distinguishes two classes of integral membrane proteins; the globular ones such as band 3 with a substantial mass embedded in the bilayer and the fibrous ones such as glycophorin A with only a strand of α-helix entering the bilayer.

PAS 3 is a sialoglycopeptide with a carbohydrate composition similar to glycophorin A, but with a different polypeptide chain. The carbohydrate is attached to the NH_2-terminal region of the protein which is uniquely located on the extracytoplasmic surface. No protein is labelled in inverted, sealed ghosts suggesting that little if any of the polypeptide chain is situated on the cytoplasmic face of the membrane. The protein is presumably anchored in the membrane by a hydrophobic COOH-terminal region.

PAS 4 is thought to be a heterodimer of glycophorin A and PAS 3 (see Figure 4.9).

The minor membrane components acetylcholinesterase and the Na^+/K^+-ATPase are both integral proteins. The Na^+/K^+-ATPase presumably spans the bilayer as it is responsible for the translocation of ions across the membrane, and antibodies directed against it bind to both sides of the membrane indicating that residues are exposed on either side. The Na^+/K^+-ATPase is in fact asymmetrically orientated within the bilayer. This allows it to pump Na^+ out of the cell and K^+ into it and to interact with its substrate, ATP, which provides the driving force for the pump by binding to a catalytic site on the enzyme at cytosol surface of the membrane. Interestingly, the cardiac glycoside, oubain, which potently inhibits this enzyme, binds to a site on the pump which is only exposed at the extracytoplasmic surface. The active site of acetylcholinesterase is localized at the extracytoplasmic surface of the bilayer as the activity in intact cells can be abolished by proteases and non-penetrant inhibitors. There is evidence suggesting that it exists as oligomers in the membrane, but none suggesting a transmembrane disposition.

4.4.2 Properties of the peripheral proteins

Bands 1, 2, 4, 5 and 6 on SDS–gels (Figures 4.9) comprise the major peripheral proteins (see section 1.4.2) of the erythrocyte membrane. These account for approximately 40 per cent of the total membrane protein. In the intact mammalian erythrocyte, exposure to proteases or to non-penetrating reagents fails to modify the peripheral proteins. Hence, all of the peripheral proteins are located uniquely on the cytoplasmic surface of the membrane. There is no evidence for any peripheral protein associated with the extracytoplasmic face of the erythrocyte membrane.

Peripheral proteins may interact with both phospholipid headgroups and regions of the integral membrane proteins that protrude on the cytoplasmic surface. In the erythrocyte it appears that all the peripheral proteins are either directly bound to an integral membrane protein or are linked to one via a third molecule. This suggests an interesting interdependence between the arrangement of peripheral proteins on the cytoplasmic surface of the membrane and the arrangement of integral proteins in the bilayer. Since some of these span the bilayer membrane it raises the intriguing possibility that information could be transmitted across the membrane in this manner.

Bands 1 and 2 are the polypeptide chains of the polymer *spectrin*, $M_r = 240,000$ and $220,000$ respectively. These form a chain-like structure which underlies the membrane and is responsible in part for the mechanical strength of the cell (Figure 4.11). Spectrin is the major cytoskeletal protein (60–70 per cent by weight) and whilst the two subunits that form this polymer are similar, they are not identical. Furthermore, spectrin appears to be a unique feature of the mammalian erythrocyte for the cytoskeleton of other mammalian cells is formed from a variety of other components (see section 6.7.3). The location of spectrin within the cell can be directly visualized in the electron microscope using anti-spectrin antibodies that have been conjugated with the iron-containing protein, ferritin, to make them electron dense. These bind only to the cytoplasmic side of the membrane and appear as dark blobs in electron micrographs due to the attached ferritin. Bands 1 and 2 may be cross-linked using bifunctional reagents. Such cross-linking may be markedly facilitated if the cells are pretreated with lectins which aggregate the carbohydrate moieties displayed at the cell surface. This suggests an interaction between spectrin and the cytoplasmic portion of either or both the transmembrane proteins band 3 or glycophorin A, which display carbohydrate residues on their extracytoplasmic portions.

Heterodimers of band 1 plus band 2 have been isolated as long (100 nm), slender (5 nm), worm-like molecules with the two subunits aligned in parallel and coiled around each other. *In vivo* the important species appears to be the heterotetramer (band 1 plus band 2)$_2$ where the heterodimers are joined end to end to form a molecule about 400 nm long. Oligomeric species are apparently not formed, presumably because the two heterodimers interact by a 'head-to-head' interaction rather than in a 'head-to-tail' fashion. The

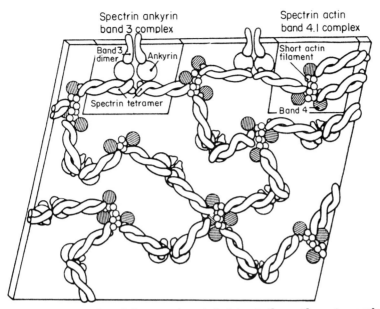

Figure 4.11 The anchoring of the spectrin cytoskeleton to the erythrocyte membrane. This diagram shows a scheme proposed to account for the formation and stabilization of a cell cytoskeleton. The cytoskeleton is a polymer composed of spectrin tetramers formed from spectrin dimers joined 'head to head'. The spectrin tetramers are polymerized at the tail end to form a network by short actin filaments (band 5) and band 4.1. This network is then anchored, at the 'head-to-head' contact region of the spectrin dimers, to the cytoplasmic surface of the membrane by molecules of ankyrin which bind tightly to dimers of the globular integral protein, band 3. (Reprinted by permission from Lux, S. E. (1979) *Nature*, 281, 427, Figure 1, copyright © 1982. Macmillan Journals Limited)

heterotetramers then bind to high-affinity sites (K_D of $10^{-7}-10^{-8}$ M), of which there are about 100,000 per cell, on the cytosol surface of the erythrocyte membrane.

These high-affinity sites for spectrin are provided by the minor peripheral proteins identified as *band 2.1* and its progeny (see Figure 4.11), which are collectively called *ankyrin* or *syndein*. Ankyrin is a pyramid-shaped molecule about 8 nm across and 10 nm long that binds at the midpoint of the heterotetramer, at the point where the heterodimers make 'head-to-head' contact, and serves to link spectrin to the transmembrane, anion transport protein, band 3. Band 3 thus acts as the true anchor to the membrane for the cytoskeleton, although only about 10–20 per cent of the total band 3 molecules are involved in this process. These band 3 molecules appear to be identical to the rest and are associated with another peripheral protein, *band 4.2*, which may have a role in the linkage of ankyrin to band 3. Unlike spectrin, ankyrin appears to be present in many cell types and may serve as a 'universal joint' for linking cytoskeletal structures to transmembrane anchoring proteins.

Two other peripheral proteins, *band 5*, which is known to be *actin*, and *band 4.1*, a globular protein about 6 nm in diameter and with a molecular weight of 82,000, have been demonstrated to interact with spectrin. These components cause the polymerization of spectrin heterotetramers to form the network of cytoskeleton underlying the red cell membrane. Electron micrographs of the polymerized structure show that short actin filaments or 'protofilaments' link the exposed tail ends of the spectrin heterotetramers together to form complex networks. The formation of such structures is dependent upon band 4.1 which, like actin, binds to the tail end of the spectrin heterotetramer. Interestingly, this reaction is maximal at physiological calcium concentrations (10^{-7}–10^{-8} M).

The other major polypeptide is *band 6* which has been identified as the monomeric form of *glyceraldehyde-3-phosphate dehydrogenase* (GAPDH). There are about 10^6 sites per cell of high affinity (K_A 10^7 M^{-1}) for band 6 displayed at the cytosol surface of the membrane of which approximately 10^5 are occupied. These sites appear to be provided by the transmembrane protein, band 3, which can bind band 6 with a 1 : 1 stoichiometry. Cross-linking and immunological studies indicate an interaction of band 6 with spectrin, showing that it binds to band 3 at a site close to the interaction of spectrin with band 3 through the ankyrin complex. As yet there has been no physiological reason proposed for why GAPDH should interact with band 3 in this way; indeed, it is likely that under *in vivo* conditions of pH and ionic strength a significant fraction of the erythrocytes GAPDH will be free in the cytosol.

Summary

All known proteins associated with the erythrocyte plasma membrane have a unique asymmetric location.

Integral proteins may have portions exposed at either or both sides of the bilayer and show a wide variety of structure.

All known peripheral proteins of the erythrocyte plasma membrane are situated on the cytoplasmic face of the membrane and are associated either directly or indirectly with the cytoplasmic portions of transmembrane proteins.

4.5 LIPID ASYMMETRY IN THE ERYTHROCYTE MEMBRANE

Forty per cent by weight of the erythrocyte membrane is lipid, which is composed of approximately equimolar proportions of phospholipids and cholesterol. Of the phospholipids, phosphatidyl choline (PC) and sphingomyelin (Sp) account for some 46–60 per cent of the total in a wide variety of species. The remainder consists principally of the aminophosphatides, PS and PE. These are present in approximately constant proportions in all species, but the Sp : PC ratio varies widely. The idea that these phospholipids were asymmetrically distributed across the bilayer came first from experiments that showed that phosphatidyl ethanolamine reacted far more readily with formylmethionylsulphone methyl phosphate (FMMP) in erythrocyte ghosts than

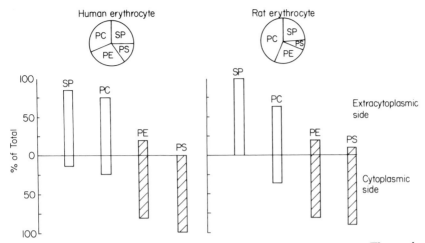

Figure 4.12   Phospholipid asymmetry in rat and human erythrocytes. The various species of phospholipid comprise markedly different proportions of the erythrocyte plasma membrane; in each case the more reactive aminophospholipids occur almost exclusively at the cytoplasmic side of the bilayer

in intact erythrocytes. It was suggested that in general the more reactive aminophosphatides were located principally on the cytoplasmic face of the membranes with the choline derivatives, phosphatidyl choline and sphingomyelin on the extracytoplasmic face (Figure 4.12). This concept has been substantiated using other non-penetrating reagents and with phospholipases. Indeed sphingomyelinase is capable of hydrolysing over 85 per cent of cell sphingomyelin in intact erythrocytes, whereas the phosphatidyl serine content is unaffected by exposure to a mixture of sphingomyelinase plus the broad specificity phospholipase A$_2$ from *Naja naja*, unless the cells are broken. Phosphatidyl serine (inner) and sphingomyelin (outer) demonstrate the most marked asymmetry of the phospholipid species, although the distribution of phosphatidyl choline (outer) and phosphatidyl ethanolamine (inner) is highly asymmetric (Figure 4.12). It is interesting to note that translocation of phospholipids in the erythrocyte membrane is sufficiently slow to allow these measurements to be made even during prolonged incubation with one half of the bilayer virtually removed by phospholipase action.

Cholesterol is present in large amounts in mammalian plasma membranes, attaining almost equimolar proportions with the phospholipids of the erythrocyte membrane. Attempts have been made to assess its distribution by exchange studies with cholesterol-free liposomes. From these it would appear that there are at least two pools of cholesterol having different exchange rates which may correspond to the inner and outer halves of the bilayer. However, cholesterol binds strongly to sphingomyelin and to neutral phospholipids with highly unsaturated fatty acyl chains, so it is possible that these bound species form a more slowly exchanging pool than non-interacting species.

An ingenious method of assessing cholesterol asymmetry has been the cleavage of the membrane bilayer by freeze–fracture, followed by analysis of the two halves for cholesterol. Despite the technical difficulties, this method has confirmed that cholesterol is present in both halves of the bilayer of the erythrocyte membrane, but with the greatest proportion associated with the outer half. This might to some extent be caused by the corresponding asymmetric distribution of sphingomyelin for which cholesterol shows a distinct preference for interaction.

Summary

Each phospholipid is found in both halves of the bilayer, but the distribution is highly asymmetric.

The more reactive aminophosphatides, phosphatidyl serine and phosphatidyl ethanolamine, are found principally in the cytoplasmic half.

The choline-derived phospholipids, phosphatidyl choline and sphingomyelin, predominate in the external half of the bilayer.

Cholesterol is found in both halves of the membrane, but a greater quantity is associated with the outer half of the bilayer.

4.6 CARBOHYDRATE ASYMMETRY IN THE ERYTHROCYTE MEMBRANE

Treatment of intact erythrocytes with the enzyme neuraminidase leads to the removal of all the sialic acid from the cell without causing lysis, indicating that it is located purely on the extracytoplasmic surface of the cell. Indeed, it can be further demonstrated that the major fraction of the sialic acid of the erythrocyte is associated with glycophorin A and that all of this is bound to the NH_2-terminal portion which extends from the cell surface. Using galactose oxidase and ^3H-NaBH$_4$ to label terminal galactose and N-acetyl galactosamine residues, it has been shown that all the carbohydrate associated with both the glycoproteins and the glycolipids of the cell are exclusively located on the external side of the cell membrane. No increase in labelling ensues if sealed inverted ghosts are used as source material. This also demonstrates that the rate of translocation (flip-flop) of the components is negligible. Further evidence supporting the purely external location of membrane-associated carbohydrate comes from studies using ferritin-labelled lectins: these bind only to the extracytoplasmic surface of the erythrocyte membrane.

Summary

The asymmetry of carbohydrate is absolute in the erythrocyte membrane.

4.7 PROTEIN ASYMMETRY IN OTHER MEMBRANES

Most eukaryotic cells consist of a variety of very different membranous structures which are not easily separated. Besides the problem of purity, definitions of leakiness and sidedness have to be made and furthermore it is not easy, nor in many instances possible, to prepare sealed vesicles of right-side-out and inside-out orientations as one can with erythrocytes. The principles and criteria discussed in section 4.3.1 require rigorous application. Consequently, most details concerning membrane structure are available for the plasma membrane, which is the only membrane exposed in intact cells, and for subcellular organelles like mitochondria which are relatively easy to prepare in a high state of purity. A further complication is that the plasma membrane of some cells, e.g. the liver parenchymal cell and cells of the brush border, have been shown to contain specialized domains with different structures and protein composition (see section 1.2.1).

4.7.1 Plasma membranes

The data available so far are rather rudimentary when one considers the multitude of proteins associated with this membrane. Three problems have inhibited these studies: (i) the variety of protein species is very large, (ii) methods are generally not available for making sealed vesicles of either orientation, and (iii) plasma membrane preparations are easily contaminated with other intracellular membranes (it constitutes less than 10 per cent of the total membrane protein or lipid).

A degree of success in making sealed plasma membrane vesicles has been achieved using either homogenization, sonication (exposure to ultrasound) or nitrogen bomb cavitation (exposure to high pressure followed by rapid depressurization). Attempts to sort these into vesicles of one or other orientation utilize the absolute asymmetry of protein or carbohydrate species. The preparations are passed down a column consisting of an inert support, agarose beads, to which either lectins, which bind carbohydrate, or antibodies specific to extracytoplasmic proteins, are immobilized. Right-side-out vesicles stick to the column and inside-out ones are found in the eluate (Figure 4.13). With a lectin column the right-side-out vesicles can be eluted from the column with a strong solution of a sugar able to compete for the lectin binding site. Using lymphocytes this procedure has been shown to give a genuine separation of such vesicles. This was determined by fixing isolated inside-out vesicles with glutaraldehyde, raising antisera against the exposed cytoplasmic portions of membrane proteins and showing that using a column made with this antibody attached to agarose beads only right-side-out vesicles could pass through. This is a difficult technique in practice but may become more feasible using molecularly pure monoclonal antibodies directed to different regions of membrane proteins.

An ingenious method for obtaining inside-out vesicles of the plasma membrane of phagocytic cells is to feed the cells with minute latex beads (1 μm diameter). These are taken up into the cell with a layer of membrane around them until the cell is full of latex beads, which cannot of course be digested. Gently disrupting the cells followed by centrifugation causes all the low-density beads, together with their associated inside-out membrane, to collect

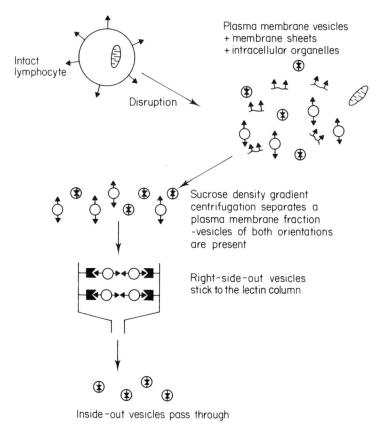

Plasma membrane vesicles
+ membrane sheets
+ intracellular organelles

Intact lymphocyte

Disruption

Sucrose density gradient centrifugation separates a plasma membrane fraction -vesicles of both orientations are present

Right-side-out vesicles stick to the lectin column.

Inside-out vesicles pass through

Figure 4.13 Preparation of orientated plasma membrane vesicles from lymphocytes

on the surface (Figure 4.14). Unfortunately, these membranes also contain contaminating structures from fused lysosomal membranes.

The wealth of enzyme activity and binding activity of receptors associated with the plasma membrane has facilitated the investigation of asymmetry amongst minor membrane components. A good example of such studies is given by the *parenchymal plasma membrane*. To obtain a preparation of pure parenchymal cells, a liver is perfused with a solution of the enzyme collagenase from *Clostridium histolyticum*. This enzyme degrades the collagen connections holding these cells together, and a pure preparation of parenchymal cells can be obtained after a simple centrifugation procedure of the released cell suspension. In the parenchymal cell the binding of insulin and glucagon is localized to distinct and highly specific glycoprotein receptors which are uniquely orientated in the membrane with their binding sites at the cell suface. This can readily be demonstrated by the fact that hormone covalently linked to agarose beads promotes a cellular response and that trypsin treatment of the cell surface under non-lytic conditions can abolish hormone binding.

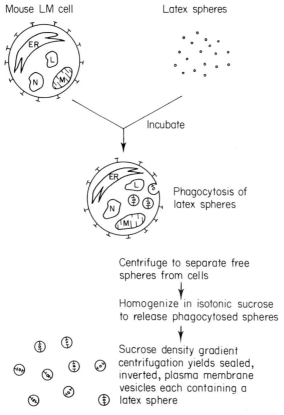

Mouse LM cell

Latex spheres

Incubate

Phagocytosis of
latex spheres

Centrifuge to separate free
spheres from cells

Homogenize in isotonic sucrose
to release phagocytosed spheres

Sucrose density gradient
centrifugation yields sealed,
inverted, plasma membrane
vesicles each containing a
latex sphere

Figure 4.14 Preparation of inverted, sealed plasma-membrane vesicles from a phagocytic cell line using latex beads

Furthermore, sealed inverted plasma-membrane vesicles do not bind these hormones. On the other hand, adenylate cyclase, which is the catalytic unit for the glucagon receptor and many other types of externally orientated receptors in various cell types, has its active site exposed at the cytoplasmic side of the membrane. This orientation of the catalytic unit was deduced from observations that in the intact cell the enzyme is resistant to proteases, yet it is destroyed by them in leaky cells; that cyclic AMP cannot be made from externally added ATP; and that cyclic AMP levels rise intracellularly in response to extracellular addition of hormone. Such an asymmetric juxtaposition of hormone receptor and catalytic unit is clearly of immense physiological importance in achieving a vectorial flow of information across the cell membrane, from the hormone binding outside to cyclic AMP production intracellularly. The catalytic unit of adenylate cyclase is apparently localized in the cytoplasmic half of the bilayer as it only senses lipid phase separations in that half in its basal state. However, when it physically couples to the glucagon receptor to form a transmembrane complex (see section 7.7) it becomes sensitive to the lipid environment of both halves of the bilayer (see section 4.3.2).

About 50 per cent of the total 5'-nucleotidase activity is expressed in whole parenchymal cells incubated with the non-penetrating substrate, AMP, and a corresponding fraction of activity can be immediately neutralized by externally added antibody. The remaining activity is found in the extracytoplasmic face of intracellular vesicles which are probably involved in membrane recycling (see section 6.7). So, far from breaking the absolute asymmetry rule, 5'-nucleotidase has the same orientation with respect to the membrane and cytoplasm in both locations. Interestingly, in collagenase-isolated adipocytes about 75 per cent of the 5'-nucleotidase activity is expressed in whole cells; seemingly the fraction of internalized, vesicular plasma membrane varies from cell type to cell type. 5'-Nucleotidase is believed to span the bilayer as it senses lipid phase separations occurring in both halves of the bilayer, and it can be iodinated at both sides of the membrane. The proposed orientation of these and other plasma membrane proteins is summarized in Table 4.1.

A particularly elegant investigation into the organization of a plasma-membrane protein has been carried out with the amino-peptidase N of the *intestinal brush border*. This enzyme forms 8 per cent of the total membrane protein. A hydrophilic, globular region comprises the major part of the enzyme; it contains the active site and attached carbohydrate moieties and is located on the extracytoplasmic surface of the membrane. It is attached to the membrane by a short hydrophobic region, which represents only about 2–6 per cent of the protein, spanning the bilayer to yield a small exposed region on the other side. Unlike glycophorin A and virus coat proteins it is the NH_2-terminal end which is situated at the cytoplasmic face of the membrane. The extent of penetration of this protein has been assessed in two ways. In the first instance it was found that the protein extracted by detergents was slightly larger than that removed by the action of the protease, papain. It also exhibited a marked tendency to aggregate, unlike the papain-produced enzyme, and appeared to have a slightly higher molecular weight. Secondly, insertion into the membrane masks 4 out of the 12 antigenic sites available on the solubilized enzyme. In order to assess whether the protein was transmembranous the following scheme was formulated (Figure 4.15). Right-side-out vesicles were prepared from brush borders in the dark in the presence of a photosensitive reagent, 4-fluoro-3-nitrophenylazide (NAP) coupled to a Fab fragment of a human IgG myeloma protein. This allowed the isolation of sealed right-side-out vesicles with the photoaffinity label coupled to the Fab fragment, to make it both non-diffusable and recognizable by a specific peroxidase-labelled antibody, trapped inside. After irradiation to link the NAP–Fab fragment to the enzyme at the cytosol side of the membrane the membrane vesicle was dissolved in detergent. It was found that one and the same molecule could bind both anti-leucine amino-peptidase antibody and peroxidase-labelled antibody directed against the NAP–Fab fragment. This complex could be readily cleaved with papain into the two fragments, where

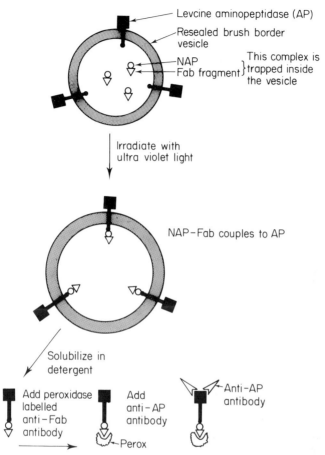

Figure 4.15 **Aminopeptidase N is a transmembrane protein. A photosensitive reagent, 4-fluoro-3-nitrophenylazide (NAP) coupled to a Fab fragment derived from a myeloma protein to make it impermeable, was trapped inside brush border vesicles. Upon irradiation, this reagent was shown to label the same aminopeptidase N enzyme associated with the vesicle as did antiaminopeptidase antibodies. The bulk of the protein of the aminopeptidase is known to be extracytoplasmic as it can readily be cleaved by externally added papain to yield a soluble, active molecule. It would appear that this globular protein is anchored to the membrane by a fibrous transmembrane tail. (Adapted from Louvard, D., Semeriva, M., and Maroux, S. (1976)** *J. Mol. Biol.,* **106, 1023–1035)**

the peroxidase-labelled anti-Fab fragment can be shown to bind to the hydrophobic fragment released, indicating the transmembrane disposition of this asymmetrically orientated aminopeptidase.

The plasma membrane of *lymphocytes* is of great interest because of the central role of these cells in the immune response. They exist in the body as two types, B-lymphocytes which upon activation produce antibody, and T-lymphocytes which do not, but amongst other things act to stimulate B-cells and kill foreign cells. These species can be separated using columns consisting of inert beads, agarose, coupled to antibodies directed against specific determinants expressed by either T- or B-cells. Anti-immunoglobulin antibodies bind only to the extracytoplasmic surface of these cells. It is believed that this surface immunoglobulin, which on a particular cell has a unique specificity, acts as a receptor for a particular antigen which, once bound, activates the cell and triggers its proliferation. The histocompatibility antigens (see section 1.1.4) of a human lymphoblastoid cell line have been studied in some detail by the lactoperoxidase-catalysed iodination technique. In this cloned cell line all cells are identical, overcoming the problems of molecular heterogeneity. The histocompatibility antigens are integral proteins, binding significant amounts of detergent, exposed at the external surface of the plasma membrane. Populations of inside-out, sealed vesicles prepared using lectin columns as described at the beginning of this section, were iodinated and the pattern of labelled proteins compared with those found using iodinated whole cells as a source. Ten major bands were iodinated in both instances, indicating their transmembrane disposition. In order to identify the bands on the SDS–gels, specific antisera were raised to purified membrane proteins and incubated with detergent extracts of iodinated, inside-out vesicles. Each immunoprecipitate was then separated and analysed by SDS–PAGE. In this manner it was established that the glycoprotein histocompatibility antigens HLA-A, B, C and Ia all spanned the bilayer. This experiment was also performed for the immunoglobulin, IgM which apparently was not iodinated in inside-out vesicles. Like all surface immunoglobulin receptors it is an integral membrane protein, uniquely orientated with its antigen binding (Fab) sites exposed at the cell surface.

4.7.2 Mitochondria

From the viewpoint of experimental studies on membrane asymmetry, mitochondria have the advantage of being a major cell constituent (20–50 per cent of the cell volume) and are also easy to purify by differential centrifugation. The outer membrane may be removed by treatment with bile salts or digitonin, which are mild detergents that preferentially attack the cholesterol-rich outer membrane but leave the inner membrane intact, sealed and functional. Alternatively, the outer membrane may be selectively removed by osmotic means followed by extremely mild sonication. Strong but controlled sonication or treatment with alkali generates sealed inside-out vesicles of the inner membrane. It is usual to designate the faces of the inner membrane by M for the side which normally faces the mitochondrial matrix and C for the cytosolic side.

The mitochondrial inner membrane contains some 60 different proteins of which about 40 can be classed as integral membrane proteins. These occupy about one-third of the area of the membrane, the rest being lipid bilayer formed of phosphatidyl choline, phosphatidyl ethanolamine and a characteristic mitochondrial phospholipid, cardiolipin, in a ratio of 4 : 3 : 2. Interest has focused principally on the organization of the active sites of the electron transport chain in the membrane since this is of primary importance to the pathway of oxidative phosphorylation. Histochemical labelling, substrate and inhibitor access, specific antibody binding and labelling with non-penetrating reagents have been used in this study. The results are shown in Table 4.3. Less is known about the overall disposition of these proteins in the membrane, e.g. whether they span the membrane. Labelling of mitochondria denuded of their outer membrane, and of inverted vesicles with the non-penetrating reagents DABS and PMBS suggests that a larger number of reactive groups (and hence more polypeptide chains) are exposed at the M-side of the mitochondria. SDS–gels of DABS-labelled right-side out and inside-out mitochondrial vesicles show that a M_r = 26,000 peptide spans the membrane, but its identity remains obscure.

Cytochrome c is a peripheral membrane protein (M_r = 13,000) situated on the C-side of the membrane. It is bound by electrostatic interactions with the headgroups of phospholipids and with the membrane protein cytochrome oxidase which spans the bilayer. There is one cytochrome c-molecule for each cytochrome oxidase molecule.

Cytochrome oxidase is a multimeric protein that contains two cytochrome active sites, a and a_3. Antibody and non-penetrating label studies have firmly established that cytochrome a is situated on the C-side and cytochrome a_3 on the M-side. The protein functions as a proton pump, accepting electrons from the peripheral protein cytochrome c on the C-side of the membrane and reacting with molecular oxygen at the M-side. Labelling studies using DABS, iodination and specific antibodies in isolated, sealed inner membranes, inverted vesicles and in a purified enzyme complex show differential labelling

Table 4.3   The asymmetric organization of the components of the electron transport system and the F_1-ATPase

Enzyme complex	No. of subunits	Mol. wt.	Type	Side exposed at	
				M	C
NADH-ubiquinone reductase (complex I)	16	850 K	Integral	Yes	Yes
Succinate-ubiquinone reductase (complex II)	2	97 K	Integral	Yes	?
Ubiquinone-cytochrome c reductase (complex III) (cytochrome-b, c_1 complex)	7	280 K	Integral	Yes (b)	Yes (c_1)
Cytochrome c	1	13 K	Peripheral	No	Yes
Cytochrome oxidase (complex IV)	7	200 K	Integral	Yes (a_3)	Yes (a)
F_1-ATPase	10	448 K	Integral	Yes	Yes

of constituent polypeptide chains, suggesting an asymmetric organisation within the membrane. This has recently been confirmed using arylazido-phospholipids where the photoreactive label can be placed at different positions on the acyl chain so as to effect 'shallow' or 'deep' labelling of peptides penetrating the hydrophobic core of the bilayer. Cross-linking experiments with cytochrome c have been performed in an attempt to define the subunits of cytochrome oxidase interacting with it. These have proved inconclusive, perhaps because of the different lengths of spacers used in the cross-linking reagent. All of these results can be largely summarized in the hypothetical model shown in Figure 4.16 where the location of subunit VII is still a matter of controversy.

Several years ago it was found that extraction of mitochondria with detergents resulted in a green membrane fraction whose predominant protein is cytochrome oxidase. Relatively large areas of this membrane consist of cytochrome oxidase ordered in a two-dimensional crystalline assay which diffracts incident electromagnetic radiation. If the specimen is examined in the electron microscope using a low dosage of electrons so as not to damage the tissue, and recording the diffraction pattern rather than the transmitted image, a pattern of diffraction spots may be obtained. The approximate shape of the molecules may be discerned from this 1–1.2 nm resolution study which also identifies the three-dimensional space group (P2, 2, 2). The crystals are envisaged as flattened, closed vesicles in which the cytochrome oxidase molecules from the two parallel membranes interact with each other at the centre of the vesicle. The molecules of cytochrome oxidase exhibit a markedly asymmetric orientation in the bilayer, where each molecule protrudes some 7 nm from the bilayer on the inside of the crystals, whereas little protein protrudes on the external surface. From antibody binding studies it would appear that these vesicles display an inside-out configuration with respect to the native mitochondrion. Thus we might expect the transmembrane cytochrome oxidase to protrude asymmetrically from the C-surface of the intact mitochondrion.

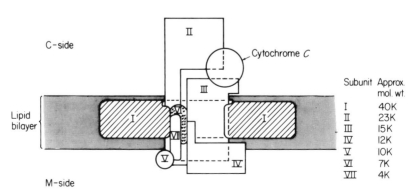

Figure 4.16   A hypothetical model of cytochrome oxidase

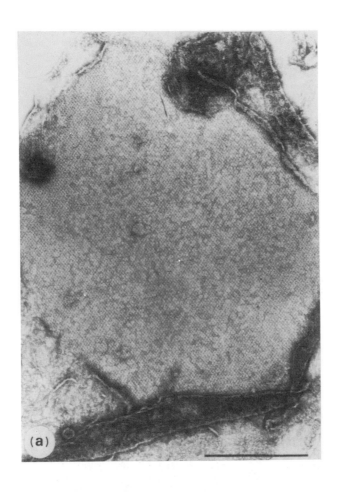

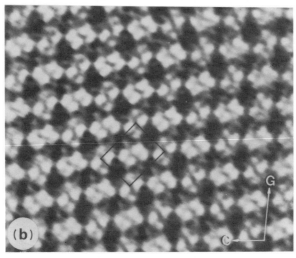

190

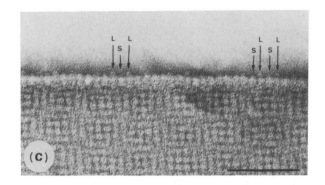

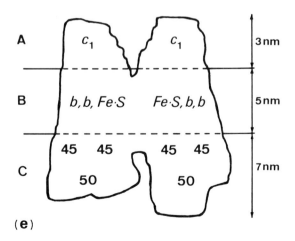

Cytochrome reductase (ubiquinol : cytochrome c reductase) forms 10 per cent of the protein of the mitochondrial inner membrane. It is a major enzyme of the mitochondrial oxidative phosphorylation system (see section 7.6.5) and can be isolated as a pure detergent–protein complex from mitochondria. After sonicating the complex with micelles of appropriate phospholipids and a non-ionic detergent, the detergent can subsequently be removed allowing the formation of 'membrane crystals' (Figure 4.17a). These crystals extend to a few micrometers in the two dimensions of the bilayer and are only one molecule thick in the third dimension, making them suitable for image analysis by electron microscope (Figure 4.17b,c). From this a three-dimensional model of cytochrome reductase can be constructed (Figure 4.17). The protein is dimeric and asymmetrically disposed about the plane of the bilayer. The smaller exposed (polar) domain (S), seen at the top of the molecule, accounts for about 20 per cent of the total protein and is normally exposed on the cytoplasmic side of the mitochondrial inner membrane. The other polar section (L) extends about 7 nm into the mitochondrial matrix and accounts for some 50 per cent of the total protein. These two molecules interact solely at their central region which lies within the membrane itself, and as the monomers are skewed relative to the dimer axis it is possible that they can exist in two enantiomorphic forms. Each of the monomers of this dimeric enzyme ($M_r = 550,000$) consists of at least eight different subunits. Five of these do not appear to have prosethetic groups but three of them, the cytochromes b ($M_r = 30,000$), C_1 ($M_r = 31,000$) and the Fe–S subunit ($M_r = 25,000$) carry the redox centres. The possible arrangements of these subunits are shown in Figure 4.18e which attempts to relate biochemical data with the structural study.

A characteristic morphological feature of the mitochondrial inner membrane seen in negative contrast is an array of 9 nm spheres on the M-side of the membrane. These were called F_1 factors and were later identified as the ATPase which generates ATP in dissipating the proton gradient across the

---

Figure 4.17 The structure of ubiquinol : cytochrome *C* reductase. Cytochrome reductase is a major enzyme of the mitochondrial oxidative phosphorylation system. It can be isolated in a pure form from *Neurospora crassa* mitochondria and reconstituted into lipid bilayers under conditions where it forms 'membrane crystals' (a) that can be visualized by negatively stained electron microscopy (scale bar is 0.5 μm). These consist of a single layer of membrane crystals, one molecule thick yielding a unit cell (box) containing two dimeric molecules (b). In the reconstitution procedure the molecules are packed alternately in an up and down manner in the bilayer, and as the molecules are asymmetric then a view of the edge of the crystal (c) shows the alternating molecules extending 7 nm (L) and 3 nm (S). In the native membrane, of course, all of the molecules will be orientated identically. From this data a model can be constructed (d) showing the dimeric nature, membrane domain and polar domains of this molecule. The biochemical data is then fitted schematically to this structure (e). (From Leonard, K., Wingfield, P., Arad, T., and Weiss, H. (1981) *J. Mol. Biol.*, 149, 259–274. Reprinted with permission of the *Journal of Molecular Biology*)

membrane (see Chapter 7). The ATPase can be divided into three sectors: (i) the spherical ATPase portion that can be released from the membrane by mechanical agitation; (ii) the oligomycin sensitivity conferring factor (the stalk) released by dilute ammonia treatment; (iii) the membrane sector which is an integral protein presumably spanning the bilayer and forming a proton channel. In this case the asymmetry is so marked that it may be observed with the electron microscope.

4.7.3 The endoplasmic reticulum

The endoplasmic reticulum membrane is impermeable to charged molecules and to uncharged molecules with a molecular weight exceeding 1000 daltons. Microsomes retain this impermeability and are thus suited to asymmetry studies, although techniques for producing inverted vesicles have not yet been described.

At least 30 different polypeptides of different molecular weight are associated with the membrane, but no evidence is available as to whether any of them span the bilayer. Selective iodination catalysed by lactoperoxidase has demonstrated that a number of SDS–gel bands are protected from label in intact vesicles, which presumably correspond to both soluble proteins trapped in the lumen and to proteins exclusively associated with the lumenal surface. The lumenal proteins may be labelled when the vesicles are made permeable by treatment with low concentrations of sodium deoxycholate. Most proteins, however, are exposed at the cytoplasmic surface of the membrane, indicating that there are few proteins located entirely in the extracytoplasmic half of the bilayer. Enzymes lining the membrane of the lumen of the endoplasmic reticulum are responsible for the addition of the 'core' sugar residues during glycoprotein synthesis (see section 6.3). Not surprisingly, therefore, there is a marked asymmetry of carbohydrate across the membrane which can easily be visualized in the electron microscope using ferritin-conjugated lectins as labels.

The location of the active sites of enzymes has been inferred from the latency of their substrates, from specific antibody interaction and by access to protease enzymes. Glucose 6-phosphatase is a characteristic enzyme of the endoplasmic reticulum. Latency studies with this enzyme show that it is situated on the lumenal face of the reticulum membrane, but has a carrier protein associated with it in the membrane which gives it access to glucose 6-phosphate in the cytosol. A summary of the activities and transport systems that have been studied in the endoplasmic reticulum is given in Table 4.4.

The endoplasmic reticulum is rich in cytochromes which are used in the synthesis of sterols. Cytochrome b_5 and NADH-cytochrome b_5-reductase constitute one electron transport system which has been closely studied. Both of these components are located on the cytoplasmic side of the membrane and they consist of a hydrophilic portion containing the active site which is anchored in the membrane by a short hydrophobic pedicle. Proteolytic diges-

Table 4.4 Asymmetry of the active sites of some enzymes associated with endoplasmic reticulum

Exposed at cytoplasmic surface	Exposed at lumenal surface
Nucleoside pyrophosphatase	β-Glucuronidase
Cytochrome b_5	Glucose-6-phosphatase
NADH-cytochrome b_5-reductase	Nucleoside diphosphatase
NADPH-cytochrome c-reductase	Acetanilide-hydrolysing esterase
GDP mannosyl transferase	(glycoproteins)
ATPase	
Cytochrome P-$_{450}$	

tion releases a protein with a slightly smaller molecular weight than detergent solubilization. The dynamic interaction between these components is discussed in section 7.8. Cytochrome P_{450} is the most abundant protein of the membrane. This probably has a similar dynamic relationship to NADPH-cytochrome P_{450}-reductase as the b_5 system. Once again both proteins are located on the cytoplasmic face.

One of the most interesting structural features of the endoplasmic reticulum is that of the ribosome itself. Each ribosome is a complex of nucleic acids and proteins which are responsible for protein synthesis. The ribosomes attached to the endoplasmic reticulum are not only asymmetrically oriented on the cytoplasmic face but are engaged in the synthesis of secretory proteins, which pass through the bilayer into the lumen, and membrane proteins which are themselves inserted asymmetrically in the membrane. Indeed it is during synthesis that the absolute asymmetry of membrane proteins is thought to be achieved. Consequently, the mechanism of ribosome attachment to the rough endoplasmic reticulum and subsequent insertion of nascent membrane proteins is a subject of great current interest (see section 6.2). Ribosomes bind to specific binding proteins, called ribophoryns, which have been identified by their ability to be cross-linked to ribosomal proteins and by their co-migration with the ribosomes to ribosome-rich areas of the membrane when the ribosomes are aggregated with bivalent antibodies. There are two ribophoryns of M_r = 63,000 and 65,000 which span the membrane and are glycosylated on the lumental side. Ribophoryn I and not ribophoryn II is sensitive to trypsin placed on the cytoplasmic surface whereas both species are sensitive to trypsin in permeable microsomal vesicles. They must therefore by asymmetrically opposed across the membrane.

4.7.4 Viral membranes

Some viruses, like the Semliki Forest virus (SFV) and vesicular stomatis virus (VSV), have a membrane envelope which is formed by the budding out of the virus from the host cell. The phospholipid profile of this membrane is

194

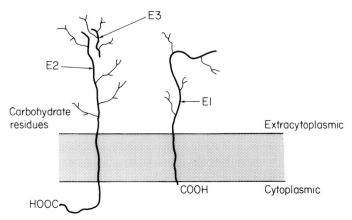

Figure 4.18   A schematic representation of the spike proteins of Semliki Forest virus

roughly like that of the host cell plasma membrane, but the protein composition is almost exclusively coded for by viral nucleic acid. Since there is only one membrane in a virus particle it provides a convenient system for the study of protein asymmetry.

The protein content of the SFV particle is particularly simple. The core of the particle is formed from nucleic acid and capsid proteins of which there are 240 copies per particle. Each capsid protein (M_r = 30,000), which may be regarded as a peripheral membrane protein on the inner face of the membrane interacts with one 'spike' protein. This 'spike' protein is in fact a complex of three polypeptide chains, two of which, E2 and E1, span the membrane (Figure 4.18). E_2 has a small segment of about 3000 daltons of its C-terminal end located on the cytoplasmic side of the membrane where it is accessible to proteolytic enzymes but both E1 and E2 have the major part of their polypeptide chain on the extracytoplasmic surface. Only this portion of the molecule is glycosylated and it forms the characteristic 'spikes' seen in electron micrographs. E1 and E2 are indeed very similar topologically to glycophorin A; fibrous, transmembrane proteins spanning the bilayer with a thin twist of α-helix.

Summary

   Protein asymmetry is a general feature of all membranes and is absolute for all copies of a particular polypeptide.

   In many cases asymmetry is directly related to protein function.

4.8   PROTEIN ASYMMETRY AND CELL FUNCTION

This description of protein asymmetry in membranes, although by no means complete, is we hope sufficient to convince the reader that protein asymmetry is both ubiquitous and absolute. From the simple virus envelope to the complex heterogeneous membranes of mammalian cells the same law applies. This high degree of conservation suggests that it is a basic property of living organisms. Our concept of a primordial cell not only involves the genetic capacity coding for biological macromolecules and for self-duplication, but also for a cell plasma membrane. This membrane is not simply a permeability barrier to isolate the cell from its environment, but as a result of its asymmetry, it confers vectorial properties on the cell which allow the maintenance of the steady-state process which we call life. We have seen examples of this in the complex mitochondrial membrane which couples vectorial transport of protons to ATP formation, and in other membranes in the asymmetric orientation of transport and carrier proteins (see Chapter 7). The vectorial ATPase pumps play an especially important role since the large free energy of hydrolysis which is coupled to transport ensures a virtually unidirectional flow.

A key question then is to ask what determines membrane protein asymmetry. It is probable that asymmetry originates in the co-translational insertion of the protein in the membrane (see section 6.2). Appropriately enough this is controlled ultimately in the genome of the cell since the signal which determines the orientation of a protein in the membrane is thought to reside in the primary sequence of the polypeptide chain. We have also discussed the entropic forces which maintain this asymmetry once it has been created (see section 1.5).

It is possible for us to ask a further question, namely whether protein asymmetry is thermodynamically stable. This would not help to create asymmetry *in vivo* since there is a kinetic barrier to the translocation of amphiphilic protein molecules through the bilayer. We may reduce this kinetic barrier *in vitro*, however, by the addition of detergents and then reconstitute by removal of added detergent. For most protein species this leads to a random orientation, but for a very few, under-defined conditions, reconstitution can be achieved to yield an apparent absolute protein asymmetry (see Chapter 5). A spontaneous process means, of course, that there is a net decrease in free energy associated with reassembly. In these examples it is a lipoprotein complex which reconstitutes to form a membrane vesicle rather than a newly synthesised polypeptide inserting in a biological membrane. At one time it was thought that glycosylation was an important mechanism for locking proteins on the extracytoplasmic face of membranes. However, asymmetry is still maintained in viral membranes if glycosylation is blocked with the inhibitor tunicamycin. The membranes of mammalian cells are in a dynamic relationship with each other. Membrane vesicles, for instance, take newly synthesized proteins from the endoplasmic reticulum via the Golgi apparatus to the plasma membrane (see section 6.7). Pinocytosis and phagocytosis involve the pinching off of plasma membrane which is directed to the lysosomes or recycled to the cell surface. Since asymmetry is observed in all these membranes we may conclude that membrane asymmetry is conserved during membrane fusion and vesiculation.

196

Summary

Protein asymmetry is responisble for the vectorial properties of biological membranes which are characteristic of a living cell.

4.9 PHOSPHOLIPID ASYMMETRY IN MEMBRANES FROM OTHER CELLS

Comparatively little is known about phospholipid asymmetry in cells other than the erythrocyte. Where such information does exist it usually relates to the external membrane of the cell since the intact cell is the most easily prepared sealed membrane preparation. Viral and bacterial membranes have therefore been studied in more detail than cells which have to be isolated from whole tissues by digestion with collagenase. Gram-positive bacteria also have the advantage of containing only one membrane system, although they also are surrounded by a cell wall consisting of polysaccharides which must first be removed. We shall also discuss the data concerning intracellular membranes of eukaryotic cells and the problems that such investigations involve.

4.9.1 Viral membranes

The mature virion of the *influenza virus* is produced from replicated particles in the host cytoplasm by a process of budding through the host plasma membrane. During this process a lipid bilayer derived from the host cell membrane forms an envelope around the viral nucleocapsid. Host cell membrane proteins are excluded, however, and replaced by those translated from the viral genome.

Influenza virus particles derived from cultured kidney cells contain almost twice as much phospholipid in the internal side of the bilayer as in the external side. This is radically different from the erythrocyte or bacterial cell plasma membrane. The composition of phospholipids in the membrane, however, are close to those of the host cell. In order to assess the phospholipid asymmetry of these membranes the infected host cell culture was grown on a medium containing ^{32}P-inorganic phosphate so that all the phospholipids became radio-labelled. The isolated viral particles were then incubated with unlabelled phosphatidyl choline liposomes and a pure phospholipid exchange protein from rat liver cytoplasm. From measurements of the incorporation of label into the liposomes and loss of label from the virus particle (see Figure 4.6) the percentage of phosphatidyl choline in the outer half of the envelope membrane was assessed. Phospholipid exchange proteins of wider substrate specificity and specific phospholipases were then used to build up a picture of the lipid organization in the virion. Sphingomyelin was located predominantly in the inner monolayer, whereas phosphatidyl choline and phosphatidyl inositol were enriched in the outer monolayer. The aminophosphatides, phosphatidyl serine and phosphatidyl ethanolamine, together with the neutral lipid cholesterol were equally distributed between the two sides of the bilayer. It was found, however, that the glycolipids, which represent a large fraction of the total membrane lipid, were exclusively located on the external side. This fraction balances the asymmetric distribution of phospholipids in favour of the inner monolayer.

In two other enveloped viruses which have been studied, VSV and SFV, the aminophosphatides were found preferentially in the inner side of the bilayer and the choline-containing ones in the outer layer. The rate of translocation of phospholipids is evidently rather small in these membranes since the fatty acyl chains of phosphatidyl ethanolamine in the VSV envelope show asymmetry, being more unsaturated on the inner side of the membrane. In fact the positioning of unsaturated acyl chains on the glycerol backbone was also different, both C1 and C2 having a similar ratio of saturated : unsaturated chains in lipid from the inner side, but C1 being predominately saturated and C2 unsaturated in the outer half of the bilayer. Cholesterol also appears to be asymmetrically distributed in VSV with the inner side containing twice as much as the external side.

Although viral particles present a convenient membrane to study, suggestions that they can be used to assess the asymmetry of the host cell membrane should be treated with caution as the viral protein content bears no relation to the host cell. If specific lipids were selected by these proteins during their segregation, prior to budding, then the lipid composition of the envelope membrane need not reflect that of the host plasma membrane. Bacteriophage PM2 provides an example of this phenomenon. When obtained from a marine pseudomonad it was shown by chemical labelling studies that most of the phosphatidyl ethanolamine was localized in the inner half of the bilayer. The other major phospholipid was phosphatidyl glycerol which was located principally in the external half of the bilayer. The composition of this bilayer envelope was markedly different from that of the host membrane.

4.9.2 Bacteria

Bacteria have a much simpler phospholipid composition than that of mammalian cell membranes. There are frequently only two or three major species present, although certain species can contain glycolipids and sulpholipids.

The polysaccharide cell wall of Gram-positive bacteria may be removed by digestion under iso-osmotic conditions using the enzyme lysozyme. This yields a population of intact protoplasts with identical sidedness. The lipid distribution in the membrane of these protoplasts has been investigated using phospholipid exchange proteins and phospholipases in a similar manner to that used for virus envelopes. The membrane of *Micrococcus lysodeikticus* contains only three phospholipids. Phosphatidyl inositol is found predominantly on the cytoplasmic side of the membrane (80 per cent), phosphatidyl glycerol on the extracytoplasmic side (80 per cent), and diphosphatidyl glycerol (cardiolipin) is equally distributed between the two halves.

Another Gram-positive bacterium, *Bacillus megaterium*, also yields protoplasts upon lysozyme treatment. In this instance only 33 per cent of the total phosphatidyl ethanolamine was available for modification in intact protoplasts by the non-penetrating reagents IAI and TNBS, whereas 67 per cent was modified in inside-out sealed vesicles. Thus, phosphatidyl ethanolamine exists predominantly in the cytoplasmic half of the bilayer and by implication the only other major phospholipid present in this particular strain of *B. megaterium*, phosphatidyl glycerol, must be located principally in the extracytoplasmic monolayer. The variation in lipid asymmetry between even closely related cells is emphasized by the observation that another strain of *B. megaterium*, MK10D, not only has a more complex lipid composition but exhibits a markedly different lipid asymmetry from that described above. In this study, which used both TNBS and phospholipase C as probes, phosphatidyl glycerol was found principally at the cytoplasmic surface of the membrane and phosphatidyl ethanolamine was symmetrically distributed. Another lipid, 3'-glucosaminyl phosphatidyl glycerol, was predominantly localized to the external half, whereas the location of other lipid components was not assessed.

Bacillus subtilis spheroplasts have also been investigated with TNBS and phospholipase C, but significant fractions of lipid species remain inaccessible to these agents even in broken membranes. It was evident, however, that the constituent lipids were asymmetrically distributed with at least 60 per cent of the phosphatidyl ethanolamine and 60 per cent of the lysophosphatidyl glycerol in the outer half of the membrane. Cardiolipin and phosphatidyl glycerol predominated in the inner half of the membrane.

4.9.3 The plasma membrane of eukaryotic cells

The problems involved in studying phospholipid asymmetry in the plasma membrane of eukaryotic cells possessing subcellular organelles are formidable. Few cells have a plasma membrane as rigid as the erythrocyte membrane and consequently they are easily damaged during preparation or spontaneously rupture during experimentation. In addition there are no generally available techniques for making sealed vesicles of unique orientation derived from just one membrane. The plasma membrane of many mammalian cells is also known to be heterogeneous, and highly purified preparations may contain only one or a mixture of vesicles or fragments reflecting this situation. Furthermore, such preparations are easily contaminated with fragments of other intracellular organelles, especially endoplasmic reticulum. This all adds to the difficulty in assessing lipid asymmetry (see section 4.1).

Figure 4.19 shows the distribution of phospholipids between the extracytoplasmic and cytoplasmic faces of a number of plasma membranes. While each membrane differs in its composition and phospholipid distribution, they all comply with the basic pattern of the erythrocyte, having the aminophosphatides and PI principally on the cytoplasmic side while Sp and PC predominate on the outer face. A variety of techniques were used to collect this data. The rat liver domains were studied in right-side-out vesicles, the platelet membrane in intact platelets, the LM cell in latex phagosomes (see sections 4.7.1) and the kidney cells in the envelope of VSV (see section 4.9.1).

4.9.4 The endoplasmic reticulum

Unfortunately the complexity of this membrane and the inability to isolate the membrane without disruption make studies on lipid asymmetry very difficult and unlikely to yield meaningful results. What data there are on phospholipase accessibility to aminophosphatides suggest diametrically opposed models. Since phospholipids, cholesterol and triglyceride are all synthesized by the endoplasmic reticulum it is possible that some areas of membrane do not exist in normal bilayer form. It has been suggested, for example, that cytochrome P_{450} may be associated with a lipid micelle in the centre of a lipid bilayer. It is almost certain that the lipid composition varies between points at or near the site of insertion of new lipid and points at which material is transferred to the Golgi apparatus for segregation and directed export.

4.9.5 Mitochondria, Golgi apparatus and lysosomes

Phospholipase A_2 and sphingomyelinase have been used to assess the phospholipid asymmetry in the Golgi apparatus, lysosomal and inner mitochondrial membrane. The results are given in Figure 4.20. It is evident that an asymmetrical distribution of the phospholipid is a general feature of these membranes. These studies, however, were only carried out using phospholipase attack on intact right-side-out organelles. It is possible, therefore, that phospholipid pools remained on the external face, but for some reason were inaccessible to enzyme attack. Golgi apparatus and lysosomes are also notoriously difficult to isolate in a pure form. It would seem that intracellular organelles exhibit a similar asymmetry to plasma membranes with the aminophosphatides located at the cytoplasmic surface.

Summary

Phospholipid classes, neutral lipids and acyl chain distribution all display asymmetry in a wide variety of biological membranes.

The plasma membrane of the eukaryotic cell has the aminophosphatides, phosphatidyl ethanolamine and phosphatidyl serine, as well as phosphatidyl inositol, predominantly localized at the cytoplasmic surface.

The choline-based phosphatides predominate at the extracytoplasmic surface.

4.10 PHOSPHOLIPID ASYMMETRY IN LIPOSOMES

Unilamellar liposomes consisting of a drop of aqueous medium surrounded by a single spherical lipid bilayer may be prepared by adding water or buffer to a thin film of phospholipid deposited on the side of a flask and sonicating

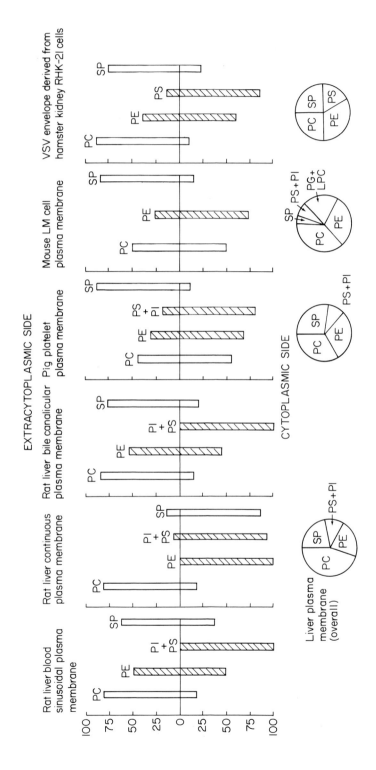

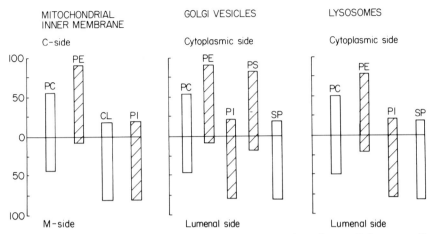

Figure 4.20   Phospholipid asymmetry in the membranes from intracellular organelles of rat liver. Phospholipases have been used as tools to probe the asymmetry of intracellular organelles. However, even in leaky vesicles not all of the lipid was available for hydrolysis by phospholipases. As such this proposed asymmetry should be regarded as tentative. (Adapted from De Pierre, J. W., and Ernster, L. (1977) *Annu. Rev. Biochem.*, 46, 201–262. Reproduced with permission from the *Annual Review of Biochemistry*, Volume 46. © 1977 by Annual Reviews Inc.). An extremely detailed study on bovine heart mitochondria (Krebs, J. J. R. and co-workers, *J. Biol. Chem.*, 254, 5308–5316) gave similar results to those shown here, except that phosphatidyl ethanolamine was found predominantly on the M-side rather than the C-side

(exposing to ultrasound) under an inert atmosphere of nitrogen. Prolonged sonication clarifies the suspension and yields vesicles with a radius of approximately 10 nm. These liposomes contain more lipid on the outside of the bilayer than on the inside, the ratio being about 2 for 10 nm vesicles, but decreasing towards 1 as the vesicle radius increases. Thus the packing of lipids in these vesicles is crucial to their formation and stability. Liposomes made with mixtures of phosphatidyl choline, which is a zwitterion and neutral at pH 7, and negatively charged phosphatidyl serine have phosphatidyl choline principally on the outside. It is surprising that the intermolecular repulsion of

Figure 4.19   Phospholipid asymmetry in plasma membranes. Asymmetry was investigated using phospholipases for the liver and platelet membranes; exchange proteins plus TNBS labelling for the LM-cell membrane; phospholipases + TNBS on the VSV envelope. The distribution of individual lipid species is given above and the proportions of phospholipid present in the membrane are given in the Venn diagrams below. For phospholipid nomenclature see section 1.4. (The data are adapted from Higgins, J. A. and Evans, W. H. (1978) *Biochem. J.*, 174, 563–567; Zwaal, R. F. A. (1978) *Biochim. Biophys. Acta*, 515, 163–205; Sandra, A., and Pagano, R. E. (1978) *Biochemistry*, 17, 332–338; Patzer, E. J., Moore, N. F., Barenholz, Y., Shaw, J. M., and Wagner, R. R. (1978) *J. Biol. Chem.*, 253, 4544–4550; Fontaine, R. N., and Schroeder, F. (1979) *Biochim. Biophys. Acta*, 558, 1–12

charged phosphatidyl serine molecules does not favour the opposite orientation. This illustrates the importance of the relative size of the headgroup, the smaller serine headgroup being more easily accommodated in the tightly packed inner leaflet. At a more alkaline pH the quaternary ammonium ion of the serine headgroup loses its positive charge and the increased intermolecular repulsion now favours an asymmetric liposome bilayer with phosphatidyl serine on the outside. Sphingomyelin, which has a larger headgroup than phosphatidyl choline, is found predominantly in the external half of the 10 nm liposomes formed from a mixture of these two species. Thus the distribution will be determined by packing constraints which place the lipids with the larger headgroups at the external surface rather than in the tightly packed inner surface. Such considerations will be balanced against charge effects due to intermolecular repulsion between acidic phospholipid species which will favour their localization in the external side where they can achieve a greater separation. As the mixtures of phosphatidyl serine and phosphatidyl choline demonstrate, due to considerations of size and charge, the distribution will be markedly pH-dependent.

Phospholipid packing is also determined by the degree of unsaturation of the fatty acyl chains. The introduction of unsaturated fatty acids, especially *cis*-double bonds, favours the location of that phospholipid in the less compact outer monolayer. Thus the distribution of phospholipids in small liposomes is governed by three factors: the size of headgroup, the charge on the phospholipids, and the packing of the fatty acyl chains.

Summary

Liposomes display asymmetry which is dependent on the thermodynamic stability of the constituent phospholipids.

4.11 THE FUNCTION OF LIPID ASYMMETRY

Lipid asymmetry is probably a general feature of all membranes, although it displays a remarkable variability. No clear pattern in the distribution of lipid classes can be derived from the available data, although it would appear that for the eukaryotic plasma membrane the more reactive aminophosphatides and phosphatidyl inositol predominate at the extracytoplasmic side of the membrane. The asymmetry of a phospholipid class is peculiar to a particular membrane from a particular cell of a particular species.

There is, however, clearly not a random selection of phospholipids for a membrane, and sequential modifications may be observed in the lipid composition of the membranes involved in the pathway of secretion (Table 4.5). This suggests that the phospholipids of a membrane do have a function and that certain sets of phospholipids are selected for each membrane to give it unique properties suited to its physiological role. These might be in terms of membrane fluidity, permeability, charge, reactivity or to support enzyme activity. These are not properties characteristic of individual phospholipid

Table 4.5   Lipid composition of some cellular membranes

	Percentage of total						
	Rat liver			Lactating cell			
Lipid	ER	Golgi	PM	ER	Golgi	PM	MFGM
Phosphatidyl choline	48	15	19	57	47	33	36
Sphingomyelin	5	7	12	6	13	24	22
Phosphatidyl ethanolamine	19	9	12	24	27	25	28
Phosphatidyl serine	4	3	7	4	6	4	4
Phosphatidyl inositol	8	5	3	6	7	12	11
Cholesterol	6	8	20	—	—	—	—

Data for rat liver adapted from Van Hoeven, R. P., and Emmelot, P. (1972) *J. Memb. Biol.*, **9**, 105–126, and that for the lactating cell from Patton, S., and Keenan, T. W. (1975) *Biochem. Biophys. Acta*, **415**, 273–309. ER—endoplasmic reticulum; PM—plasma membrane; MFGM—milk fat globule membrane, this is a membrane derived from the plasma membrane which surrounds the milk globule secreted from the cell.

classes and it is likely that there is a degree of flexibility in the types of phospholipid chosen by the cell provided that the properties of the bilayer are maintained. The same could be true for lipid asymmetry. In this case the distribution of lipids across the bilayer may not just be a quirk of the specificity of translocase enzymes but may be designed to give a membrane in which the two halves have different properties that are important for the function of the membrane as a whole.

Headgroup asymmetry, for instance, may be related to surface reactivity, the more reactive aminophosphatides being located on the cytoplasmic face of the plasma membrane in order to prevent them from interacting with the environment.

Surface charge will also be determined by the headgroup asymmetry: phosphatidyl serine and phosphatidyl inositol being strongly anionic, phosphatidyl ethanolamine slightly anionic and phosphatidyl choline and sphingomyelin zwitterionic at neutral pH. The ratio of the strongly anionic to zwitterionic phospholipids varies widely between cell types, but is usually constant for a particular cell type in different species. Thus, in erythrocytes from various species the net charge of polar headgroups is strikingly similar although the proportions of individual headgroups varies markedly. To some extent this ratio is conserved in the two halves of the bilayer, as shown by studies on human and rat erythrocytes (Table 4.6). This asymmetric conservation of charge distribution observed in the erythrocyte membrane may have biological significance since it is known that certain densities of anionic phospholipids can promote clot formation. The predominant localization of negatively charged phospholipids to the cytosol side of the bilayer may therefore be important for avoiding a spontaneous thrombosis. In support of this idea, inverted sealed erythrocyte ghosts and broken erythrocytes and platelets have

Table 4.6   The proportions of zwitterionic and strongly ionic phospholipid species in the cytoplasmic and extracytoplasmic sides of erythrocyte plasma membranes

RATIO	Zwitterionic species (SP + PC + PE) per cent of total lipid	
	Strongly anionic species (PS + PI) per cent of total lipid	
	Extra-cytoplasmic side	Cytoplasmic side
Rat erythrocyte	47 : 2	35 : 15
Human erythrocyte	49 : 0	35 : 10

SP—sphingomyelin, PC—phosphatidyl choline, PE—phosphatidyl ethanolamine, PS—phosphatidyl serine and PI—phosphatidyl inositol.

been found to be effective procoagulants. In both these cell types the strongly anionic phospholipid phosphatidyl serine is almost exclusively localized on the cytoplasmic side of the plasma membrane.

The headgroup content of cell membranes is extremely resistant to manipulation by dietary means. However, this has been successfully achieved with *Neurospora* mutants which showed a conservation of the ratio of strongly anionic to zwitterionic lipids despite extremely large variation in chemical composition of the membrane. It would be interesting to see if this was true for both halves of the membrane. The charge distribution on the cell surface may affect cell–cell interactions and is also likely to be important for the binding of peripheral proteins to the membrane.

Calcium and magnesium ions also bind to the anionic headgroups on the cytoplasmic face of the membrane where they increase bilayer rigidity and can cause lateral segregation of phospholipid species (see Chapter 3). The membrane may therefore act as a store of these cations which could be released as a result of membrane perturbation. This could serve to modulate certain cellular processes and has been proposed as one of the actions of certain polypeptide hormones, e.g. insulin, on the plasma membrane of mammalian cells.

Phospholipid asymmetry may also be required for the activation or structural requirements of asymmetric membrane enzymes. β-Hydroxybutyrate dehydrogenase, Na^+/K^+ activated ATPase and adenylate cyclase, for example, each require specific phospholipids for activity and have different asymmetric locations.

One of the most interesting concepts for the function of lipid asymmetry is the formation of biological membranes with different fluidities in each half of the bilayer. Support for such a proposal comes from studies on rat and hamster liver plasma membranes, where Arrhenius plots of the activities of a number of membrane-bound enzymes whose orientation in the membrane is known can be used to identify different lipid phase separations occurring in the inner and outer halves of the bilayer (Figure 4.7). During hibernation of the hamster, induced by the dark and cold (4 °C), its body temperature drops from 38 °C to about 5 °C and under such conditions only the lipid phase separation occurring in the outer half of the liver plasma membrane bilayer is affected. This lipid phase separation is lowered to the temperature at which the animal maintained homeothermy, implying a selective increase in the fluidity of the external half of the bilayer. These fascinating observations imply that the cell can independently manipulate the composition of each half of the bilayer. It is interesting, therefore, that in a number of plasma membranes, e.g. from liver, LM-fibroblasts, synptosomes and erythrocytes, there is evidence for a difference in the degree of unsaturation of the fatty acyl chain between the two halves of the bilayer.

Summary

Lipid asymmetry may have a functional role in supporting the activity and regulating the properties of penetrant proteins, maintaining a charge distribution, and providing an intracellular 'sink' for cations.

There appear to be cellular adaptation mechanisms available for modulating lipid asymmetry.

FURTHER READING

Bretscher, M. S. (1973) Membrane structure: some general principles, *Science*, **181**, 622–629.

Bergelson, L. D. and Barsukov, L. I. (1977) Topological asymmetry of phospholipids in membranes, *Science*, **197**, 224–230.

Etemadi, A.-H. (1980) Membrane asymmetry: a critical appraisal of methodology I. Methods for assessing the asymmetrical orientation and distribution of proteins, *Biochim. Biophys. Acta*, **604**, 347–422.

Etemadi, A.-H. (1980) Membrane asymmetry: a critical appraisal of methodology. II. Methods for assessing the unequal distribution of lipids, *Biochim. Biophys. Acta*, **604**, 423–476.

Kenny, A. J., and Booth, A. G. (1978) Microvilli: their ultrastructure, enzymology and molecular organisation, *Essays in Biochemistry*, **14**, 12–44.

Marchesi, V. T. (1975) in Molecular orientation of proteins in membranes, *Membranes of Cell Walls and Membranes*, Vol. 2, Fox, C. F., ed., MTP International Review of Science, Biochemistry Series One, Butterworths/University Park Press, pp. 123–154.

Op den Kamp, J. A. F. (1979) Membrane asymmetry, *Annu. Rev. Biochem.*, **48**, 47–71.

Rothman, J. E., and Lenard, J. (1977) Membrane asymmetry, *Science*, **195**, 743–753.

Chapter 5

Reconstitution of defined membrane functions

5.1 INTRODUCTION

As we have seen in Chapter 1, biological membranes consist of a multitude of components which together carry out the diverse jobs that characterize a particular membrane. The reconstitution of a particular membrane function allows us to define the *minimum* number of these components that are absolutely necessary to enable a particular job to be carried out. It also satisfies the very basic desire to take apart a functioning system and reassemble it from its constituent pieces. Once such a system has been assembled we can investi-

206

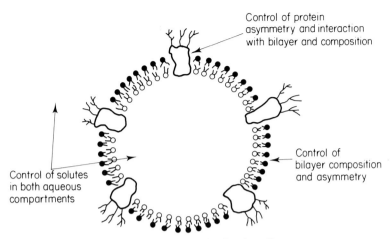

Control of protein
asymmetry and interaction
with bilayer and composition

Control of
bilayer composition
and asymmetry

Control of solutes
in both aqueous
compartments

Figure 5.1   Aims of reconstitution into liposomes

gate its properties in detail, free from the influence of any other membrane processes. The constituents of its aqueous and liquid environment can be manipulated, the asymmetry of the lipid environment can be controlled and ligands such as substrates or effector molecules can be presented to the inserted protein asymmetrically (Figure 5.1).

In this chapter we will discuss the various methods by which functional reconstitutions have been achieved and also ways of manipulating the lipid environment of integral proteins. The complete control of protein and lipid composition in reconstituted systems has been achieved, but control over the absolute orientation of these components remains as yet out of reach.

5.2   METHODS OF RECONSTITUTION

The origins of current procedures can be traced back to the investigators responsible for 'dissecting-out' components of the mitochondrial electron transport chain and reconstituting the various segments. Most information can be obtained when pure proteins are reconstituted into vesicles of defined lipid composition. Unfortunately, since the proteins responsible for many of the functions we are interested in constitute a relatively minor fraction of the total membrane protein and are difficult to purify, the majority of defined reconstitutions utilize major membrane proteins for which simple purification procedures are available.

In general the only means of assaying transport proteins during purification is that of functional reconstitution. This is often a lengthy process in itself, making the purification extremely tedious. Fortunately the ATP-driven transport pumps can easily be identified by virtue of their ATPase activity, making them particularly easy subjects for study.

5.2.1 Detergent removal

This is a highly successful procedure that has been used to reconstitute a wide variety of membrane functions. In this method, phospholipids and proteins are added to a detergent solution to yield a soluble mixture. This is then slowly dialysed, often for between one and two days against large volumes of detergent-free buffer. During the removal of detergent, the phospholipids and proteins aggregate to form vesicles with the integral proteins inserted within the bilayer. These vesicles then have membrane properties characteristic of the proteins inserted in them, providing a reconstituted membrane function.

The detergent used most often is Na cholate (2 per cent) because it can be obtained in a pure form, it is dialysable, many proteins survive exposure to it, its structure is defined, and it can be obtained radio-labelled, which makes assessment of its removal extremely easy. More recently octyl-β-D-glucoside, a dialysable non-ionic detergent, has found favour with proteins that are easily denatured by charged detergents. Again, this detergent can be obtained as a defined, pure and radio-labelled species. Unlike other non-ionic detergents it has a high critical micelle concentration of about 20 mM which facilitates its use in solubilization and subsequent removal (see section 1.6).

The need to dialyse either overnight or more often for a few days, in order to reconstitute and remove residual detergent, can lead to problems, especially with easily denatured proteins. To some extent this period can be shortened by adding inert polystyrene beads which act to adsorb the detergent. Stabilizing reagents like β-mercaptoethanol or dithiothreitol may be added to the dialysis buffer in order to protect essential thiol groups on the proteins, and protease inhibitors can also be added. Continuous-flow dialysis procedures can be used to speed up this method. This alleviates the need to move dialysis bags from one buffer container to another. It also allows detergent to be removed at a controlled rate. In addition, by adjusting the rate of removal of detergent the size of the reconstituted liposomes may be manipulated.

To obtain a rapid removal of detergent the detergent–lipid–protein solution may be gel-filtered on Sephadex G-50 equilibrated with detergent-free buffer. The detergent is removed as the mixture passes down the column, leading to the formation of essentially detergent-free liposome vesicles which elute in the void volume.

The use of any detergent-based reconstitution procedure requires a sensitive means of assay for residual detergent. For whilst its complete removal may not be absolutely necessary to observe a reconstituted activity, it is necessary if any serious study on lipid–protein interactions is to be pursued.

5.2.2 Detergent dilution

A solution of phospholipids in a detergent (cholate)-containing buffer is briefly sonicated before it is added to a membrane protein solution where the overall detergent concentration is kept low (0.5–1 per cent). After a short

Table 5.1 Some membrane functions that have been successfully reconstituted

Method	Detergent dialysis	Detergent dilution	Sonication	Detergent catalysed incorporation	Planar membrane
	Ca^{2+}-pump (sarcoplasmic reticulum)	Ca^{2+}-pump	Ca^{2+}-pump	Ca^{2+}-pump	Monosaccharide transport (erythrocyte)
	Cytochrome oxidase	Acetylcholine receptor	Cytochrome oxidase	Cytochrome oxidase	Acetylcholine receptor
	H^+-pump (bacteriorhodopsin)		H^+-pump (bacteriorhodopsin)		Ca^{2+}-channel proteolipid
	Na^+/K^+-ATPase		Na^+-pump (Na^+/K^+-ATPase)		
	Acetylcholine receptor		Na^+-channel (lobster nerve)		
	Proline carrier (E. coli)		Adenine-nucleotide carrier (mitochondria)		
	Glutamate carrier (B. subtilis)		Proline carrier (M. phlei)		
	Glutamate carrier (rat brain)		γ-Glutamyl transpeptidase (amino acid carrier)		
	γ-Aminobutyrate carrier (rat brain)		Phosphate-carrier (mitochondria)		
	Catecholamine carrier (chromaffin granules)				
	Glucose carrier (adipocyte)				
	Glucose carrier (erythrocyte)				

period of incubation this mixture is rapidly diluted about 20-fold into the assay medium.

This offers a quick and usually quite effective means of reconstitution that appears to work for a number of membrane functions (see Table 5.1), although it fails to work for others, e.g. bacteriorhodopsin and the Na^+/K^+-ATP-driven pump. Because of its virtue of rapidity it offers a convenient method of assay during the process of purification of proteins that have transport or ion-gating roles.

The residual cholate may well be the reason why certain systems cannot be suitably reconstituted by this method and its presence mitigates against the use of this procedure for preparing systems to study lipid–protein interactions.

5.2.3 Sonication

A solution of phospholipids in $CHCl_2$: MeOH (2 : 1) is dried on to the side of a glass vial and the residual solvent is removed under reduced pressure. Then a suspension of membrane proteins ($0.5-2.0$ mg ml^{-1}) in 0.15 M KCl is added to the tube containing the phospholipids (25 μmol ml^{-1}) and the mixture is exposed to ultrasound, in a bath-type sonicator, for periods of 3–10 min in an inert atmosphere.

This method has proved extremely successful for the reconstitution of the Ca^{2+}-pump from sarcoplasmic reticulum and the light-driven H^+-pump from bacteriorhodopsin and others (see Table 5.1). It has the distinct advantage that it is quick and that the addition of detergents is not necessary. This can be particularly useful for sensitive proteins and also for transport systems where residual detergent may produce leaky vesicles. Major disadvantages are the lack of reproducibility that often occurs using sonication equipment, as it is difficult to control precisely both the output of power and the power that is absorbed by the sample. Furthermore, some membrane proteins are particularly susceptible to inactivation by exposure to ultrasound.

In the two methods outlined above, dialysis or dilution of the detergent allow the formation of a bilayer around the integral proteins to yield a vesicle. However, in this sonication procedure the bilayer of a pre-existing liposome has to be perturbed to such an extent that the integral proteins can insert themselves. The facility with which this process can be achieved is likely to be dependent upon the protein in question, the lipid constituents of the bilayer and the experimental conditions. Presumably, for transmembrane proteins, it is likely to be an energetically unfavourable process involving the movement of hydrophilic residues through the bilayer core. All these factors may well explain the very variable effectiveness and reproducibility of this technique.

5.2.4 Detergent-catalysed incorporation

In an analogous fashion to the previous procedure which relies on ultrasound to perturb a pre-existing bilayer to such an extent that it accepts an integral membrane protein, extremely low concentrations of detergent may be used to catalyse the insertion of integral membrane proteins. Lysolecithin and Na cholate (<0.1 per cent) have both been used successfully to reconstitute cytochrome oxidase and the Ca^{2+}-pump into pre-formed phospholipid vesicles.

Again, this is a technique which is likely to be more suitable for some proteins than for others. It is potentially of value in achieving the asymmetric reconstitution of certain membrane proteins. Studies with cytochrome oxidase have shown that whilst detergent–dialysis techniques have yielded a random orientation in the membrane, the detergent-catalysed method can produce an asymmetric system. Here the cytochrome c binding face of cytochrome oxidase is exposed on the outside of the vesicle, as is normally seen in the mitochondrion. This phenomenon is presumably related to the relative ease with which cytochrome oxidase in different orientations can insert itself into a pre-formed lipid bilayer in the presence of low detergent concentrations. Constraints upon insertion may also be imposed by packing criteria within the small curved vesicles (see section 4.10).

5.2.5 Black lipid films

Phospholipids in a hydrocarbon solvent can be applied across a small hole (approx. 2 mm diam.) separating two buffer-filled compartments. After a short period of time the solvent disappears leaving a thin bilayer stretched across the aperture. If this film is looked at under reflected light then the light reflected from the front and back surface of the bilayer are very nearly in counterphase: it thus appears to be black. The electrical capacitance across this black lipid bilayer can be measured provided that the compartments have been suitably insulated. This allows the permeability of the film to solutes to be assessed. Changes in the permeability of this bilayer caused by alterations in phospholipid, cholesterol and of course protein content, can now be gauged. To incorporate proteins into the black lipid film they need first of all to be 'dissolved' in the hydrocarbon solvent, usually n-decane, containing the phospholipids. This of course limits the method to those proteins that can withstand such a treatment without being denatured; furthermore, single defined phospholipids will not form black lipid films.

Reconstitution into black lipid films is a particularly useful means of investigating ion fluxes through transport proteins and has been used successfully to investigate the properties of the monosaccharide transport system of the human erythrocyte. This system is especially flexible in that solutes can easily be added to either of the compartments that are separated by the black lipid film. This means that solutes can readily be presented asymmetrically to the incorporated proteins.

Summary

Reconstitution allows the minimum number of components necessary for a particular membrane function to be defined.

It may be achieved by either the removal of detergent from detergent–lipid–protein complexes or by the perturbation of pre-existing liposomes so as to accept the integral membrane proteins.

5.3   ASPECTS OF RECONSTITUTION STUDIES

5.3.1   Purity of components

If the aim of the reconstitution procedure is to define the minimum number of components necessary to carry out a particular membrane function, then these components need to be pure. Protein purity can readily be checked by non-denaturing and SDS–PAGE, and lipid purity by tlc and glc (see section 1.7).

5.3.2   Asymmetry

If a transport function is to be reconstituted and shown to work, then it is essential that both the solutes and integral membrane proteins should be positioned asymmetrically about the plane of the vesicle membrane. Substances can be conveniently trapped inside the reconstituted vesicles during their formation, either at the detergent dilution or dialysis stage, or during the addition of buffer to dried phospholipid films when making vesicles for sonicated or detergent-catalysed incorporation. To create a solute gradient across the vesicle membrane, vesicles can be gel-filtered on Sephadex G25 into buffer free of the solute trapped inside the vesicles.

A membrane protein may require an asymmetric lipid bilayer in order to function. However, as we have discussed in section 3.3, there are only a few examples of enzymes which exhibit a clear-cut specificity for a particular headgroup phospholipid in order to function, so this is unlikely to pose a major problem. It is possible, though, that the efficiency of asymmetric transmembrane proteins may be sensitive to asymmetric changes in their lipid environment. This will only become apparent when chemical techniques and phospholipid exchange proteins (see section 4.3) are used to create asymmetry in reconstituted vesicles. For in most circumstances the phospholipid vesicles formed during reconstitution processes do not exhibit any marked asymmetry unless their dimensions are particularly small (see section 4.10). This is only likely to occur to any particular extent in procedures using prolonged sonication which splits up large vesicles to yield small unilamellar ones.

For vectorial processes such as those carried out by transport proteins and receptors transmitting information across bilayers, there is ideally a requirement for an asymmetric incorporation of proteins into the reconstituted vesicles. In practice, proteins usually exhibit a random orientation about the reconstituted vesicle membrane. So, in order for the system to be functionally assessed it is necessary to add the substrate or effector ligands asymmetrically about the vesicle membrane. For example, in reconstitutions of the ATP-driven Ca^{2+}-pump, proteins are randomly orientated in the membrane (see Figure 5.2). However by adding both the energy source, MgATP, and the solute to be translocated, Ca^{2+}, to the exterior of the sealed vesicle, Ca^{2+} accumulates inside the vesicle at the expense of ATP hydrolysis. This accumulation is only effected by those proteins correctly orientated in the membrane. The others, which exhibit the opposite orientation, are unable to pump Ca^{2+} out of the vesicle because the energy source, MgATP, is present only at the external face of the vesicle.

There are, however, a few instances in which asymmetric reconstitutions have been achieved. These are exceptions, confined to specific proteins under specific conditions, for in a standard cholate dialysis procedure self-assembly yields a random reconstitution. However, in those instances where integration of a membrane protein has to be made into a pre-formed lipid vesicle, then a number of rather different considerations need to be made. For example the vesicle's dimensions are often of such a size that packing constraints become an important consideration. Furthermore, insertion in one particular orientation may be easier than in another. This is especially true if bulky, charged sialic acid residues are attached to one face of the protein, as is the case with glycophorin (see section 4.4). These residues would also exhibit considerable charge-repulsion if packed inside small vesicles. Taking this into consideration it is not perhaps too surprising that in sonication or detergent-catalysed reconstitutions of glycophorin A the major fraction of the molecules are found with their sialic acid residues orientated at the vesicle surface.

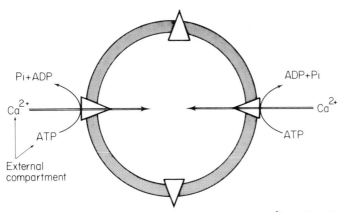

Figure 5.2 Reconstitution of the calcium pump. When the Ca^{2+}-ATPase from rabbit muscle sarcoplasmic reticulum is reconstituted into sealed vesicles there is a random orientation of the protein. However, by adding Ca^{2+} and an energy source, ATP, to the exterior of the sealed vesicle only, those carriers able to interact with both Ca^{2+} and ATP drive the net uptake of Ca^{2+} into the vesicle

5.3.3 Insertion

It is extremely important that the reconstitution procedure used achieves an integration of the membrane protein into the lipid bilayer itself. In this way the protein can interact with the bilayer in a similar fashion to that achieved *in vivo*. This is of course essential in order for transmembrane proteins effecting ion- or information-translocation across the bilayer to function. It is also necessary that other integral proteins insert correctly in order that they reflect their native sensitivity to a bilayer environment. Problems of this nature usually arise only when the membrane protein is presented with a pre-formed lipid vesicle with which to interact. Integration is then usually assisted using sonication or catalytic amounts of detergent. The failure to ascertain whether true incorporation has occurred can lead to rather dramatic differences in the properties of an integral protein. A good example of this is shown by the enzyme β-hydroxybutyrate dehydrogenase, an enzyme whose activity is totally dependent upon interaction with the choline-headgroup region of a phospholipid (see section 3.3). This enzyme can be studied in a defined lipid environment either by mixing the purified apoprotein with pre-formed vesicles of synthetic phosphatidyl choline which reactivate the enzyme, or by using a cholate–lipid substitution procedure (see section 5.4) somewhat akin to detergent-dilution reconstitution (section 5.2.2). Figure 5.3 shows the strikingly different temperature activity profiles exhibited by the same enzyme, interacting with dipalmitoyl phosphatidyl choline, but 'reconstituted' using these two very different procedures. The cholate-mediated lipid substitution procedure ensures integration into the bilayer where the enzyme is sensitive to both the headgroup and acyl chain region of the bilayer. As the fluidity of the bilayer decreases with decreasing temperature, then the rigidity of the acyl chains surrounding the enzyme progressively inhibits its activity. The completely different picture seen when the apoenzyme is merely mixed with pre-formed lipid vesicles is due to lack of integration of the enzyme into the bilayer. So high activities are seen at the low temperatures because the enzyme adsorbs itself to the exterior of the vesicle, allowing it to interact functionally with the headgroup region without being constrained by the rigidity of the bilayer. However, as the temperature increases, molecular motion within the bilayer increases and dipalmitoyl phosphatidyl choline undergoes a pre-transition (see section 2.8) which correlates with the onset of a progressive temperature-dependent decrease in activity of the absorbed enzyme. This continues until the main transition (see section 2.6) is reached when the activity of the enzyme in the pre-formed liposome system is nearly zero, whereas that in the lipid-substituted system is progressively increased. Presumably the greatly increased molecular motion in the bilayer, together with the change in orientation of the headgroup as phosphatidyl choline goes through its phase transition, does not allow for a functional absorption of the enzyme to the vesicle exterior.

It is essential to ensure that the protein is inserted correctly into the bilayer.

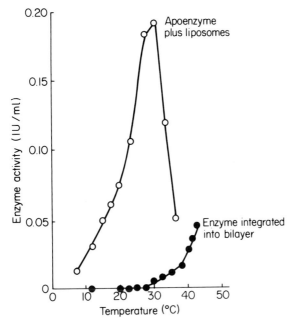

Figure 5.3 The temperature dependence of β-hydroxybutyrate dehydrogenase in different environments. There is a dramatic difference in the temperature dependence of the activity of this mitochondrial integral enzyme between the apoenzyme reactivated by the addition of pre-formed liposomes of dipalmitoyl phosphatidyl choline ($\bigcirc$) and the enzyme reconstituted by lipid substitution/detergent removal techniques into dipalmitoyl phosphatidyl choline vesicles ($\bullet$). These data are adapted from Houslay, M. D., Warren, G. B., Birdsall, N. J. M., and Metcalfe, J. C. (1975) *FEBS Lett.*, 51, 146–151, and Gazzotti, P., Bock, H.-G., and Fleischer, S. (1975) *J. Biol. Chem.*, 250, 5782–5790

This is easy to ascertain for proteins effecting a flux of solute across the bilayer, as failure to achieve the correct integration will not result in a net vectorial flux. However, for other types of integral proteins the studies similar to those described above or chemical labelling experiments need to be carried out.

5.3.4 Integration

Transport proteins require a sealed vesicle in order to function. This is necessary both for the accumulation of solute and for the asymmetric presentation of an energy source, be it ATP, a membrane potential or a solute gradient. To this end it is essential that a spectrum of phospholipids are offered so that an impermeable liposome can be formed with the functioning protein sealed and integrated into the bilayer. After a functional reconstitution has been achieved, the phospholipid composition of the vesicles can be manipulated.

216

This allows the lipid requirements for the functioning of the protein to be separated out from those required both to seal the protein into the bilayer and to form impermeable liposomes. The problems of using a single, defined phospholipid species for reconstitution studies is that they may be unable either to produce sealed vesicles or to interact sufficiently well with the transmembrane protein, in which case leaks will also occur at the lipid–protein interface.

In some instances it is necessary to trap solutes inside vesicles in order to provide a driving force for the accumulation of another species. For example, to study the isolated bacterial proline transport system it must be reconstituted into liposomes containing KCl which are subsequently transferred to a KCl-free medium. On addition of the cyclic peptide ionophore valinomycin (see section 7.3) a membrane potential is formed by the movement of K^+ ions across the membrane. This membrane potential is used to drive proline uptake into the vesicles. In such cases it is essential that the vesicles prepared during the reconstitution procedure retain the entrapped KCl until it can be expended in driving the accumulation of proline into the vesicles. This is not

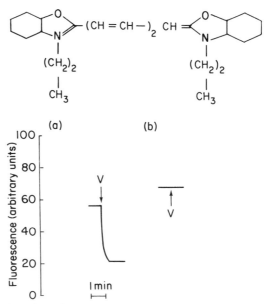

Figure 5.4 The detection of a membrane potential across a phospholipid vesicle using a fluorescent cationic dye. Liposomes generated from pigeon erythrocyte lipids were formed containing 110 mM K^+ phosphate, pH 7.4. When diluted into a K^+-free medium (a) a membrane potential could be generated upon the addition of the ionophore, valinomycin(V). This was readily detected by a change in the fluorescence of the carbocyanine dye, 3′,3′-dipropyloxadicarbocyanine. If the liposomes were diluted into 110 mM K^+ phosphate, pH 7.4, then no change in fluorescence occurred (b). (Adapted from Kimmich, G. A., Philo, D., and Eddy, A. A. (1977) *Biochem. J.*, **168**, 81–90. Reprinted with permission of the Biochemical Society of London)

always so easily accomplished as the 'leakiness' of the vesicles is dependent upon their lipid composition, the type and concentration of integral protein(s) inserted, temperature and residual detergent. It is therefore essential that the integrity of the reconstituted vesicles should be ascertained in order to ensure that a 'driving force' of sufficient magnitude is present to power the transport process. The presence of entrapped solute and its rate of leakage out of the membrane can be measured. However, the generation of a membrane potential, achieved for example by expending a K^+ gradient using the ionophore valinomycin, can easily and rapidly be quantitated by a fluorescence method. This technique utilizes certain fluorescent carbocyanine dyes which carry a delocalized positive charge. They penetrate lipid bilayers easily and are concentrated within the lipid boundary in response to the membrane potential. This process is apparently amplified by the subsequent binding of the dye to the membrane itself. The addition of valinomycin to vesicles containing K^+ hyperpolarizes them, resulting in a decrease in the fluorescence intensity (Figure 5.4), the magnitude of which can be used to calculate the membrane potential. By following the rate of change of fluorescence with time it is possible to assess the loss of the membrane potential caused by leakage and the rate of transfer of charged solutes across the membrane by the inserted protein.

Summary

In order to study integral proteins in reconstituted systems it is necessary to ensure that protein is integrated into the bilayer and solute concentration on either side of the membrane selectively controlled.

Reconstitution usually results in a random orientation of components in the bilayer. Transport proteins need to be reconstituted into sealed vesicles.

5.4 METHODS OF MANIPULATING THE LIPID ENVIRONMENT OF INTEGRAL PROTEINS

For both physical and biochemical studies aimed at investigating lipid–protein interactions, and in order to achieve defined reconstitutions, it is necessary to be able to change at will the lipid environment of any particular membrane protein.

5.4.1 Lipid substitution

The entire lipid environment of a membrane protein may be substituted by incubating the protein and its endogenous lipid, or a complete biological membrane, with an excess of a defined lipid in the presence of the detergent sodium cholate. This allows equilibration between the exogenous and endogenous lipid pools. Because the exogeneous lipid is added in excess it dominates the resultant common lipid pool. By centrifuging this soluble mixture into a detergent-free sucrose density gradient vesicles are formed with similar ratios of lipid to protein as found in native biological membranes

218

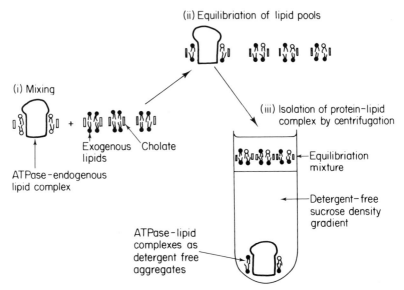

Figure 5.5 **Lipid substitution. This scheme outlines the lipid substitution process for Ca^{2+}-ATPase, but it is appropriate to any biological membrane or lipid–protein complex. (Adapted and reprinted with permission from Warren, G. B., Toon, P. A., Birdsall, N. J. M., Lee, A. G., and Metcalfe, J. C. (1974)** *Biochemistry***, 13, 5501–5507. Copyright (1974) American Chemical Society)**

(Figure 5.5). This procedure may be repeated until at least 99 per cent of the original lipid has been substituted with the defined exogenous species. The degree of substitution achieved at each stage is simply proportional to the molar ratio of exogenous lipid to endogenous lipid provided that equilibrium is attained. At the end of this procedure it is usual to resuspend the pellet and remove any residual detergent by recentrifugation and dialysis against detergent-absorbing polystyrene beads (see section 5.2.1). This can be extremely important as residual detergent can affect the functioning of membrane proteins, it can perturb annular lipid–protein interactions and can induce leakage in phospholipid vesicles.

Many proteins can be manipulated in this way provided that they are not denatured by the detergent used to achieve equilibration of the lipid pools. Usually Na cholate is used, but Na deoxycholate and the non-ionic detergent, octyl glucoside, can both be used effectively. It is important, however, that the detergent used should be easily identified, should not bind tightly to proteins and should be easily removed after lipid substitution. These detergents satisfy such criteria whereas many others do not.

Lipid substitution yields membrane proteins fully inserted into a defined lipid bilayer forming membrane sheets or vesicles. These proteins are sensitive to the physical and physicochemical properties of the lipids that surround them and can be used for both biochemical and physical studies of lipid–protein interactions. As the proteins are never stripped free of lipid before insertion into a defined environment, the chances of denaturation or incorrect insertion into the defined bilayer (see section 5.3.3) is much reduced, thus providing us with a useful model system that should reflect the *in vivo* lipid–protein interactions. It also offers the opportunity of isolating lipid–protein complexes of various stoichiometries. This can be achieved by varying either the detergent concentration, the nature of the sucrose density gradient and by fractionating the equilibrating mixture using ammonium sulphate to precipitate out proteins complexed with various amounts of lipid. For true substitution, however, the lipid : protein ratio in the isolated complex should never exceed that seen in the original membrane. Although it is usual to use defined phospholipids in this procedure, fully substituted complexes have been obtained using cholesterol.

5.4.2 Lipid titration

This procedure provides a rapid and extremely useful means of screening lipids for their effects on membrane protein function. The membrane protein, together with its associated endogenous lipids, is incubated with excess exogenous lipid in the presence of an equilibrating detergent like sodium cholate, as in the lipid substitution procedure (see section 5.4). Then in a similar fashion to the detergent-dilution reconstitution method (see section 5.2.2) this mixture is rapidly diluted into either the assay or a detergent-free buffer solution. The rapid and extensive dilution of the detergent yields membranous particles consisting of integral proteins inserted in a lipid bilayer whose composition is dominated by the exogenous lipid. It is important to show that any changes in protein function that occur upon lipid substitution or titration are fully reversible. Lipid titration offers a means of doing just this. Complexes of protein with lipid A can be titrated with lipid B and then retitrated with lipid A. This should yield a complex with an identical activity to that observed at the beginning. Such a procedure has been used successfully with β-hydroxybutyrate dehydrogenase and Ca^{2+}-ATPase where the inhibitory effect of dimyristoyl phosphatidyl choline and the activating effect of dioleoyl phosphatidyl choline can be shown to be fully reversible (Figure 5.6). In this case the actual changes in activity are due to the very different physical constraints placed upon the enzymes in bilayers of very different lipid fluidities (see section 3.3).

5.4.3 Lipid fusion

The phospholipid composition of biological membranes can be altered by incubating them with unilamellar liposomes. Depending upon their composition, charge, size and the incubation temperature, the liposomes can fuse with and be incorporated into the membranes (see section 3.5). This procedure causes a net increase in the size of the lipid pool, altering the dilution of the

220

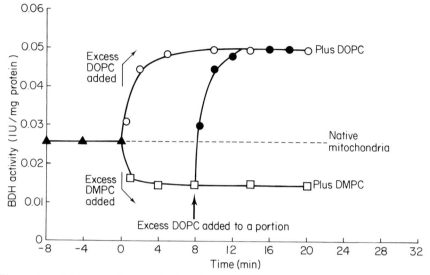

Figure 5.6   **Lipid titration of β-hydroxybutyrate dehydrogenase.** Mitochondrial inner membranes containing this integral enzyme can be treated with a tenfold excess of exogenous defined lipid in the presence of cholate (2 : 1 lipid/cholate). Incubation with the highly fluid dioleoyl phosphatidyl choline (DOPC) leads to the activation of the enzyme, whereas incubation with the less fluid dimyristoyl phosphatidyl choline (DMPC) leads to the inhibition of enzyme activity. The inhibition by DMPC is due to the increased rigidity of the membrane containing this lipid compared with that of the native mitochondria. Complete reversal of the inhibitory effect of DMPC can be achieved by adding a tenfold excess of DOPC

proteins in it (lipid : protein ratio). It also does not remove exogenous phospholipids and so is of little use in defined studies. However, it may be of use in investigating the relative effects of protein concentration in a defined lipid pool created by substitution or titration techniques. It is also of use in assessing the importance of minor phospholipid species, e.g. acidic phospholipids, in the functioning of integral proteins. For in this instance the fusion of relatively little of an exogenous minor lipid species into the membrane will dramatically alter its proportion of the total lipid pool.

5.4.4   Other methods

Any reagent that can interact with lipids is a potential tool for modulating the composition of the bilayer. Thus, cholesterol oxidase, phospholipases, phospholipid exchange proteins, phosphatidyl serine decarboxylase and many others have been used. However, none of these reagents react with a biological membrane to produce a well-defined system for study. The only serious alternatives to the lipid fusion and substitution studies are those methods which yield a lipid-free protein that can subsequently be inserted into a defined lipid environment. Such lipid-free proteins can be obtained by treating them extensively with detergent, organic solvent or phospholipase A_2. These reagents strip away the endogenous lipid, yielding a protein that can subsequently be purified. This has been successful for very few integral proteins. Presumably the removal of lipid exposes hydrophobic sites on the protein to the aqueous environment. This results in conformational changes which can alter the properties of the protein or lead to its denaturation and loss of function.

Summary

Lipid substitution allows the complete control of the lipid composition of isolated lipid–protein complexes.

Lipid titration offers a simple and rapid method of assessing reversible, lipid-mediated effects on the functioning of integral proteins.

5.5 EXAMPLES OF RECONSTITUTED SYSTEMS OF PARTICULAR INTEREST

All the techniques that we have described have been used successfully to reconstitute a variety of membrane functions, examples of which are listed in Table 5.1. It is worth discussing a few selected examples of reconstitutions in some detail, both because of their intrinsic interest and because they provide an insight into the type of information that can be gleaned from reconstitution studies. They also highlight points of methodology.

5.5.1 Bacteriorhodopsin

The only protein present in the purple membrane of *Halobacteria* is bacteriorhodopsin. This protein (M_r = 26,000) with covalently attached 11-*cis*-retinal functions as a light-driven H^+-pump (see section 7.6.3). It can readily be purified to homogeneity and then functionally incorporated into lipid vesicles by detergent removal (see section 5.2.1) or sonication (see section 5.2.3) procedures. If small lipid vesicles are formed then a predominantly asymmetric incorporation results during reconstitution. This is governed presumably by packing considerations due to the increased curvature in small vesicles, imposing constraints on the incorporation and orientation of the asymmetric bacteriorhodopsin. Thus in reconstituted vesicles, illumination results in the net translocation of protons into the vesicles. This is opposite to that found *in vivo* where illumination results in H^+ extrusion, highlighting the fact that insertion of proteins during biosynthesis proceeds by a completely different process with a different set of constraints (see section 6.2).

A fully defined reconstitution can be made using a single protein species, bacteriorhodopsin, and a single lipid species, dimyristoyl phosphatidyl choline. These two components can be assembled into a light-driven H^+-pump.

The ability to produce a H^+-gradient across sealed vesicles was used in an elegant experiment to test the hypothesis that the mitochondrial oliyomycin seritive, F_1-ATPase could utilize such a H^+-gradient to synthesize ATP from ADP + inorganic phosphate. Pure samples of the F_1-ATPase and bacteriorhodopsin were co-reconstituted into sealed lipid vesicles using either a sonication or cholate dialysis procedure and then mixed with externally added Mg^{2+}, ADP and inorganic phosphate. Upon illumination this system synthesized ATP. Thus the H^+-gradient formed upon illumination of the inserted bacteriorhodopsin was expended by the F_1-ATPase in synthesizing ATP (see Figure 5.7). For although the F_1-ATPase was randomly orientated within the bilayer an overall vectorial process was achieved by virtue of the asymmetric incorporation of bacteriorhodopsin into the membrane and the asymmetric addition of ADP and inorganic phosphate to the system (see Figure 5.7). So only those F_1-ATPase molecules with access to substrates would allow H^+-translocation to occur along with ATP synthesis.

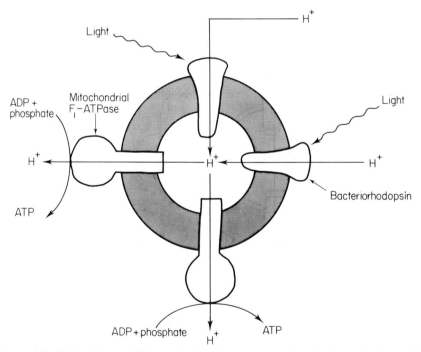

Figure 5.7 Light-driven ATP synthesis. Co-reconstitution of the mitochondrial F_1-ATPase and bacteriorhodopsin into small phospholipid vesicles using a sonication procedure allows the proton gradient formed by the action of bacteriorhodopsin to be expended by the F_1-ATPase in a process which effects the synthesis of ATP. Bacteriorhodopsin is reconstituted asymmetrically. The addition of ADP and inorganic phosphate to the exterior of the vesicle ensures that only those F_1-ATPase orientations with their binding site at the vesicle surface will function as H^+-channels coupled to ATP synthesis

In a similar fashion, co-reconstitution of bacteriorhodopsin with cyto-chrome oxidase can yield a functioning system which supports the continued oxidation of externally added reduced cytochrome c. Although the cyto-chrome oxidase is randomly orientated within the vesicle bilayer membrane, an overall vectorial process is obtained due both to the asymmetric incorpora-tion of bacteriorhodopsin and to the addition of cytochrome c to the exterior of the vesicle only. Thus, this light- and oxygen-dependent process supports a continuous flow of electrons through cytochrome oxidase across the vesicle membrane.

Bacteriorhodopsin vesicles can be incorporated into millipore filters. Here, millipore filters are impregnated with phospholipids by dipping them in soya-bean phospholipids dissolved in hexadecane. The filter is then used to separate two compartments containing Ca^{2+}-loaded buffers. Into one of these compartments is added bacteriorhodopsin-containing vesicles which then fuse, in a Ca^{2+}-dependent process, with the phospholipid-impregnated filters. This yields a system which generates a photopotential across the millipore filter upon illumination. Indeed, this immobilized system has all the advan-tages of a black lipid film in that the potential across the membrane can be measured, yet it is created by gentle techniques and is remarkably stable. Its application to other membrane processes may well be rewarding.

5.5.2 Calcium pump

The major protein of rabbit muscle sarcoplasmic reticulum is a Ca^{2+} and Mg^{2+}-dependent ATPase. This enzyme is responsible for pumping calcium back into the sarcoplasmic reticulum after it has been released during muscle contraction. It is easily purified to apparent homogeneity and shown to con-sist of peptide of $M_r = 110,000$, although in the membrane it is believed to function either as a dimer or in oligomeric units. If an ionic detergent, like sodium cholate, is used to solubilize the enzyme during purification it is essential that the detergent concentration is not so high that annular lipid is stripped from the protein. Sufficient lipid to form a single ring of bilayer around the protein must remain if it is to remain active (see section 3.3). It is possible, however, fully to strip the enzyme of lipid using the non-ionic detergent, Triton X-100. This detergent, unlike cholate, is not easily removed, making it unsuitable for reconstitution studies. Thus, it has been found best to purify the enzyme together with those annular lipids derived from the bilayer of the sarcoplasmic reticulum and then replace them by lipid substitution in order to provide suitable starting material for a defined recon-stitution.

Using lipid-substitution and cholate dialysis procedures this protein has been functionally reconstituted into vesicles containing dioleoyl phosphatidyl choline as the only phospholipid. However, in order to achieve high levels of Ca^{2+} accumulation, oxalate must be trapped inside the vesicles during the detergent–dialysis reconstitution step. This allows it to complex the Ca^{2+}, as the oxalate salt, inside the vesicles and prevent Ca^{2+} leakage. Thus, by adding the energy source, MgATP to the outside and by trapping the Ca^{2+} inside the vesicles using a complexing agent, a net flow of Ca^{2+} into vesicles is achieved. This is despite the protein molecules being randomly orientated at either side of the bilayer.

The ability to work with such a defined system has demonstrated that whilst a fluid lipid bilayer is essential for the pump to function, a too fluid bilayer is too leaky to allow net Ca^{2+} accumulation. This problem is apparently overcome in natural membranes by the use of cholesterol to diminish leaks in the bilayer and annular phosphatidyl ethanolamine to seal the protein into the lipid bilayer (see section 3.3).

It would seem from reconstitution studies that a single protein species and a single lipid species are sufficient entities to construct an ion-pump. However, a certain degree of controversy still surrounds the contention that an additional 'coupling factor' or 'ionophore' is also necessary. The suggestion is that this is a low-molecular-weight proteolipid, that is a peptide with lipid attached covalently to it. The proteolipid forms a minor component of the sarcoplasmic reticulum proteins but appears to contaminate, albeit to a small and variable extent, many so-called pure preparations of the Ca^{2+}-pump. Reconstitution of the proteolipid into either black lipid films or liposomes shows that it can act as a Ca^{2+} ionophore. Indeed, when it is added to reconstitution systems containing the Ca^{2+}-pump it actually increases the efficiency of the pump; that is, more Ca^{2+} ions are pumped into the vesicles per molecule of ATP hydrolysed than in its absence. However, as one might expect of a molecule possessing ionophoric properties, the incorporation of high concentrations into vesicles containing the Ca^{2+}-pump actually leads to a decrease in the efficiency of the pump because it creates a leak. The inference is that only a certain number of proteolipid molecules can interact with the pump and once this has been achieved the excess proteolipids merely promote a leak. So for the proteolipid to function efficiently it must occur in defined stoichiometric amounts with the Ca^{2+}-pump. At the present time there is no evidence suggesting this; indeed there are reports that preparations of the Ca^{2+}-ATPase which are free of proteolipid can be used to reconstitute functionally a Ca^{2+}-pump. It therefore remains to be seen whether the proteolipid is either essential for the functioning of the pump or merely acts to increase its efficiency.

5.5.3 Nicotinic acetylcholine receptor

All copies of this integral, transmembrane protein are inserted asymmetrically within the membrane. When acetylcholine or a suitable agonist, e.g. nicotine, occupies a specific recognition site on the molecule, an ion flux ensues. This results in the depolarization of the membrane. The incorporation of such molecules into a synapse provides the receiver for chemical neurotransmission.

Acetylcholine receptors form the basis of the neuromuscular junction (see section 13.2), but a particularly rich source of receptor is found in the electric tissue (electroplax) of fish such as *Torpedo* species, where it constitutes a major protein. Treatment with either ionic detergents, such as Na cholate, or non-ionic detergents, such as Triton X-100 or octyl glucoside, can be used to solubilize the protein. Afterwards it can be purified by affinity chromatography using immobilized ligands that are based either on agonists or on toxins. Although there has been a long-standing controversy on the structure of the nicotinic receptor, it is accepted that it is a glycoprotein ($M_r = 270,000$) consisting of a number of non-identical subunits. These provide both the recognition or binding site for acetylcholine and the ion-channel that is opened upon agonist occupancy of the recognition site. The most recent investigations indicate that four distinct types of polypeptides form the intact functional molecule. These are an α-subunit ($M_r = 40,000$) which contains the acetylcholine binding site, a β-subunit ($M_r = 50,000$), a γ-subunit ($M_r = 60,000$) and δ-subunit ($M_r = 64,000$) whose functions have yet to be elucidated. Presumably some or all are involved in both the formation of the ion-channel and in allowing the formation of the receptor array at the post-synaptic junction (see section 2.14). These subunits have been demonstrated to be asymmetrically positioned in the membrane and the suggestion is that they form a complex of stoichiometry ($\alpha_2\beta\gamma\delta$).

Over the past few years attempts to reconstitute a functional acetylcholine receptor from a pure protein have met with considerable difficulty. For whilst the delipidation that occurs during the purification has little effect on acetylcholine binding, the functioning of the ion-channel is easily lost due to denaturation of the protein. Furthermore, the process of affinity chromatography and the use of competing ligands to release the bound receptor can lead to changes in function. This may be due to changes in receptor conformation and also to loss of the integrity of the complex. Recently, however, a number of laboratories using different methods of purification and reconstitution have been successful in placing a functioning receptor into receptor-containing vesicles loaded with ^{22}Na. Subsequent treatment with agonist leads to an extremely rapid efflux of ^{22}Na which can be completely inhibited by pretreatment with the antagonists, α-bungarotoxin or *d*-tubocurarine.

This pure receptor protein can be integrated into planar (black lipid) membranes constructed by a technique which does not involve organic solvents (Figure 5.8). In this case two electrically isolated compartments are separated by a Teflon barrier in which there is an aperture. The planar membrane containing the integrated acetylcholine receptor extends across this aperture, making it possible to investigate conductance changes across the membrane that occur on agonist addition. Succinyl or carbamoyl choline are usually chosen because they are stable to enzymatic or spontaneous degradation. When they are added to one compartment, they cause the development of a large increase in conductance over a short period of time (1–2 min), which can be blocked either by α-bungarotoxin or *d*-tubocurarine. Reconstitution into planar membranes also allows a measurement of the single-channel con-

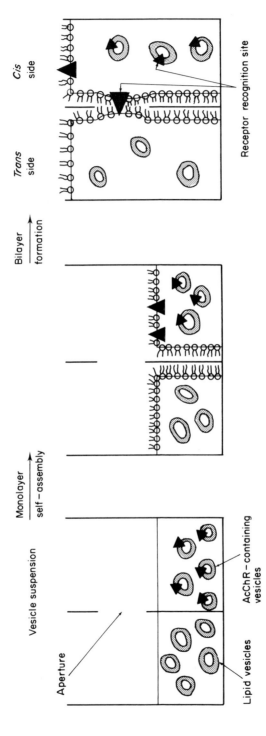

Figure 5.8 Integration of the acetylcholine receptor into planar membranes. Vesicles from *Torpedo* electroplax, where the acetylcholine receptor is the major protein species, are incubated in a compartment of a membrane cell filled to just below the aperture. Under appropriate conditions some of the vesicles can be induced to form monolayers. By raising the water levels the monolayers easily combine to form a planar bilayer across the aperture. (Adapted from Schlindler, H., and Quast, U. (1980) *Proc. Natl. Acad. Sci., U.S.A.*, **77**, 3052–3056)

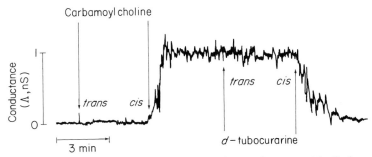

Figure 5.9 **A functioning acetylcholine receptor inserted asymmetrically in a planar membrane. The acetylcholine receptor inserted into a planar membrane as described in the legend to Figure 5.8 is asymmetrically orientated, as agonist or antagonist addition to the *cis*-side of the membrane only results in conductance changes. (Adapted from Schlindler, H., and Quast, U. (1980) *Proc. Natl. Acad. Sci., U.S.A.*, 77, 3052–3056)**

ductance and of the mean channel open time to be made, and these values are very similar to those obtained with the native receptor in its natural environment. Interestingly, this method of reconstituting the acetylcholine in planar membranes results in a functionally asymmetric orientation of the protein (see Figure 5.9).

5.5.4 Amino-acid transporting systems

Studies with intact eukaryotic and prokaryotic cells have demonstrated that the major pathways for the accumulation of amino acids are electrogenic. The proteins that achieve the transport of amino acids across the membrane are powered by the membrane potential (see section 7.5.4). Indeed a pure protein, $M_r = 20,000$, from *Mycobacterium phlei*, when inserted into pre-formed soya-bean phospholipid (azolectin) liposomes, can effect proline uptake into the vesicles. In order for the protein to facilitate proline uptake a membrane potential must first be generated. In this case it can be achieved by allowing K^+-exit from the KCl-loaded vesicles using the ionophore valinomycin.

The presence of a low (10 mM) external Na^+ concentration is essential (Figure 5.10) in order for the transport protein to work. Such a requirement has been noted before for the uptake of certain types of amino acids into both eukaryotic and prokaryotic cells. In these instances it is thought that Na^+ together with the amino acid move into the cell, simultaneously collapsing the proton gradient (see section 7.5.4). This reconstitution demonstrates that a single protein suffices to allow the electrogenic movement of proline into the vesicle. The uptake activity shown by this protein can be abolished by sulphydryl reagents or by adding proton-conducting uncouplers to remove the power source.

228

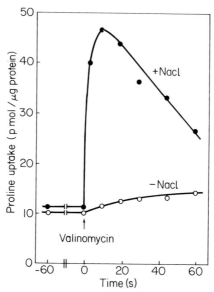

Figure 5.10 **Reconstitution of a proline carrier.** A pure protein from *Mycobacterium phlei* was reconstituted using a sonication procedure into KCl-loaded liposomes. Uptake of proline was triggered by the addition of the cyclic pentapeptide ionophore valinomycin. This allowed K^+ exit with the formation of a membrane potential that was used to drive proline accumulation. For the protein to function, Na^+ (10 mM) was required in the extra-vesicular medium. (Reprinted with permission from Lee, S.-H., Cohen, N. S., Jacobs, A. J., and Brodie, A. F. (1979) *Biochemistry*, **18**, 2232–2239)

Summary

A variety of individual proteins have been successfully reconstituted into defined lipid vesicles with the recovery of a membrane function.

This method has proved to be a powerful tool in defining the minimum number of components necessary to describe particular membrane functions.

FURTHER READING

Blok, M. C., and Van Dam, K. (1978) Association of bacteriorhodopsin containing phospholipid vesicles with phospholipid impregnated millipore filters, *Biochem. Biophys. Acta*, **507**, 48–61.

Darson, A., Vandenberg, C. A., Schonfeld, M., Ellisman, M. H., Spitzer, N. C., and Montal, M. (1980) Aspects of membrane reconstitution procedures, *Proc. Natl. Acad. Sci., U.S.A.*, **77**, 239–243.

Drachev, L. A. *et al.* (1974) Direct measurement of electric current generation by cytochrome oxidase, H^+-ATPase and bacteriorhodopsin, *Nature (Lond.)*, **249**, 321–324.

Jones, N. M., and Nickson, J. K. (1978) Electrical properties and glucose permeability of planar bilayer membranes on incorporation of erythrocyte membrane protein extracts, *Biochem Biophys. Acta*, **509**, 260–271.

Kagawa, Y. (1976) Reconstitution of the mitochondrial inner membrane in *The enzymes of Biological Membranes* Vol. 4, Martonosi, A., ed., John Wiley, London, pp. 125–142.

Kramer, R., and Klingenberg, M. (1980) Modulation of the reconstituted adenine nucleotide exchange by membrane potential, *Biochemistry*, **19**, 556–560.

Nelson, N., Anholt, R., Lindstrom, J., and Montal, M. (1980) Reconstitution of purified acetylcholine receptors with functional ion channels in planar lipid bilayers, *Proc. Natl. Acad. Sci., U.S.A.*, **77**, 3057–3061.

Racker, E. (1975) Reconstitution of membranes, *Proceedings of the Tenth FEBS Meeting*, Vol. 41, Montreuil, J. and Mandel, P. eds., North-Holland/Elsevier, pp. 25–34.

Shanahan, M. F., and Czech, M. P. (1977) Purification and reconstitution of the adipocyte plasma membrane D-glucose transport system, *J. Biol. Chem.*, **252**, 8341–8343.

Sikka, S. C., and Kalra, V. K. (1980) α-Glutamyl transpeptidase-mediated transport of amino acid in lecithin vesicles, *J. Biol. Chem.*, **255**, 4399–4402.

Warren, G. B., Toon, P. A., Birdsall, N. J. M., Lee, A. G., and Metcalfe, J. C. (1974) Reconstitution of a calcium pump using defined membrane components, *Proc. Natl. Acad. Sci. U.S.A.*, **71**, 622–626.

Wohlrab, H. (1980) Purification of a reconstitutively active mitochondrial phosphate transport protein, *J. Biol. Chem.*, **255**, 8170–8173.

Chapter 6

Membrane turnover and circulation

6.1 INTRODUCTION

In Chapter 2 we discussed the most rapid dynamic properties of membrane molecules which give the membrane its characteristic fluidity. Superimposed on top of these movements of individual molecules is the movement of the membrane as a whole. Membranes are not static structures but are continually pinching off and moving as vesicles to other parts of the cell. In this chapter we shall describe how this circulation of membrane may be associated with a variety of physiological processes such as nutrition, hormone action and locomotion, and may also have an important function of its own—the inspection and repair of the otherwise inaccessible outer face of the plasma membrane. Intracellular transport via vesicles which bud off from one membrane and fuse with another, has recently become a subject of great interest following the isolation of a new organelle, the coated vesicle, which is widely regarded as the vehicle for these movements. Circulation of membranes round the cell can be a surprisingly rapid phenomenon with time constants for the complete replacement of a membrane in the order of minutes. Coated vesicle movement between adjacent membranes can be even faster, taking place in seconds.

We shall also be concerned in this chapter with another aspect of membrane turnover, namely the continual degradation and resynthesis of membrane components. These processes are somewhat slower than membrane circulation, having a time constant in hours or days, but still contribute to the ever-changing nature of biological membranes. The need for turnover is evident: proteins are gradually denatured at physiological temperatures, receptors become bound to their ligands, and membrane components on the outside of the plasma membrane face the possibly adverse conditions of the environment. Denatured membrane components must, however, be removed from the membrane and new components inserted whilst the membranes continue their normal function. The mechanisms for the synthesis and degradation of membrane proteins and lipids must therefore account not only for the synthesis of the membrane macromolecules but also for their correct insertion into the membrane.

6.2   THE SYNTHESIS AND INSERTION OF MEMBRANE PROTEINS

There is no reason to believe that the mechanism by which the polypeptide chain of membrane proteins is synthesized is any different from that of cytoplasmic proteins. The characteristic feature of membrane proteins is not a different type of amino acid or peptide bond, but their unique orientation with apolar residues interacting with lipid in the core of the bilayer. Therefore, we shall not describe the reactions that occur on the surface of the ribosome during peptide bond formation but focus on the events that immediately follow.

The synthesis of integral membrane proteins which span the bilayer presents an interesting paradox for the cell. These proteins contain charged amino-acid residues on both sides of the bilayer, yet their synthesis occurs only on the cytoplasmic face where the machinery for protein synthesis is situated. This must mean that charged amino acids pass across the apolar core of the membrane, yet it is precisely the inability to do this which is responsible for the maintenance of the absolute asymmetry of membrane proteins. Before we descibe how the cell overcomes this problem, we must introduce the experimental techniques which have made its investigation possible.

6.2.1   Protein synthesis *in vitro*

Our knowledge of membrane protein synthesis is derived entirely from the study of protein synthesis *in vitro*. This is achieved by combining mRNA, ribosomes, soluble factors and aminoacyl-tRNA activating enzymes from various sources. The addition of amino acids then allows the synthesis of proteins to occur which may be subsequently analysed by SDS–polyacrylamide gel electrophoresis. If membrane vesicles purified from the endoplasmic reticulum are now added to the *in vitro* system it should be possible to observe the insertion of proteins into the vesicle membranes. This may be tested by adding proteolytic enzymes. If the protein is embedded in the membrane, little if any, degradation of the protein will occur, if however it is attached to the outside of the membrane or in solution, it will be completely degraded (Figure 6.1). Using this system it should be possible to study the factors which determine the correct insertion of membrane proteins into biological membranes.

Unfortunately if this experiment is attempted using a total mRNA extract

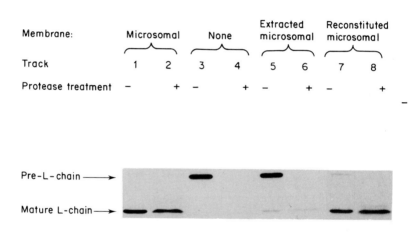

Figure 6.1   Assay of membrane protein insertion. The correct insertion of a membrane or secretory protein into membrane vesicles may be tested by adding proteases to the product of an *in vitro* protein synthesis system. In this figure mRNA from myeloma cells which codes for the secretory protein immunoglobulin light chain, was used. In the absence of membranes no insertion of the protein can occur and the proteolytic enzyme degrades the protein into small fragments which run near the dye front on an SDS–PAGE gel (tracks 3 and 4). When membranes are added to the *in vitro* protein synthesis system two effects are noticed: (1) the protein synthesized is of a lower molecular weight (track 1) and (2) it is no longer degraded by the exogenous protease (track 2). After a mild proteolytic treatment the ability of the membranes to translocate immunoglobulin light chain is lost by washing in 0.5 M KCl (tracks 5 and 6). Simply adding back the salt extract is sufficient, however to reconstitute translocation of the protein (tracks 7 and 8). (Reprinted by permission from Warren, G., and Dobberstein, B. (1978) *Nature*, 273, 569–571. Copyright © 1978, Macmillan Journals Limited)

from mammalian, or even bacterial, cells it is found that the large number of proteins synthesized makes interpretation of the SDS–PAGE patterns impossible. The initial advances in this field therefore followed the development of experimental situations in which only a small number of proteins are synthesized. It was in fact the study of the synthesis of a secretory protein that gave the first clues as to how membrane proteins are inserted into the bilayer. Messenger-RNA extracted from myeloma cells codes predominantly for a single protein product (immunoglobulin light chain) which is secreted by these tumour cells in large quantities. The results of *in vitro* studies using this system could be extrapolated to membrane protein synthesis and insertion because secretory proteins, like membrane proteins, must be translocated across the membrane from the site of their synthesis on the cytoplasmic face to the lumen of the endoplasmic reticulum. In this case, however, the whole protein is discharged into the extracytoplasmic space rather than remaining with part of the polypeptide chain embedded in the membrane.

The first reports of the synthesis and assembly of a membrane protein used mRNA extracted from virus-infected cells. When virus infection occurs, host protein synthesis is shut off, and only virus DNA encoded proteins are synthesized. The membrane-encapsulated viruses (e.g. vesicular stomatitis virus, VSV, and Semliki Forest virus, SFV) produce only one or two species of membrane glycoprotein which gives rise to the characteristic 'spikes' seen in electron micrographs of virus particles. *In vitro* experiments using mRNA extracted from infected cells have therefore been extensively used to study membrane assembly. It must be remembered, however, that in this case it is the assembly of viral coat proteins rather than host membrane proteins which is being studied.

More recently it has been possible to produce copies of a single species of mRNA molecule by using recombinant DNA techniques. In principle this involves the formation of a double-stranded DNA molecule complementary in base sequence to a mRNA molecule using the reverse transcriptase and DNA polymerase. This segment is then inserted into a bacteriophage or a small circular self-replicating piece of nucleic acid called a plasmid (see Figure 6.2). Infection of an *Escherichia coli* culture with this material then produces many copies of the foreign DNA segment as the cells divide. If a mixture of mRNA molecules is used initially then the cells may be plated out and individual colonies, or clones, tested for the presence of the required foreign DNA. Isolation of the plasmids from the cell culture now provides a pure source of the foreign DNA segment to which the RNA may be hybridised. The foreign DNA segment may also be sequenced, and the primary structure of its protein product determined using the genetic code. This procedure is much more rapid than protein sequencing.

In some cases it has been possible to synthesize an oligonucleotide complementary to the DNA which codes for a particular protein and then to use this as a hybridisation probe for screening a cDNA library. These tech-

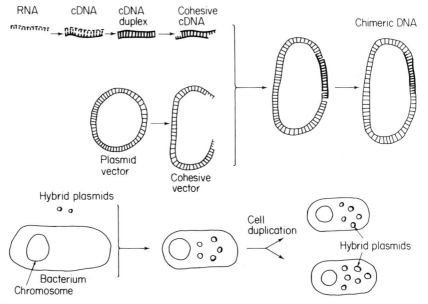

Figure 6.2 Cloning of an individual gene. This simplified scheme shows how to proceed from a purified mRNA in order to produce many copies of its corresponding DNA for sequencing or for expression studies

niques have enabled mRNA species which represent less than 0.1 per cent of the total to be purified.

An *in vitro* protein synthesis system also requires the addition of ribosomes, which are the complex, ribonucleic-acid-containing particles responsible for catalysing peptide bond formation. These may be observed *in vivo* in clusters, called polysomes, in which several ribosomes are attached to one mRNA molecule. Two classes of these polysomes may be distinguished in the cell by their location and protein products. Cytoplasmic polysomes are responsible for the synthesis of soluble proteins and peripheral membrane proteins that are attached to the cytoplasmic face of membranes within the cell. Integral membrane proteins, on the other hand, are synthesized by membrane-bound polysomes which are found on the endoplasmic reticulum, thus giving it its 'rough' appearance. Both classes of ribosome are, however, identical and will both translate mRNA molecules coding for either soluble or membrane proteins. The difference is only observed if membrane vesicles are added to the *in vitro* protein synthesis system. It is then seen that the membrane proteins are segregated into the vesicles whereas the soluble proteins are not. These experiments show that the information specifying the insertion of a membrane protein is contained in the mRNA molecule rather than in the protein synthesis machinery that is used to translate it.

6.2.2 The signal hypothesis

When the mRNA for a secretory or membrane protein is translated *in vitro* in the absence of membranes the product is larger than that normally found *in vivo*. This 'pre-protein' is also devoid of any carbohydrate that may normally be found in the mature protein. An example is shown in Fig. 6.1. This shows the *in vitro* synthesis of immunoglobulin light chain by mRNA extracted from myeloma cells. In track 1 the authentic product ($M_r = 21,000$) is produced *in vitro* in the presence of microsomal membranes. In the absence of membrane vesicles (track 3) the product of translation is some 4000 daltons larger. This is not due to false initiation of synthesis *in vitro* at some point to the left of the initiator codon since, under conditions where no new initiation is occurring, polypeptide chains commenced in the absence of membranes, but completed in the presence of membranes show the normal molecular weight. It would therefore appear that the mRNA for secreted and membrane proteins contains more information than is required to specify the mature protein. Care was taken in these experiments, of course, to ensure that no endogenous mRNA was introduced into the *in vitro* protein synthesis system on the surface of the membrane vesicles.

From results like these it was suggested that the extra amino acids in the pre-protein are the signal which directs the ribosome to the membrane surface. Since this signal is absent in soluble cytoplasmic proteins, ribosomes translating mRNA from soluble proteins do not bind to membranes. Sequencing of precursor secretory and membrane proteins has revealed that the additional amino-acid residues are always located at the NH_2-terminus of the polypeptide chain. This 'signal sequence' is therefore the first part of the polypeptide chain to emerge from the ribosome and is well placed to direct the ribosome to the endoplasmic reticulum surface.

Naturally a great deal of interest has been focused on the structure of these signal sequences and the mechanism by which they interact with biological membranes. One approach has been to compare the NH_2-extensions of a variety of pre-proteins and to look for homologies. Most signal sequences are about 15–30 amino-acid residues long and have a predominantly hydrophobic character (Figure 6.3). Some similarities may be observed between sequences although no unique sequence, which might be a binding site for a membrane receptor, emerges. At the NH_2-terminus is a region of charged amino acids which usually contains lysine and/or arginine. Then in the central portion is an extremely hydrophobic segment containing a high proportion of leucine. It is usually about 10 amino acids long and stretches between residues -7 and -17 of the signal sequence. This is frequently followed by a proline residue and then finally at the cleavage site is situated an alanine or glycine residue. From this analysis it seems likely that the interaction of the signal sequence with the membrane (or membrane receptor) could involve both electrostatic and hydrophobic interactions. Evidently the enzyme which cleaves the signal sequence from the polypeptide chain (the signal peptidase)

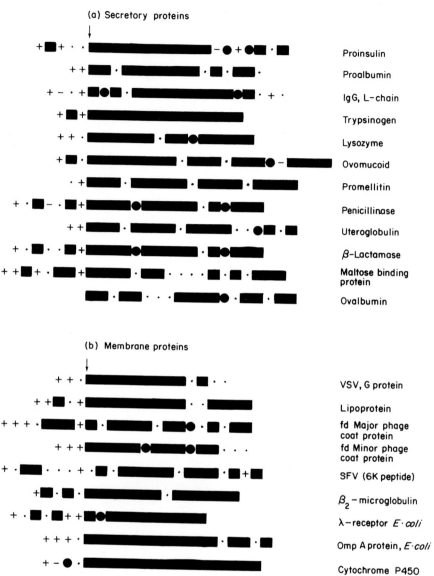

Figure 6.3 Homologies between signal sequences. This figure shows some of the known signal sequences of secretory and membrane proteins which have been aligned so that the similarities are clearly visible. On the left are the NH$_2$-termini of each protein. These regions are hydrophilic and contain several positive charges at neutral pH. Adjacent to these hydrophilic regions are regions of extreme hydrophobicity (starting at the arrow). The hydrophobic region is frequently terminated with a proline residue. Cleavage of each peptide occurs next to the amino acid on the right-hand side of the sequences shown here. This amino acid is usually an alanine or glycine.

■ hydrophobic amino acids	: Ala Gly Leu Val Phe Trp
●	: Pro
· uncharged amino acids	: Ser Thr Cys Met His Tyr
+ positive charged amino acids	: Lys Arg
− negatively charged amino acids	: Glu Asp

has a specificity for amino acids with a small hydrophobic side chain. The significance of the proline may lie in its ability to interrupt an α-helix. The hydrophobic region followed by the proline is predicted by computer studies on protein folding to assume an α-helix followed by a β-turn. It is also possible that the proline interrupts a β-pleated sheet structure formed between the signal sequence and a membrane receptor. In the absence of a unique primary structure in the signal sequence these similarities of secondary structure assume a greater importance.

Another approach to examining the function of the signal sequence is to screen bacterial secreted and membrane proteins for mutations in this region. It has been found for instance that any charged amino acid or proline residue introduced into the central hydrophobic region of the signal sequence prevents the secretion or insertion of the protein and causes it to accumulate in the bacterial cytoplasm in a high-molecular-weight precursor form. Evidently the hydrophobicity of the signal sequence plays the key role, therefore, in the translocation of a protein through the membrane. The ability to block the processing of secretory proteins by the exchange of a hydrophobic residue in the signal sequence with an equally hydrophobic proline residue adds further support to the view that some secondary structure in this hydrophobic region is also important.

In a mutant strain of E. coli a single amino-acid change in the signal sequence of membrane lipoprotein from glycine to asparagine results in the insertion of unprocessed lipoprotein molecules into the bacterial membrane. In this case the mutation appears to have affected a second site in the signal sequence, which is the recognition site for the signal peptidase. This result also shows that cleavage of the signal peptide is not a necessary requirement for translocation of a membrane protein.

The nature of the molecular interaction of the signal peptide with the membrane remains obscure. Originally it was proposed that the hydrophobicity of the signal sequence caused the growing polypeptide chains to simply 'dissolve' in the apolar core of the bilayer and thus direct the synthesis of secretory and membrane proteins to membrane surfaces. This explanation is inadequate however, to explain the specificity of protein synthesis which only occurs on the membranes of the endoplasmic reticulum. Furthermore, it cannot account for the competition that may be observed between different secretory proteins for translocation across the membrane when the concentration of membranes in the in vitro protein synthesis system is made limiting. This competition cannot be explained by a competition for ribosome binding sites (see section 4.7) since the cleaved signal peptide itself is a competitive inhibitor of secretion showing that a receptor must also exist for the signal sequence. The fact that secretory and membrane proteins also display competition for receptors in the membrane is the most powerful evidence for a similar mechanism of translocation of these proteins through the bilayer.

When microsomal membranes are subject to extremely mild proteolysis and then washed in solutions of high ionic strength their ability to translocate

238

immunoglobulin light chain precursor is destroyed (Figure 6.1, tracks 5 and 6). Addition of the salt extract back to the membranes restores their processing ability. The active fragment in the salt extract is a polypeptide of $M_r = 60,000$. It seems probable, therefore, that the receptor for the signal sequence is a membrane protein which may be cleaved on mild hydrolysis to give this 60 000 dalton fragment.

Two models have been proposed to account for the translocation of the signal sequence to the site of the signal peptidase in the lumen of the endoplasmic reticulum (Figure 6.4). The earlier model proposed that the protein went 'head first' through the membrane, the charged NH_2-terminus being important in binding to the receptor. More recently it has been suggested that the charged NH_2-terminus remains on the cytoplasmic side of the membrane whilst the apolar residues loop across the membrane maintaining β-pleated sheet interactions with the receptor protein. There is insufficient evidence at the moment to distinguish between these two models.

So far we have implied, but not formally stated, an important concept of the signal hypothesis, namely that a membrane protein is inserted as an extended polypeptide chain during, rather than after, its synthesis. In order for a membrane protein to be correctly inserted in the membrane it is necessary for membranes to be present during a short 'window' in time after synthesis is

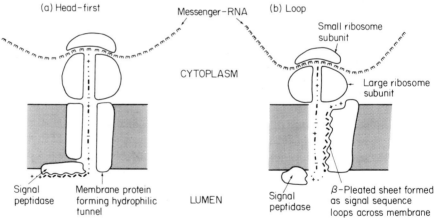

Figure 6.4 **Translocation of the signal sequence through the membrane. Two methods have been proposed for the insertion of the signal through the membrane of the endoplasmic reticulum. In (a) the NH_2-terminus of the protein is envisaged as travelling 'head first' through the bilayer. Binding of the signal sequence to the signal peptidase then draws the first 20 or so amino acids across the bilayer. This may be aided by the formation of a hydrophilic tunnel formed by integral membrane proteins of the endoplasmic reticulum. The alternative (b) shows the hydrophilic portion of the signal sequence binding to a membrane receptor on the cytoplasmic surface while the hydrophobic residues loop through the membrane. The distinguishing feature between these models is which side of the membrane the signal peptide is released after cleavage by the signal peptidase**

initiated. This may be shown *in vitro* by adding membranes at various times after initiation of protein synthesis. Only if membranes are present during the synthesis of the first 70 amino-acid residues does correct insertion occur. Addition of membranes after this time does not achieve the processing of the pre-protein, and resistance to proteases is not obtained. Since the ribosome structure itself covers up the first 40 amino-acid residues the critical moment in protein synthesis occurs when the 30 amino acids of the signal sequence have just emerged from the ribosome. If no membranes are present it presumably becomes buried in the hydrophobic centre of the protein and affinity for membranes is lost.

This 'co-translational' insertion of membrane proteins is the key to how the cell achieves an absolute asymmetry of protein distribution. Instead of synthesizing a pool of folded proteins on the cytoplasmic face of the membrane and then subsequently translocating them across the membrane, the two operations are combined into one operation which occurs at one site. This automatically dictates an asymmetry for the protein since protein synthesis always proceeds in the same direction and insertion always occurs from the cytoplasmic side. The most natural arrangement is for the NH_2-terminus to pass right through the membrane and be located on the extracytoplasmic face whilst the COOH-terminus remains in the cytoplasm. Passage of the protein across the membrane occurs as an extended polypeptide chain. The emerging protein, however, will fold according to the various secondary forces which dictate the lowest free energy structure. This folding of the protein after insertion has the effect of making the whole process irreversible.

The signal hypothesis also provides some rationalization of the energetics of protein translocation across membranes. As the signal sequence becomes exposed on the surface of the ribosome the mRNA ribosome complex becomes associated with the membranes of the endoplasmic reticulum. This is a spontaneous process which must occur within about 0.1 s, since this is the time taken for the first 70 amino acid residues to be synthesized *in vivo*. Stabilization of the hydrophobic segment of the signal sequence by the exclusion of water in the core of the membrane is the principal driving force for this reaction. The ribosome then binds to the ribophoryn proteins (see section 4.7) with an energy of interaction of about -195 kJ mol^{-1} in eukaryotic cells. Extrusion of the polypeptide chain across the membrane occurs against this binding force. Calculations made from the sequence of known secretory and transmembrane proteins show that about 140 kJ mol^{-1} is required to insert the NH_2-terminus of these polypeptides (including the signal sequence) into the bilayer. This represents the energy required to take the polypeptide chain from a random coil in aqueous solution into an α-helix conformation in the apolar environment of the membrane. Thus the energy of ribosome binding is sufficient to account for insertion. Transfer of the polypeptide chain across the membrane is probably aided by the binding of the signal sequence to the signal peptidase which is located on the extracytoplasmic face of the endoplasmic reticulum. Once across the bilayer the signal peptide is cleaved by the signal peptidase, its function now being complete. Finally, the folding of the protein in the lumen of the endoplasmic reticulum prevents any tendency for a reverse translocation and may provide some pull on the emerging polypeptide chain (the energy of folding is in the order of -1 kJ per mol of residue).

This co-translational insertion has considerable energetic advantages. Instead of passing the whole protein across the membrane at one instant, the charged and polar groups are translocated one at a time. So long as the initial energy for insertion of the protein does not exceed that of the ribosome–ribophoryn interaction, translocation will be initiated. Thereafter the energetic disadvantage of one polar amino acid entering the bilayer will frequently be cancelled out by another leaving on the other side of the membrane. It has also been proposed that more integral proteins are involved at this stage to form a hydrophilic tunnel through which the polypeptide chain may pass. There is no direct evidence for these, however, and they only make it more difficult to explain why membrane proteins do not pass all the way through the bilayer like secretory proteins. Presumably they contain 'stop–transfer' sequences just like the signal sequences is a 'start transfer' sequence. If the affinity of these regions for the hydrophobic core of the membrane exceeds the binding energy of the ribosome then it is possible for the ribosome to dissociate from the membrane and a cytoplasmic tail of the protein to be synthesized.

There are now several examples of stop-transfer sequences in integral membrane proteins. At the COOH-terminus of the SFV glycoprotein E1, for instance, there is a stretch of 24 hydrophobic amino acids terminated with two positively charged arginine residues. There are therefore two energetic barriers to complete translocation of the protein. The low free energy of this hydrophobic region in the bilayer together with the large energy required to pass the terminal arginine residues through the hydrophobic core of the membrane are presumably what anchor the protein in place. The E2 glycoprotein of SFV also has a region of 35 hydrophobic amino acids set in 31 residues from the COOH-terminus. This agrees with the known orientation of the protein with a short COOH-terminal projection on the cytoplasmic side of the membrane. Similar stop–transfer sequences have also been found in glycophorin A, haemoglutinin, and immunoglobulin M.

There are now many examples of secretory and membrane proteins which contain a signal sequence which is cleaved during translation. Two exceptions to this rule are included in Figure 6.3. Cytochrome P_{450} is an integral membrane protein which is not synthesized in a larger precursor form *in vitro*. Sequencing of this protein has revealed, however, that it does contain a hydrophobic NH_2-terminus similar to signal peptides. Cytochrome P_{450} therefore resembles the *E. coli* mutant already mentioned in that it contains the information for insertion but not for signal peptidase cleavage. Ovalbumin is a secretory protein which similarly has no precursor form. In this case, however, no homology of the NH_2-terminus and signal sequences can be found.

Ovalbumin is glycosylated by enzymes in the lumen of the endoplasmic reticulum during synthesis, showing that the protein moves across the membrane as an extended polypeptide chain and it competes with other secreted proteins for transport across the membrane. The source of competing activity in this case is found at an internal site—between residues 234 and 253 (shown in Figure 6.3). It is possible, therefore, that ovalbumin is delivered into the lumen of the endoplasmic reticulum by a 'breach presentation'! If this is the case it is not known how the long NH_2-terminal region of the nascent chain could be transferred across the bilayer.

This observation of an internal signal sequence in ovalbumin opens the possibility that other internal signals may also exist which determine the final topography of the protein. We have already mentioned two of these 'topogenic' sequences: the signal and stop–transfer sequences. Arranged in this order a fibrous transmembrane protein is generated (like the viral coat proteins) with the NH_2-terminus on the extracytoplasmic face of the membrane. More complex arrangements of these signals could give rise to globular proteins with many folds in the membrane (e.g. erythrocyte band 3 and bacteriorhodopsin) or even proteins with the NH_2-terminus on the cytoplasmic side (e.g. band 3, urate oxidase and penicillinase from *Bacillus lichiformis*). This hypothesis has yet to be tested by genetic manipulation of these proteins.

6.2.3 The membrane trigger hypothesis

A fundamental tenet of the signal hypothesis is that the membrane or secretory protein is inserted at the same time as it is being synthesized. While this hypothesis appears to be correct for most secretory and fibrous membrane proteins there are some examples in which insertion of these proteins has been shown to be a post-translational event. In these cases it is necessary to propose a second mechanism for protein insertion which is called the membrane trigger hypothesis. In particular, we shall consider the insertion of proteins into chloroplasts and mitochondria, and the mechanism of entry of some toxins and filamentous virus into cells.

Although mitochondria and chloroplasts have all the machinery to be autonomous (i.e. DNA, DNA polymerase, RNA polymerase and ribosomes) every multisubunit protein in these organelles derives most of its subunits from protein synthesis on the cytoplasmic ribosomes. Many of these proteins synthesized in the cytoplasm must cross not just one, but two membranes in order to reach their location in the organelle. For example the cytoplasmically synthesized subunits of mitochondrial cytochrome oxidase must pass through the mitochondrial outer membrane and then become embedded in the inner membrane, and the soluble chloroplast protein, 3-phospho-D-glycerate carboxylase, must pass through both membranes of the chloroplast. This latter protein is of considerable interest in itself because of its role in photosynthesis and its great abundance—in fact it is the most abundant protein in nature. It consists of two subunits, the large subunit being synthesized by the chloroplast while the small subunit is synthesized by nuclear DNA on cytoplasmic polysomes. This small subunit is synthesized as a precursor which is some 5000 daltons larger than the mature product. Processing of this precursor is carried out by soluble enzymes in the stroma of the chloroplast. Unlike secretory proteins, however, this transport across the membrane and processing does not have to occur during the synthesis of the protein, indeed it can still take place in the presence of protein synthesis inhibitors. Exogenous small subunit precursor added to chloroplast under these conditions is rapidly translocated across the membrane and cleaved to its mature form. The NH_2-terminus of the polypeptide in this case is not hydrophobic, but contains many acidic amino-acid residues.

A similar post-translational insertion occurs during the insertion of cytoplasmically synthesized proteins into the membranes of mitochondria. In these organelles a novel mechanism of protein insertion appears to have evolved. A 'signal' sequence is still used, although this may or may not be hydrophobic in nature. The important difference is that in these proteins the signal sequence does not cause the immediate insertion of the protein, but instead it activates it by altering the folding pathway of the polypeptide chain. Contact with a membrane surface then triggers the refolding of the protein so that it becomes embedded in the bilayer. Most soluble proteins are micellar in structure, containing a central hydrophobic core with charged residues interacting with the aqueous environment on the outside. It is easy to see that a membrane protein could be produced simply by inverting this structure so that the hydrophobic regions interact with the apolar core of the bilayer. For most soluble proteins the energy barrier for this transition is very high, although it has been possible to invert some soluble proteins after treatment with SDS and then dialysis against non-ionic detergents. The synthesis of a signal peptide could activate a protein to a metastable state in which this transition occurred spontaneously at physiological temperatures.

The filamentous phage, M13, provides another interesting example of post-translational insertion of a protein into a membrane. This virus contains a single-stranded circular DNA molecule surrounded by 2400 copies of a coat protein. At each stage of virus infection this coat protein is found embedded in the host cell plasma membrane with its NH_2-terminus on the outer surface and its COOH-terminus exposed to the cytoplasm. Of the 50 amino acids in this protein, residues 20–40 are hydrophobic and span the membrane. When synthesized in vitro a precursor is made with an additional 23 residues on the NH_2-terminus. These have the classic arrangement of basic and hydrophobic residues found in other signal sequences and are cleaved from the protein after insertion into the membrane. Insertion in this case occurs in the absence of protein synthesis, it does, however, require energy in the form of a membrane potential, and an interesting model has therefore been proposed to account for the insertion of this protein (Figure 6.5). Once correctly inserted the signal peptide is cleaved in the normal manner. There are several more

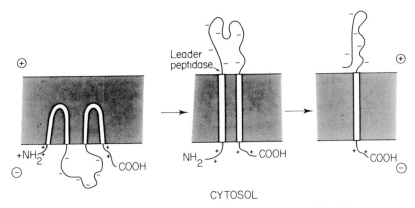

Figure 6.5 Insertion of M13 coat protein. The coat protein of the filamentous phage, M13, is synthesized with a classical signal peptide on the NH$_2$-terminus of the polypeptide. Insertion of the protein into the membrane can occur post-translationally, but requires a membrane potential. This model depicts the spontaneous association of the hydrophobic regions of the molecule with the bilayer followed by the translocation of the hydrophilic domains driven by the membrane potential. Cleavage of the signal peptide then generates the mature, asymmetrically located coat protein. (Reprinted with permission from Date, T., Goodman, J. M., and Wickner, W. T. (1980) *Proc. Natl. Acad. Sci., U.S.A.*, **77**, 4674–4678)

examples of membrane potential-dependent insertion of membrane proteins among mitochondrial proteins.

One way in which the membrane could trigger the re-folding of a protein is by proteolytic cleavage of the polypeptide chain. Several toxin molecules and the membrane attack complex of complement (see section 7.4) might be considered as examples of this mechanism. In both cases proteolytic cleavage exposes large hydrophobic regions on the molecule which direct it to the adjacent membrane surface.

Of course, membrane-triggered insertion of proteins need not necessarily be post-translational, it could equally well occur during synthesis. It becomes more difficult then to distinguish it from the signal hypothesis, which could anyway be considered as a special case of membrane-triggered insertion. In addition to the examples mentioned above, proteins inserted with their NH$_2$-terminus on the cytoplasmic face and proteins with many folds in the membrane are attractive candidates for membrane-triggered insertion since complex topogenic sequences need not then be invoked.

Summary

Nascent membrane proteins contain signal sequences which determine their segregation into membranes.

These sequences are frequently found at the NH$_2$-terminus of the protein, are largely hydrophobic in character and are usually removed during insertion of the protein into the membrane.

Some proteins are inserted post-translationally by a membrane triggered rearrangement.

In all cases asymmetry is achieved by insertion of the protein from a unique side of the membrane.

6.3 GLYCOSYLATION OF MEMBRANE PROTEINS

Many proteins in a cell, including both soluble and membrane proteins, contain mono- or oligosaccharide moieties covalently bonded to amino acids of the polypeptide chain. For plasma-membrane proteins this carbohydrate is only present on the extracellular face where it makes a very important contribution to the chemical properties of the cell surface. In mammals the immense diversity of carbohydrate moieties, which can vary in size and sugar content, is an important factor in determining the antigenic properties of individual cells. The carbohydrate of glycoproteins is also important in the interaction of a cell with its surroundings—whether it spreads on a particular surface, or how it reacts to its neighbours. We shall discuss first how asymmetric synthesis of carbohydrate is achieved. The problem is different from that of asymmetric protein synthesis since it is known that the carbohydrate is added during or after the synthesis of the polypeptide chain, and that it involves enzymes present in the lumen of the endoplasmic reticulum and Golgi apparatus. Asymmetry is therefore simply a consequence of the location of the enzymatic machinery. The problem is how the hydrophilic sugar residues themselves are passed across the endoplasmic reticulum membrane.

The first step towards our understanding of protein glycosylation was the discovery that sugar residues could be activated by nucleotide triphosphates. Successive transfer of the sugar residues to an amino-acid residue of a protein may then occur to produce mono-, di- or oligosaccharide side chains.

$$\text{UTP} + \text{mannose} = \text{UDPmannose} + \text{phosphate}$$

$$\text{UDPmannose} + \overset{|}{\underset{|}{\text{Ser}}} = \text{mannose-}\overset{|}{\underset{|}{\text{Ser}}} + \text{UDP}$$

These sugars are covalently bonded by an oxygen ester linkage to hydroxyl-containing amino acids such as serine, threonine, hydroxylysine, and hydroxyproline. Most membrane proteins, however, are glycosylated by a different mechanism which involves the addition of a large oligosaccharide in one step to an asparagine residue. The bond in this case is a nitrogen ester. Once again the oligosaccharide is activated by a co-factor which in this case is a polyprenol with the general structure:

$$\underset{\text{H(CH}_2\text{CH}=\text{CH}-\text{CH}_2)_n\text{OH}}{\overset{\text{CH}_3}{\overset{|}{}}}$$

In eukaryotic cells dolichol is the most abundant polyprenol with 14–24 isoprene units of which the α-unit is saturated. Retinol (vitamin A) may also act as an acceptor for oligosaccharide intermediates in mammals. Dolichol is a large lipid-soluble molecule which partitions into the apolar core of the bilayer. For this reason dolichol sugars are usually referred to as lipid-linked. Dolichol is present in many tissues, mainly esterified to fatty acids, but a small part is found esterified to sugars. These dolichol-sugars are formed by transfer of the sugar from nucleotide sugars to dolichol phosphate using enzymes in the endoplasmic reticulum:

$$\text{GDPmannose} + \text{dol-P} \rightleftharpoons \text{dol-P-mannose} + \text{GDP}$$

The synthesis of a complex lipid-linked oligosaccharide, however, starts with the formation of a pyrophosphate sugar ester:

$$\text{UDP-}N\text{-acetylglucosamine} + \text{dol-P} \rightarrow \text{dol-P-P-}N\text{-acetylglucosamine} + \text{UMP}$$

This reaction is specifically blocked by the glucosamine-containing antibiotic, tunicamycin, produced by *Streptomyces lysosuperficus*. The dol-P-P-*N*-acetylglucosamine once formed is built up by a series of reactions in which sugar residues are added from both nucleotide and lipid-linked precursors to form a complex lipid-linked oligosaccharide. The structure and synthesis of the commonly occurring glucose containing oligosaccharide, dolichol-diphosphate-G-oligosaccharide, is shown in Figure 6.6.

The similarity of this lipid-linked oligosaccharide with the oligosaccharide core of most carbohydrate moieties of glycoproteins suggests that this is a universal donor which may be subsequently modified by addition or deletion of side chains. The details of this process have been most elaborately studied using VSV, although a similar mechanism is thought to occur for other proteins. By using a synchronous *in vitro* protein synthesis system with VSV mRNA it was shown that glycosylation is dependent on the presence of vesicles from the endoplasmic reticulum and occurs during the synthesis of the polypeptide chain. Two core oligosaccharides are normally added at positions about 150 and 400 residues from the NH_2-terminus. If the membranes are destroyed by adding detergent after various times of translocation, protein products with none, one or two carbohydrate moieties may be formed. All the core oligosaccharides are added before synthesis is complete.

Processing of this core oligosaccharide may be tested by its resistance to hydrolytic enzymes and by structural investigations of the isolated protein. The enzymes which trim and elaborate the core oligosaccharides are found in the lumen of the endoplasmic reticulum and Golgi apparatus. Thus, the state of the oligosaccharide processing provides a biochemical marker for the transport of nascent protein through these organelles. Vesicular stomatitis virus glycoprotein is processed in two steps. The first two glucose residues of the dol-P-P-G-oligosaccharide (Figure 6.6) are removed *in vivo* after times of approximately 2 and 8 min. These reactions probably occur in the smooth elements of the endoplasmic reticulum. After 30 min the third glucose residue is removed and the core immediately elaborated with *N*-acetylglucosamine, galactose and sialic acid. This results in a complex or

246

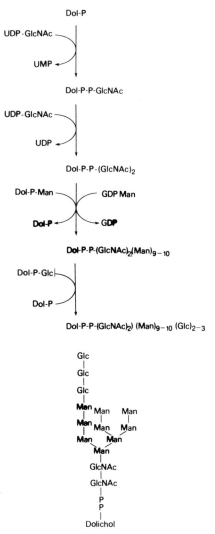

Figure 6.6 The synthesis and structure of dol-P-P-G-oligosaccharide. Oligosaccharides are added to the nascent polypeptide chain in the lumen of the rough endoplasmic reticulum as one unit. This is commonly the glucose-rich oligosaccharide shown here. This oligosaccharide is synthesized on the polyrenol called dolichol by a step-wise procedure from nucleotide and dolichol-linked sugars (see Figure 6.7 for key to sugars)

lactosamine type of oligosaccharide, so called because of the presence of the disaccharide galactosyl $\beta(1{\rightarrow}4)N$-acetylglucosamine. Other glycoproteins contain different oligosaccharides which have the same core structure but side chains of mannose (Figure 6.7). These are referred to as high mannose or simple oligosaccharides. The burst of reactions occurring after a 30 min delay probably occurs as vesicles containing nascent protein fuse with the Golgi apparatus. The mechanism by which secreted membrane proteins move between the endoplasmic reticulum, Golgi apparatus and plasma membrane is not, however, very clear. We shall discuss in section 6.7 the possibility that coated vesicles are involved.

Not all asparagine residues on a polypeptide are glycosylated. The sequence which is common to those sites which are glycosylated is ASN–X–SER–(or THR), although other factors must also be involved since some asparagine residues situated in this sequence do not react. Specificity for the type of oligosaccharide processing is determined by the nature of the residue X. If X is polar a complex oligosaccharide is formed, if X is apolar, it is of the high mannose type. This finding is given more importance by the suggestion that the type of oligosaccharide linked to a protein determines its eventual location in the cell. Thus, plasma cells treated with tunicamycin do not secrete unglycosylated immunoglobulin, but accumulate it in swollen endoplasmic reticulum vesicles. The primary sequence of a protein not only determines the translocation of the protein across the membrane but may also determine its final destination in the cell.

The apparent duplication of sugar acceptor molecules is a reflection on their site of action. Nucleotide sugars cannot pass through membranes and are therefore restricted to reactions in the cytoplasm. Lipid-linked sugars, on the other hand, are anchored to the face of the membrane and are therefore well placed for glycosylating membrane and secretory proteins. It would be convenient to assume that the large dolichol moiety allowed the hydrophilic sugar residues to be transported across the apolar core of the membrane and assembled into complex lipid-linked oligosaccharides in the lumen of the endoplasmic reticulum. This would automatically generate the correct asym-

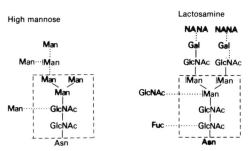

Figure 6.7 The structure of mature oligosaccharides. Once an oligosaccharide has been transferred to an asparagine residue on a protein it is trimmed and elaborated by enzymes in the smooth endoplasmic reticulum and Golgi apparatus. Two commonly occurring structures of mature oligosaccharides are the high mannose and lactosamine oligosaccharides shown here. (GlcNAc–N-acetylglucosamine, Man = mannose, Gal = galactose, Fuc = fucose, NANA = N-acetylsialic acid. Residues linked by a dotted line are present in some but not all oligosaccharides. The core structure is indicated by the dashed line)

metry when a membrane protein was glycosylated. Unfortunately, not all evidence is in accord with this hypothesis and it is also possible that glycosylation occurs on the cytoplasmic face of the endoplasmic reticulum just before translocation of the polypeptide chain.

Summary

Carbohydrate is added in an asymmetric manner to membrane proteins by enzymes which occur in the lumen of the endoplasmic reticulum and Golgi apparatus.

A core oligosaccharide is transferred from a lipid-linked intermediate to an asparagine residue on the glycoprotein during synthesis of the polypeptide chain.

The oligosaccharide may then be modified during transport to other membranes in the cell.

6.4 SYNTHESIS OF MEMBRANE LIPIDS

The synthetic pathway for the major phospholipids in bacteria and mammals is shown in Figure 6.8. Phosphatidic acid, which contains the basic elements of an amphipathic phospholipid molecule, is first synthesized from a fatty acyl co-enzyme A molecule and glycerophosphate. After this a variety of different headgroups can be added to the phospholipid using the energy of CTP hydrolysis. The phospholipids must end up in the correct orientation in the bilayer since in mature membranes the rate of flip-flop is very small (see section 2.3). Phospholipids do not, however, display an absolute asymmetry like membrane proteins.

The mechanism of insertion of nascent phospholipid molecules into the bilayer has been most thoroughly studied in the Gram-positive bacterium, *Baccillus megaterium* which has the advantage of having only one membrane which surrounds the cell, and which contains only two phospholipids, phosphatidyl ethanolamine (PE) and phosphatidyl glycerol. Here, PE, which represents 70 per cent of the total lipid in the membrane, is asymmetrically distributed between the two halves of the bilayer with about 30 per cent on the outer surface, where it reacts readily with the non-penetrating reagent TNBS, and 70 per cent on the cytoplasmic face. The enzymes of phospholipid synthesis with one exception are membrane bound and utilize soluble precursors and co-factors. It might be argued, therefore, that newly synthesized phospholipids must be assembled on the cytoplasmic face of the membrane. This was indeed found to be the case when *B. megaterium* was pulse-labelled with ^{32}P-phosphate. The radioactive phosphate was incorporated into phospholipids that were not available to exogenous TNBS and must therefore be located on the cytoplasmic face of the bilayer. If labelling of the outer surface with TNBS was delayed for 30 min after the pulse of ^{32}P-phosphate then the TNBS was found associated with radioactive PE. In fact the radioactive phospholipid now represented the same proportion of the total PE in both halves of the bilayer. Careful measurements of this sort suggested that the newly synthesized phospholipid moved across the membrane with a half-time of about 5 min. This rate of equilibration is almost 100,000 times faster than the

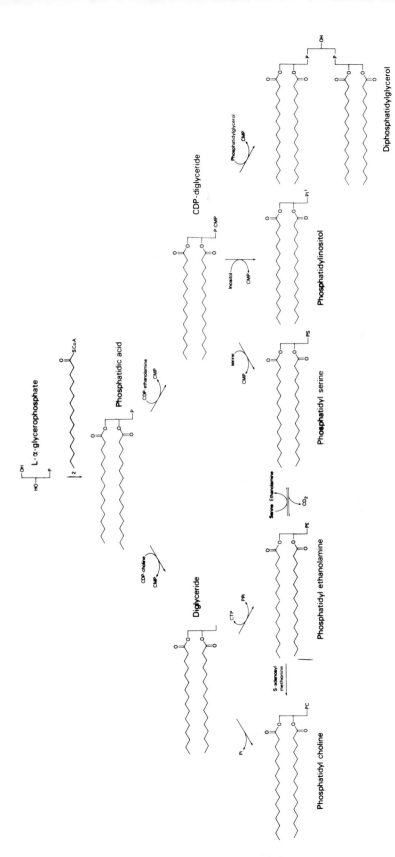

Figure 6.8 Biosynthetic pathways of glycerolipids

rate of flip-flop observed in other membrane systems, and cannot be attributed to spontaneous translocation of phospholipid molecules across the bilayer. Two possibilities are being investigated. One suggestion is that a special carrier protein exists in the membrane which facilitates phospholipid translocation. Evidently, one must also conjecture that this is only present in rapidly growing membranes. No such protein has yet been isolated. The other possibility is that the increase in size of the cytoplasmic monolayer distorts the bilayer in such a way that phospholipid translocation becomes energetically feasible.

Some light has been thrown on this problem by studies on the synthesis of phosphatidyl choline in the endoplasmic reticulum of mammalian cells. This phospholipid may be formed by either a direct transfer of phosphoryl choline on to diglyceride or by successive methylation of PE using S-adenosyl methionine (Figure 6.8). Of particular interest here is the mechanism of the latter pathway. The enzymes responsible for methylation of PE are located on the extracytoplasmic face of the microsomal membrane, but the final phosphatidyl choline product is found on both faces of the membrane, despite the fact that phosphatidyl choline is not able to cross the membrane by flip-flop. When the location of the mono- and di-methyl intermediates was tested using phospholipase C, it was found that they were not accessible in either sealed or leaky vesicles. The simplest conclusion is that the methylated intermediate is bound to an enzyme in the membrane and methylated during transfer across the bilayer. Similar observations have been made in rat liver microsomes and the erythrocyte membrane, suggesting that this modification-linked transport of PE may be a widespread phenomenon in mammalian cells.

Summary

Phospholipid molecules are inserted into the cytoplasmic face of the membrane in which they are synthesized.

Rapidly expanding membranes have a mechanism for translocating phospholipids, after their synthesis, across the hydrophobic core of the membrane.

6.5 TURNOVER OF MEMBRANE PROTEINS

The steady-state concentration of protein in a cell or whole organism is determined not only by its rate of synthesis but also by its rate of degradation. In a non-growing cell the rates of synthesis and degradation are equal and either of these may be measured to give a measurement of protein turnover. In a growing cell the rate of synthesis exceeds that of degradation, usually as a result of an increased rate of synthesis whilst the rate of degradation remains constant. However, the two processes are not independent. Cyclohexamide, which inhibits protein synthesis in eukaryotic cells, also inhibits degradation, thus maintaining a steady-state condition. In non-steady-state conditions the rate of turnover may be taken as the rate of synthesis or degradation, whichever is smaller.

Membrane turnover is a surprisingly rapid aspect of membrane dynamics.

Table 6.1   Half-lives of membrane proteins from rat liver

Membrane	Half-life (days)
Nuclear envelope	2–3
Rough endoplasmic reticulum	2–6
Smooth endoplasmic reticulum	2–5
Golgi apparatus	1–2
Plasma membrane	2–3
Mitochondrial inner membrane	7–9
Peroxisomes	3–4
Lysosomes	7–8

Proteins in membranes are broken down with half-lives between 2 and 5 d, while in some exceptional cases a soluble protein may have a half-life as short as 10 min (see Table 6.1). Degradation is also very extensive, it is not just a small fraction of the cell which turns over but every component except the DNA. In fact more protein is broken down in body tissues than in the gastrointestinal tract in any 24 h in man. This is not, however, caused by destruction of whole cells which would result in synchronous turnover of all components, but is heterogeneous in nature and occurs within each cell. A rat hepatocyte for instance replaces 70 per cent of its protein every 4–5 d. Since the life of a cell is 160–400 d, this must represent the degradation and resynthesis of components by the cell itself.

Investigation of membrane turnover has been frustrated by the problems involved in measuring protein degradation. Measurements of enzyme activity or antibody binding can be very misleading, since changes in the steady-state concentration only represent the rate of degradation if synthesis is inhibited. The most common technique is to pulse-label a whole animal or cell suspension with a radioactive amino acid and then follow the decay of radioactivity in the protein. This technique assumes that the labelled amino acids released by degradation are not reincorporated into new proteins. In fact this does happen to a large extent and gives rise to a gross overestimate of the protein half-life. Some improvements have been made by the use of poorly re-utilized labels such as (guanidino-^{14}C) arginine and $NaH^{14}CO_3$, although these have the disadvantage of being rather expensive and easily incorporated into other macromolecules. A more sophisticated method of labelling the proteins is to administer two different precursors to the animals at different times. The first might be $NaH^{14}CO_3$ for instance, followed several days later by ^3H leucine. Proteins are then separated by fractionation procedures soon after the administration of the second label. This technique gives two points on the decay curve of the protein, and the ratio of radioactivity from each label provides a sensitive measurement of the half-life. This ratio may be compared with the ratio in control cells in which both isotopes are administered together. A large ratio is indicative of a rapidly turning over protein and vice versa. Unfortunately the longer half-life of membrane proteins makes the measurement of turnover less accurate than for soluble proteins. Care must also be taken in extrapolation of results with soluble proteins to membrane proteins, since different mechanisms of degradation are probably involved.

Membrane proteins, like soluble proteins, are degraded with first-order kinetics, i.e. once a newly synthesized (radioactive) protein has entered the pool of similar (unlabelled) proteins, its chance of breakdown is identical to all molecules of the same type. Degradation is not therefore dependent on the age of the protein but occurs in a random fashion. This does not exclude the possibility that proteins synthesized in a defective manner are more rapidly degraded. The evidence for a correlation between protein size and rate of degradation, as seen for soluble proteins, is much weaker for membrane proteins. This is of particular interest because two different mechanisms could be envisaged for membrane-protein degradation. First we must consider the possibility that the degradation of membrane proteins could be heterogeneous, each protein having its characteristic rate of turnover. This could most plausibly be achieved by group-specific proteases embedded in the membrane itself. The probability of nascent or mature membrane proteins colliding with the membrane proteases would be equal, thus giving first-order kinetics, while heterogeneity could occur as the result of different mobilities in the bilayer and on the susceptibility of the protein to attack. Interestingly, proteolytic enzymes are found in the plasma membrane although their function is not clear.

Plasma-membrane vesicles taken up by endocytosis can fuse with primary lysosomes containing degradative enzymes. Some membrane proteins, as well as the contents of the vesicle, may be broken down at this stage. It is also possible that intact membrane proteins are recycled by the formation of small vesicles which bud off from the secondary lysosome and fuse with the plasma membrane (see Figure 6.9). Cross-linked or denatured membrane proteins may be segregated into areas of the lysosomal membrane which are degraded by intralysosomal membrane degradation.

Care must be taken, however, to ensure that a heterogeneous rate of membrane-protein turnover is not simply the result of studying a heterogeneous population of membranes. In rat liver, for example, there are several different cell types each of which contain membranes which could turn over at different rates. Even in the isolated hepatocyte plasma membrane it is known that several distinct domains exist and it is very likely that these would be endocytosed at different rates. Finally, it is important to separate peripheral proteins from integral membrane proteins and wherever possible to examine the degradation of individual types of protein.

In the alternative, co-ordinate model, of membrane degradation large areas of membrane are removed (as in endocytosis) and degraded at one time, thus giving an identical half-life for all proteins and lipids in the membrane. Support for a co-ordinate turnover of plasma-membrane protein has come from

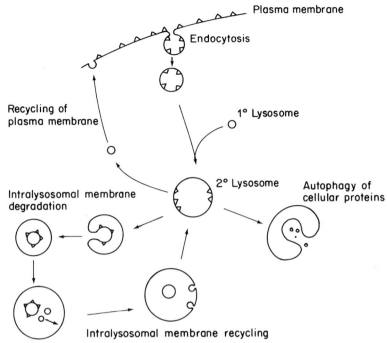

Figure 6.9 **Pathways of membrane uptake into lysosomes. Primary lysosomes fuse with plasma-membrane vesicles formed by endocytosis to generate secondary lysosomes in which both the contents and some membrane components may be degraded. Soluble proteins are taken up into lysosomes by autophagy. It is also possible that membrane is selected for degradation by intralysosomal degradation or recycled to the cell surface.**

some experiments in which the lactoperoxidase iodination of intact rat hepatoma cells was studied in culture. The culture was divided into two and labelled with either ^{125}I or ^{131}I. The ^{131}I-labelled culture was then frozen while the ^{125}I-labelled cells were allowed to metabolize the label at 37 °C for various periods. Performing the experiment this way ensured that both labels were added to the culture in the same growth phase. Plasma-membrane proteins were then separated by SDS–gel electrophoresis and similar ratios of ^{125}I to ^{131}I were found. A series of elegant experiments have also been performed on mitochondrial inner membrane turnover. This organelle may be purified in a reasonably homogeneous state, allowing accurate measurements of turnover to be made. The mitochondria were derived from the liver of rats double labelled with $NaH^{14}CO_3$ and ^3H-leucine and fractionated into water-soluble (mitochondrial matrix proteins), salt extracts (extrinsic proteins), and detergent solubilized material (intrinsic membrane proteins). Although heterogeneity of turnover was observed in the matrix and extrinsic proteins the intrinsic membrane proteins showed very little heterogeneity. It is not known whether intracellular organelles like the mitochondria are degraded by

254

the lysosomes or whether they contain their own proteases and are capable of autophagy.

There are examples of membrane proteins for which the rate of turnover cannot be explained by this hypothesis in its simplest form. For instance the receptors for lectins, which are capped to one pole of the cell when bivalent lectin is added, are subsequently internalized and degraded at a high rate. Similarly, the rate of recovery of HLA-2 histocompatability antigen of lymphocytes after digestion by papain and the recovery of $Na^+K^+ATPase$ of HeLa cells after ouabain blockade suggests a half-life of about 6 h which is much faster than other proteins in the same membranes. Drugs like phenobarbital are also known to induce differential levels of proteins in rat liver endoplasmic reticulum, although it is difficult in this case to distinguish changes in the rates of synthesis and degradation. Evidently, additional mechanisms must be postulated to account for these abnormal situations.

Early cell biologists considered the cell to be a stable structure, its components having a similar half-life to the cell. The observation that cells, on the contrary, engage in such a rapid rate of protein turnover suggests that this must be an important aspect of cell metabolism, especially considering the large quantity of energy required to resynthesize proteins. One benefit is the ability to control the concentration of regulatory enzymes and receptor molecules. Rapid synthesis of these proteins must be followed by rapid destruction if sensitive control is to be achieved. Another idea is that protein degradation may play an important role in destroying aberrant forms of proteins and have a function in slowing the onset of senility. Interest has focused, therefore, on the mechanism of protein degradation in senescent cell cultures and in patients with progeria syndrome who suffer premature ageing and early death. Abnormal proteins are found in these patients which have increased heat lability and reduced enzyme activity per molecule. Ageing cultures also have increased numbers of lysosomes and higher rates of protein degradation. In bacteria it has been shown that higher rates of degradation (up to 20 times) occur with proteins containing incorrect amino acids or polypeptides prematurely terminated with puromycin. It is uncertain, however, whether the accumulation of malfunctional proteins in senility is caused by random errors in translation, post-translational modifications, or a failure of the degradative process to remove abnormal forms.

Summary

Membrane proteins are degraded more slowly than soluble proteins with a half life of 2–5 d.

It is not yet clear whether intrinsic membrane proteins are degraded in a heterogeneous or a co-ordinate fashion.

6.6   TURNOVER OF PHOSPHOLIPIDS

Measurement of the degradation of membrane phospholipids is considerably more complex than that of protein. The decay curve of radioactively labelled phospholipids may be different when different isotopes are used or even non-exponential, making the measurement of a half-life impossible. The immense diversity of phospholipid structure also makes generalizations difficult. Simply changing the fatty acid side chains can effect the turnover rate. Phospholipids containing arachidonic acid, for instance, are particularly inert. Individual membranes often have a characteristic rate of phospholipid turnover, although this can be confused by the action of phospholipid exchange proteins which equilibrate phospholipids between adjacent membranes. Thus, in the rat liver the half-life of phospholipids is 5–10 d and is similar to the half-life of membrane proteins. In the myelin sheath, however, the half-life of phosphatidyl choline is between 40 and 90 d. This is a consequence of the compact multilamellar structure and low metabolic activity of this membrane.

The phospholipase enzymes, which break down phospholipids are frequently membrane-bound enzymes and are abundant in bacterial and animal cells. They are classified according to the bonds that they break (Figure 6.10). In mammalian cells a high concentration of A-type phospholipases with acid pH optima are found in the lysosomes where most phospholipid degradation occurs. No phospholipase C or D enzymes, which are characteristic of snake venum and bacteria, are found in mammalian lysosomes, although a sphingomyelin phosphodiesterase, which attacks sphingomyelin in a similar manner to phospholipase C, is found in the lysosomes of many tissues including brain, liver and kidney. Phosphodiesterases are also present in the lysosomes which can complete the degradation of the phospholipid structure. Some phospholipases with neutral or alkaline pH optima are also found in other subcellular organelles like the endoplasmic reticulum.

The turnover of one phospholipid has aroused considerably more interest than any other. That is the phospholipid, phosphatidyl inositol (PI), which occurs in most membranes and can contribute up to 10 per cent of the total lipid. Several different phosphatidyl inositides may be found in biological membranes (Figure 6.11) all of which have the hexahydroxycyclohexane headgroup but which contain a variable number of phosphate groups. The interest in PI turnover stems from the remarkable observation that the degra-

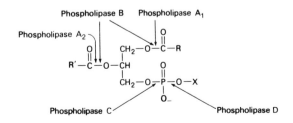

X = phospholipid head group

Figure 6.10 Enzymes of phospholipid degradation

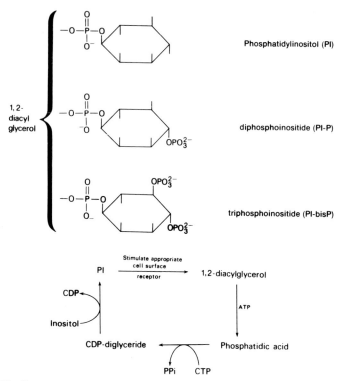

Figure 6.11 Interconversions of phosphatidyl inositol. Addition of ^{32}P-ATP to many cells during stimulation with a variety of hormones, etc. causes the ^{32}P to become incorporated into phosphatidyl inositol. This has been shown to occur as the result of cyclic turnover of the inositol lipids. Poly-phosphoinositides are also found in biological membranes. These membranes have a high affinity for divalent cations

dation and then resynthesis of PI is simulated by a wide variety of hormones, neurotransmitters and changes in growth conditions. No other phospholipid shows a similar effect. The response was initially detected by the incorporation of ^{32}P-ATP into PI and its immediate precursor, phosphatidate (Figure 6.14). It appears, however, that the primary response is the degradation of PI to 1,2-diacylglycerol and in fact PI levels usually fall during stimulation.

Common to each situation where a PI response is observed is the interaction of an external ligand with a cell surface receptor, although not all receptors evoke a response. Notable for their absence are the stimuli which act via changes in intracellular cyclic AMP concentration. A strong correlation is, however, observed between the PI response and stimuli which act via changes in intracellular calcium ion concentration. If stimulation of the receptor is bypassed by adding extracellular calcium and the ionophore A23187 (see section 7.3) then no PI turnover occurs, suggesting that the PI response is intimately associated with receptor stimulation and is not a secondary consequence of raised intracellular Ca^{2+} concentrations. Whether a PI response may occur when an appropriate stimulus is applied to cells in the absence of extracellular Ca^{2+} is the subject of contradictory reports in different tissues. Evidently, if the PI response is the link between receptor stimulation and opening of a 'Ca^{2+}-gate' in the plasma membrane, then it should occur before the rise in intracellular calcium is detected. Most awkward for this hypothesis is the location of the PI response in the membranes of secretory vesicles, nerve endings and endoplasmic reticulum, as well as in the plasma membrane. Clearly, if it has a role in controlling cell surface Ca^{2+} permeability then it must be located at the cell surface.

The di- and tri-phosphoinositides are powerful chelators of divalent cations and were once thought to be involved in nerve conduction due to their concentration in the myelin sheath surrounding nerve axons. Although there is little evidence to support this view it has more recently been found that they can potentiate the effects of elevating the intracellular Ca^{2+} concentration. As the intracellular Ca^{2+} concentration rises, Ca^{2+}-dependent phosphodiesterases cause the breakdown of the polyphosphoinositides to PI and the release

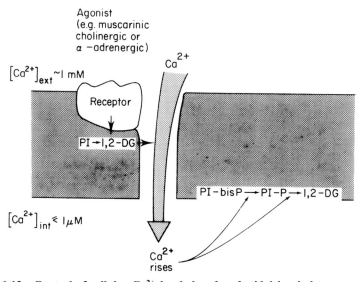

Figure 6.12 **Control of cellular Ca^{2+} levels by phosphatidyl inositol turnover. Since phosphatidyl inositol turnover increases rapidly after treatment of cells with hormones it is possible that this process is involved in the control of intracellular metabolism. In particular the PI response is associated with agonists which give rise to a change in the intracellular Ca^{2+} concentration. It has therefore been suggested that phosphatidyl inositol turnover is involved in the opening of a Ca^{2+}-gate in the plasma membrane. This Ca^{2+} in turn may stimulate Ca^{2+}-sensitive phosphodiesterases of the plasma membrane which break down the poly-phosphinositides and so deplete the membrane of high-affinity sites for divalent cations. The rise in intracellular Ca^{2+} concentration is thus potentiated. (Reprinted with permission from Michell, R. H. (1979) *Trends Biochem. Sci.*, 4, 128–131)**

of bound calcium ions. These two levels of PI metabolism may therefore have a synergistic effect on the intracellular ionized calcium concentration (Figure 6.12).

Summary

Turnover of membrane phospholipids is more complex than that of membrane proteins.

Phosphatidyl inositol turnover in response to receptor stimulation is a widespread phenomenon which may be related to changes in intracellular calcium ion concentrations.

6.7 MEMBRANE CIRCULATION

Our discussion of the synthesis and degradation of membrane components has so far ignored an important problem, namely the transport of membrane material from its site of synthesis (the endoplasmic reticulum) to specific membranes within the cell, and from those membranes to its site of degradation (the lysosomes). The rapid rate of turnover of membrane proteins and lipids means that there must be a constant flow of membrane material between these organelles. There are four mechanisms by which this transport might occur:

(1) Diffusion. Serial sections of cells viewed in the electron microscope have shown that most if not all of the membranous structures of the cell are joined together at various places. Therefore, transport of membrane components could be achieved simply by lateral diffusion in the bilayer. Specificity could be the result of barriers to diffusion such as the tight junctions between cells which prevent the diffusion of freeze–fracture particles through the plasma membrane.

(2) Carriers. Phospholipids can be transported through the cytoplasm bound to the phospholipid exchange proteins, no example, however, of a membrane-protein carrier has yet been reported.

(3) Flow-differentiation. On this model the growing endoplasmic reticulum differentiates into other organelles of the cell, principally the Golgi apparatus and the plasma membrane.

(4) Vesicle shuttles. Newly synthesized membrane components are packaged into vesicles and transported to other cell organelles. Specificity is achieved by the speoificity of fusion of the vesicles with the target membrane.

Of these possibilities it is the latter two which are of principal interest here, since they involve the movement of large areas of membrane around the cell, a process which we shall call 'membrane circulation'. Membrane circulation can also occur without synthesis and degradation of the membrane components. In particular we shall discuss the mechanism of cell locomotion and the various processes of exocytosis and endocytosis.

6.7.1 Exocytosis and endocytosis

The term 'exocytosis' is used for the release at the cell surface of secretory proteins from intracellular membrane-lined vesicles or granules. Nascent membrane proteins and phospholipids may also be present in the membrane of these vesicles so that, after fusion of the vesicle with the plasma membrane, they become part of the cell surface. Exocytosis is therefore the last step of both secretion and plasma-membrane protein synthesis.

In some cells secretion occurs continuously, while in others the secretory product accumulates in granules in the cytoplasm and is released in response to a specific stimulus. A good example of the latter is the histamine release by rat peritoneal mast cells. These cells are packed full of membrane-bound granules which contain histamine (Figure 6.13a). In cells from immunized animals the receptors on the mast cell surface pick up IgE antibodies against the specific antigen with which the animal was immunized. Addition of the antigen, antibodies to immunoglobulin, or a lectin, then causes the cross-linking of the surface receptors and within 10 s the cell responds by releasing the contents of the granules (Figure 6.13b). Degranulation is complete in about 1 min. The ease of obtaining a relatively pure homogeneous population of cells from a peritoneal exudate, and this rapid response to stimulation has made mast cell degranulation an ideal system for studying both the mechanism of exocytosis, and also the means by which secretion is coupled to stimulation. Since the effects of cross-linking the surface receptors may be mimicked by using the Ca^{2+} ionophore A23187 (see section 7.3) it appears that the immediate effect of clustering these molecules is to cause a rise in intracellular

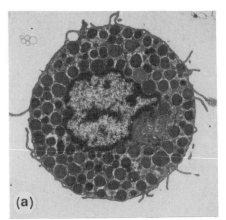 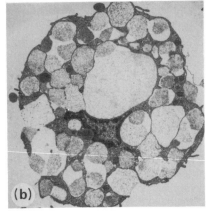

Figure 6.13 Degranulation of a rat peritoneal mast cell. A resting mast cell (a) contains many membrane-bound, histamine-containing granules. On stimulation of a sensitized cell (b) these vesicles fuse with the cell surface and release their contents. (Reprinted with permission from Lawson, D., Raff, M. C., Gomperts, B., Fewtrell, C., and Gilula, N. B. (1977) *J. Cell Biol.*, 72, 242–259)

Ca^{2+} concentration, although the molecular mechanism for this effect remains unknown. It is then observed that the cytoplasm between the granule membranes and the plasma membrane thins and then disappears as the two membranes fuse.

Most interestingly, this process is accompanied by the segregation of membrane proteins away from the area of contact, although it is not known how this is accomplished. One possibility is that the membrane proteins are attached directly or indirectly to cytoskeletal structures (see section 6.7.3) below the membrane which contract as the cytoplasmic Ca^{2+} concentration increases. What does seem clear is that in this cell type, membrane fusion occurs between areas of membrane containing few, if any, membrane proteins. Fusion of the granule membrane with the plasma membrane is followed by fission (or rupture) of the bilayer at the point of contact, which allows the release of the granule contents. In the adrenal medulla it has been shown that this process can be inhibited by compounds which block anion transport. These compounds also inhibit release of transmitter from isolated secretory vesicles prepared from the adrenal medulla, showing that the anion transport sites are present in the secretory vesicle membrane. It is probable, therefore, that a rapid flux of anions into the secretory vesicle at points where the secretory vesicle membrane has fused with the plasma membrane causes the fission of the cell surface by osmotic lysis.

During exocytosis the granule or secretory vesicle membrane is added to the plasma membrane which becomes grossly enlarged as a consequence. Excess membrane may appear in microvillae on the cell surface, be shed from the cell as small vesicles, or be internalized by endocytosis, which morphologically resembles the reverse of exocytosis. Two types of endocytosis may be distinguished, phagocytosis ('cell eating') and pinocytosis ('cell drinking'). In phagocytosis, particles of up to 1 μm are ingested in membrane-lined vacuoles which fuse with primary lysosomes to form phagocytotic vacuoles where digestion of the extracellular material occurs. Phagocytosis occurs in all classes of animals from protozoa, where it is a major means of nutrition, to mammals. In the latter case specialized cells (the mononuclear phagocytes and polymorphonuclear leucocytes) scavenge the body, engulfing foreign material. Mononuclear phagocytes have a high-affinity receptor for immunoglobulin on their cell surface which binds to foreign material coated with antibodies made by the immune system. Binding is rapidly followed by phagocytosis. Initial attempts to investigate the mechanism of phagocytosis have used a number of pharmacological compounds thought to interact with elements of the cytoskeleton (see Table 6.3). In particular the vinca-alkaloids and colchicine are used as inhibitors of the microtubules whilst the cytochalasins are used to inhibit microfilament formation. It must be stressed, however, that, like any drug, these compounds have many side effects which could also affect exocytosis and endocytosis. In general the more specific effects, which occur at the lowest concentrations, are the most credible. Using cytochalasin B two steps may be distinguished in phagocytosis. The first step which is independent of cytochalasin B is the binding of a particle to the cell surface. This may occur by non-specific hydrophobic interaction with the membrane (as when latex beads are engulfed by cells) or by binding to cell surface receptors. The plasma membrane then invaginates to enclose the particle in a manner which is inhibited by cyctochalasin B. This part of the process is energy dependent and presumably involves microfilament attachment to the membrane.

Macropinocytosis is a similar process to phagocytosis except that no particle is enclosed. Vesicles of 0.3–2 μm diameter are formed at the cell surface and migrate through the cytoplasm towards the nucleus. Cytochalasin B reversibly inhibits the uptake of vesicles in a similar manner to phagocytosis. Interestingly, if colchicine, which inhibits microtubule formation, is added, the macropinocytoic vesicles are still formed but no longer migrate towards the nucleus in a concerted manner. It has been suggested, therefore, that microtubules are used for guiding vesicles through the cell whilst microfilaments produce the contractile energy with which they are moved. Both phagocytosis and macropinocytosis are energy-dependent processess inhibited by glycolytic poisons.

6.7.2 Micropinocytosis, coated vesicles and receptor-mediated endocytosis

A substantial part of the cellular uptake of extracellular fluid takes place by a third process called micropinocytosis. This involves the formation of very small vesicles with a uniform diameter of 70–100 nm which may only be seen using the electron microscope. The formation of these vesicles does not require metabolic energy and is not inhibited by colchicine or cytochalasin B. Movement of these vesicles through the cell is thought, therefore, to occur by random diffusion. Cooling to below 4 °C is the only effective way of preventing micropinocytotic vesicle formation, probably as a consequence of the decreased membrane fluidity at that temperature.

Closely parallel to micropinocytosis is the process of absorptive endocytosis or 'receptor-mediated endocytosis'. As these names suggest, this form of endocytosis is used for the specific uptake into the cell of macromolecules bound to cell surface receptors. Proteins taken up by this mechanism have a variety of fates within the cell (Table 6.2). Complex transport particles such as the cholesterol-containing low-density lipoprotein (LDL) particle are degraded in the lysosomes so that their contents (in this case cholesterol) are released for use in the cell. Other proteins like the yolk proteins and maternal immunoglobulins are transported direct to their site of biological action without degradation. Some hormones, whose action is primarily at the cell surface are taken up, still bound to their receptor, and degraded (a process termed 'down-regulation'). The rapid turnover of receptor-bound hormones ensures a rapid response to changes in circulating hormone concentration. Other hormones may have intracellular sites of action and be delivered to these sites by receptor-mediated endocytosis. A candidate for both processes is the hor-

Table 6.2   Examples of receptor-mediated endocytosis

Protein	Cell type	Internalization via coated vesicles	Fate of protein
LDL	Fibroblasts	Yes	Degraded in lysosomes
Yolk proteins	Oocytes	Yes	Delivered to yolk granule
Transferrin	Erythroblasts, reticulocytes	Yes	?
Epidermal growth factor	Fibroblasts	Yes	Degraded in lysosomes
Insulin	Hepatocytes	?	Degraded or taken to Golgi or nucleus
α-2-Macroglobulin	Fibroblasts	Yes	Degraded in lysosomes
Maternal IgG	Foetal yolk sac	Yes	Transferred to foetal circulation
SFV particles	BHK	Yes	Transferred to lysosomes
Lysosomal enzymes	Fibroblasts	Yes	Delivered to lysosomes where they remain active

mone, insulin, which has a variety of short-term (cell surface) and long-term (intracellular) effects.

Much of the biochemical mechanism of receptor-mediated endocytosis has been deduced from experiments studying the uptake of LDL particles. These large spherical particles contain a central apolar core of cholesterol ester surrounded by a polar coat of phospholipid, unesterified cholesterol and protein. When cells require cholesterol for membrane synthesis, they synthesize a cell surface receptor that binds the protein component of LDL. About 50,000 copies of the receptor may be present on a human fibroblast surface which can bind LDL at nanomolar concentrations. The LDL, once bound is rapidly internalized and degraded in the lysosomes. Cholesterol released into the cytoplasm of the cell may then be used for membrane synthesis and cholesterol metabolism. It also has a feedback regulation on the synthesis of LDL receptors, so ensuring a homeostasis of intracellular cholesterol concentration.

The secret of the extraordinary efficiency of LDL internalization, which is complete within 10 min of binding, lies in the segregation of the bound LDL to defined regions of the plasma membrane where endocytosis is taking place. These indentations of the plasma membrane have been called 'coated pits' because of the fuzzy coat of material seen on the cytoplasmic side of the membrane in transmission electron micrographs (Figure 6.14a). Budding off from these coated pits are endocytotic vesicles coated with a similar fuzzy material. The mechanism by which the LDL receptor segregates into the coated pits has been studied in fibroblasts isolated from human patients suffering from familial hypercholesterolaemia. These patients have an elevated blood cholesterol level as a result of impaired LDL uptake. Using biochemi-

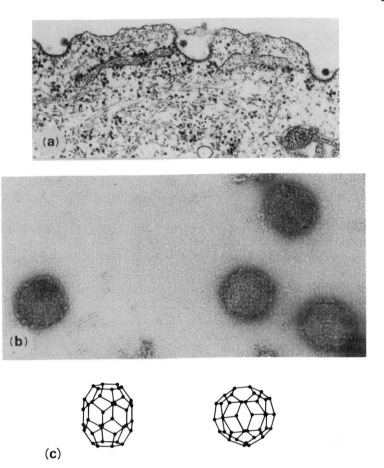

Figure 6.14   The formation and structure of coated vesicles. Indentations of the cell surface surrounded by a fuzzy coat (a) give rise to coated vesicles containing receptor-bound proteins. Shown here is the uptake of SFV particles by mammalian tissue culture cells (transmission electron micrograph × 45,000). In negatively stained electron micrographs (b) the coat appears as a basket-like lattice of protein molecules (shown here are coated vesicles from chicken oocytes × 75,000). This lattice contains a 180,000-dalton protein called clathrin. Two of a number of possible structures of the coat are shown in (c). Here 108 molecules of clathrin have been arranged in 12 pentagons and 8 hexagons to form closed basket structures. (Panel (a) was kindly supplied by Jürgen Kartenbeck and Ari Helenius, panel (b) was kindly supplied by Barbara Pearse)

cal assays for binding and internalization of LDL it was found that two different genetic defects resulted in hypercholesterolaemia. In one case no binding of the LDL to the cell surface occurred, while in the other binding occurred but the LDL was not segregated into the coated pits. The simplest interpretation of these experiments is that the LDL receptor is a transmembrane protein with a functional domain on each side of the membrane (Figure 6.15).

264

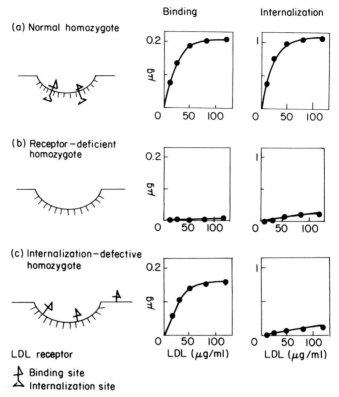

Figure 6.15 **Genetic defects of receptor-mediated endocytosis of LDL. Uptake of cholesterol in fibroblasts occurs by receptor-mediated endocytosis in coated vesicles. Binding of the cholesterol-containing LDL particle and internalization are governed by separate domains on one transmembrane receptor molecule. Fibroblasts from patients deficient in receptors (b) show both poor binding and internalization. Those from internalization-defective patients (c) can bind LDL particles, but not internalize them. The bound particles in this case are randomly dispersed on the plasma membrane rather than being aggregated into coated pits. (Reprinted with permission from Brown, M. S., and Goldstein, J. L. (1979)** *Proc. Natl. Acad. Sci., U.S.A.,* **76, 3330–3337)**

That on the extracytoplasmic side binds the LDL particle whilst that on the cytoplasmic side is responsible for signalling the LDL binding to cytoplasmic structures which cause the segregation of the protein into the coated pit. Naturally the fuzzy cytoplasmic material associated with these pits was the prime candidate for this activity.

The first observation of coated pits was in electron micrographs of the developing oocyte of the mosquito which showed a curious array of bristles over regions of the plasma membrane where micropinocytotic vesicles were forming and on the vesicles themselves. Negatively stained electron micrographs of these vesicles showed that the structure around the vesicles was a basket of protein containing lattices of regular hexagons and pentagons (Figure 6.14). The principal protein in this coat is a 180,000-dalton protein called 'clathrin' on account of its ability to associate spontaneously into lattice structures. Once the vesicles have entered the cytoplasm the clathrin dissociates and the smooth vesicles released may fuse with each other, lysosomes, or other membranes in the cell. The mechanism conferring specificity to these processes is unknown.

The interaction between a receptor protein and the clathrin coat is not understood. It is possible that it might also involve some of the more minor protein bands that are observed in SDS–PAGE of coated vesicles. There is an interesting parallel between segregation of receptors into coated vesicles and the budding of membrane-encapsulated viruses. In both cases a transmembrane protein moves from a random distribution in the bilayer to one in which it is in contact with a peripheral cytoplasmic protein. By analogy, therefore, receptors for endocytosis might contain a common domain for interaction with clathrin.

The structure of the clathrin coat has been examined by ultrastructural studies of negatively stained specimens in the electron microscope. By comparing the images of vesicles in tilted specimens with various projections of model basket structures it has been possible to identify some of the more commonly occurring structures that are found in isolated coated vesicle preparations. Two of these are shown in Figure 6.14c. It can be seen that these are composed of 12 pentagons and 8 hexagons. In fact 12 is the minimum number of pentagons required to give a closed structure, but the number of hexagons can be varied to give vesicles of a variety of sizes. Coated vesicles *in vivo* may have diameters between 60 and 250 nm. Each vertex of the basket structure is believed to contain three clathrin molecules thus giving the centre of each edge a local two-fold axis of symmetry. The structure in Figure 6.14c contains 108 molecules in all.

The ability of clathrin to form both pentagons and hexagons suggests that the excision of clathrin molecules at the cell surface might be the mechanism for pinching-off membrane vesicles. Thus, an array of hexagons would be stable on the planar face of the plasma membrane, but a process which introduced pentagons into this array would create increasing curvature until vesicle formation became a spontaneous process. Some evidence for this hypothesis has been produced by freeze–etch studies of the clathrin networks below the plasma membrane of fibroblasts. After freeze–fracture of the cells the cytoplasm is removed by extensive etching until the cytoplasmic face of the plasma membrane is revealed. On this face are found areas of clathrin coats that contain as many as 200 polygons. Although the number of pentagons in these arrays correlates with the observed curvature the absolute number of pentagons is too small. An additional complexity is the presence of heptagons in these planar areas of the clathrin network. These are found next to pentagons and have the effect of relieving the strain in the lattice caused by the pentagons. No heptagons are observed, however, in isolated coated vesicles.

Selective uptake into endocytotic vesicles is a general mechanism for the uptake of molecules too large to penetrate directly through the membrane and is used for a wide variety of cellular functions apart from nutrition. Notable in Table 6.2 is the uptake of a number of acid hydrolase enzymes by absorption to phosphomannosyl receptors in the plasma membrane which are then taken up into the lysosomes by receptor-mediated endocytosis. These enzymes are synthesized as secretory components, but are directed by their sugar-binding capacity to their ultimate lysosomal location. The low pH optima for activity and stability in the lysosomal space shows that these enzymes are genuinely lysosomal components rather than being merely taken up into the lysosomes for degradation. In fact this method for the uptake of lysosomal enzymes is only the minor route of transport for these proteins, and cells maintained in medium containing mannose-6-phosphate (which competes with the enzymes for uptake) have similar quantities of these enzymes in their lysosomes. The rest of the acid phosphatases (some 80–90 per cent of the total) are transported by a similar receptor-bound method, only directly from the Golgi apparatus to the lysosome. Once in the lysosome the acid hydrolase enzymes dissociate from their receptors in the low pH environment and have their carbohydrate binding site removed by proteolytic cleavage. This polypeptide fragment could therefore be regarded as a 'sorting sequence' since it allows the newly synthesized lysosomal enzymes to be sorted from other secretory proteins.

Receptor-mediated endocytosis is also used as a means of entry of some viruses and toxins into cells. Semliki Forest virus, for instance, binds to the HLA antigens on the cell surface of fibroblasts (which ironically are normally part of the immune defence system of the cell) and are taken up into the lysosomes. That this is a route of infection is shown by the fact that treatment with amines or ammonia which neutralize the acid environment of the lysosome also prevents infection. It is perhaps surprising that the nucleocapsid escapes the hydrolytic action of the lysosomal enzymes. Experiments *in vitro* have shown, however, that virus particles can fuse with liposomes, resulting in transfer of the viral nucleic acid into the liposome provided the pH is <6. Infection probably occurs *in vivo*, therefore, as soon as a primary lysosome fuses with an endocytotic vesicle carrying virus particles.

Coated vesicles are not only seen at the cell surface but are also observed in the region of the Golgi apparatus. This is clearly seen in Figure 6.16 in which a human fibroblast cell has been stained using a fluorescent-labelled antibody to clathrin and then viewed in the fluorescent light microscope at two focal planes. At the cell surface (Figure 6.16a) the coated pits are stained as a random array of bright dots. Deeper in the cell (Figure 6.16b) the large dark nucleus may be seen surrounded by the densely stained Golgi membranes. This clathrin at the Golgi is almost certainly concerned with another important protein transport process in the cell, that is the transport of secretory proteins from the endoplasmic reticulum to the plasma membrane. Once again it is studies with encapsulated viruses which have yielded the most

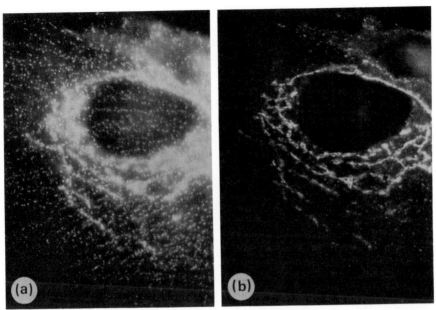

Figure 6.16 Distribution of clathrin in a human fibroblast. At least two distinct pathways of membrane transport involve clathrin-coated vesicles. This is shown here by staining a human fibroblast cell (strain WS 31) with affinity purified rabbit anti-(pig brain clathrin) and rhodamine-labelled goat anti-(rabbit IgG) antibodies. In this figure part of a single cell is shown photographed in the fluorescent light microscope. In (a) the microscope is focused on the cell surface and shows the random array of coated pits (the bright fluorescent spots). In (b) the microscope is focused in the plane of the nucleus (the large dark body) and shows the intense fluorescence associated with the surrounding Golgi apparatus. (Photograph kindly supplied by Daniel Louvard)

convincing evidence for this process. Since defined modifications to the virus glycoprotein carbohydrate residues occur at the Golgi apparatus, the transport to this organelle can be monitored by structural analysis of the sugars or by molecular weight changes. Also, in the case of SFV, the proteolytic cleavage of the p62 protein to E2 and E3 provides a biochemical assay for its arrival at the plasma membrane. Using a pulse-chase technique with radioactive amino acids it has been possible to follow the transport of immature virus particles in infected cells. Coated vesicles isolated from these cells at short times after the pulse, contain virus coat proteins similar to those synthesized in the endoplasmic reticulum. After about 20 min, however, the molecular weight changes due to a modification in the carbohydrate moiety. This suggests that the coated vesicles deliver the protein to the Golgi apparatus where it is modified and then take it up again in a second phase of transport towards the plasma membrane. Both phases involve transport in coated vesicles.

6.7.3 Cell locomotion

Many bacteria, single-celled organisms and gametes are able to swim in free solution using special organs of locomotion like cilia, flagella and tails. Some other cells are able to 'crawl' across a surface by extending regions of cytoplasm, forming new points of attachment and then hauling along the bulk of the cell. In time-lapse cinematography of mammalian fibroblasts and macrophages it is clear that the plasma membrane is very active in this process. On making contact with a solid surface these cells spread out and flatten. One edge of the cell is then extended as a delicate, sheet-like protrusion called a lammellipodium. Blebs and ruffles of membrane can rapidly form on the surface of these lammelipodia and then 'melt' back into the cytoplasm. Some blebs extend into long, thread-like filopodia which make new attachments with the substrate. As a lammellipodium extends, points of attachment at the opposite pole of the cell are released, and the nucleus and cytoplasm stream along behind the advancing margin of the cell. Sometimes long fibres attach the tail of the cell to the substrate. These fibres stretch until they are broken by tension and left behind.

The motive power behind such movements observed in the light microscope has been the subject of much research. Evidently the flexibility of the plasma membrane is essential to the locomotion of these cells. Equally clear is the fact that contractile elements of the cytoskeleton must be involved since

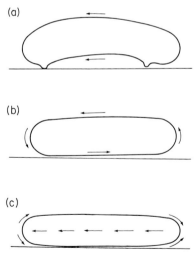

Figure 6.17 Theoretical models of amoeboid movement. During the locomotion of amoeboid cells pseudopodia are extended from the cell body in the direction of cell movement. The worm-like movement in (a) and caterpillar-track model of (b) cannot account for this movement since particles attached to the cell surface remain stationary. Most likely is the membrane circulation model of (c) in which the membrane is dismantled at the rear of the cell, taken through the cytoplasm and inserted at the growing edge of the cell

drugs which disrupt these elements have a marked effect on locomotion. It is also possible that locomotion in these cells involves the rapid disassembly and reassembly of plasma membrane.

Observations of amoeboid cells suggest three methods by which movements of the cell surface might be involved in locomotion (Figure 6.17). Of these the worm-like (Figure 6.17a) and caterpillar-track models (Figure 6.17b) are inadequate because particles adhering to the cell surface remain stationary relative to the substrate as the cell advances. This leaves the model of Figure 6.17c in which the membrane components disassembled at the rear of the cell are reinserted at the growing edge. Such a process obviously constitutes a form of membrane circulation and is interesting in terms of the dynamic properties of cell membranes. The relative rigidity of the central region of the cell is necessary to achieve a forward motion.

A somewhat similar model has been described for the locomotion of fibroblasts in tissue culture. Particles adhering to the dorsal surface of these cells migrate backwards relative to the substrate at a speed of $1-2$ μm min^{-1} and at a speed of about 3 μm min^{-1} relative to the advancing tip of the lammelipodium (Figure 6.18). Most of these particles accumulate over the nucleus. This is therefore a good candidate for the hypothetical point of internalization of the plasma membrane. Particles adhering to the ventral surface of the lamellipodium also move backwards until they become lodged between the cell and the substrate, suggesting that a net movement of membrane to the front of the cell occurs during locomotion. Although it is possible that particles could be linked by transmembrane proteins to cytoplasmic structures which pull them about the cell, the light microscopic observation of motile cells strongly suggests that the plasma membrane is dynamically involved in cell movement. Some biochemical evidence for this mechanism of locomotion has been produced by labelling surface components with ferritin-labelled concanavalin A. This label coats the cell evenly, but is rapidly aggregated to the area of membrane overlying the nucleus in a similar manner to the particles.

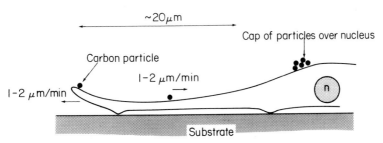

Figure 6.18 Particle movement during fibroblast locomotion. Carbon particles allowed to settle on the lammellipodia of fibroblasts in tissue culture move backwards along the dorsal surface of the cell towards a region overlying the nucleus (n). This 'cap' marks a likely site for the hypothetical internalization of membrane vesicles. Speeds are given relative to the substrate

Meanwhile a clear zone appears at the extending margin of the cell which has been interpreted as freshly inserted membrane.

Cell locomotion also involves the action of cytoskeletal elements in the cytoplasm. As the name suggests, these are filamentous structures which give the framework and contractile power for many movements within the cell and of the cell itself. Most cells contain three types of filament which are classified by their diameter. Microfilaments are the thinnest structures which appear in the electron microscope as 6 nm threads. Intermediate filaments are about 10 nm and microtubules about 25 nm in diameter. Each type of filament is formed by the reversible polymerization of small subunits into long strands, often in association with regulatory proteins. Table 6.3 summarizes some of the properties of these proteins.

Microfilaments are the main contractile element of the cytoskeleton having actin as their principal constituent. They may be observed in two forms depending on the motile activity of the cell. In quiescent fibroblasts they may be seen in bundles called 'stress fibres' that radiate from the nucleus out to the plasma membrane. During locomotion these stress fibres line up with the direction of elongation of the cell and reduce in amount in favour of a diffuse meshwork of microfilaments. In the active blebs and ruffles of the lammellipodia all cytoplasmic organelles and microfilament bundles are excluded and only the meshwork form of actin is found. These observations suggest that the stress fibres are concerned with maintaining the flattened shape of the cell whilst the actin meshwork is involved in the mechanism of membrane ruffling and locomotion.

The actin of microfilaments is associated with several other proteins. Tropomyosin, which is a regulatory protein of muscle actin, is also found associated with microfilaments but only in the bundle form. It therefore appears that in non-muscle cells this molecule stabilizes the condensation of microfilaments into the stress fibres. Myosin is also found associated with the actin of microfilaments, although in a smaller proportion than that in muscle cells. In intestinal epithelial cells the microvillae have highly ordered microfilaments in which all the actin polymers are oriented the same way and interact with myosin at the base of the structure. Here there is a clear example of a sliding filament mechanism for contraction as occurs in muscle. There is further analogy with myofibrils in the adhesion plaques where microfilaments are attached to the plasma membrane. These contain a protein which cross-reacts immunologically with α-actinin of the z-band of muscle. Bundles of microfilaments probably exert their contractile force, therefore, in a similar manner to muscle cells.

The actin found in meshworks of microfilaments is also associated with a number of ancillary proteins. The most important of these are the actin binding protein and filamin which are similar proteins serving the same function in different tissues. These proteins act as cross-linking molecules forming a gel with actin polymers. This gel may be formed *in vitro* simply by mixing actin polymers with the cross-linking protein. Furthermore, on addition of myosin

Table 6.3 Proteins of the cytoskeleton

Protein	Subunit (mol. wt)	State	Function	Structure	Size (nm)	Inhibitors
Actin	42,000	Polymer	Contractile			
Myosin (H-chain)	200,000	Dimer	Contractile			
Tropomyosin	35,000	Dimer	Stabilization of actin bundles			
Actin binding protein	220,000	Dimer	Cross-linking	Microfilaments	6	Cytochalasin B
Filamin	240,000	Dimer	Polymer in gels			
Gelsolin	91,000	Dimer	Ca^{2+} regulation of gelation			
Profilin		Monomer	Regulation of actin polymerization			
α-Actinin	102,000	Monomer	Attachment to membranes			
Vimentin	52,000	Polymer	Fixes position of nucleus			
Keratin	40–65,000	Polymer	Structural protein of epidermis	Intermediate	10	Colcemid
Desmin	55,000	Polymer	Mechanical integration of myofibril z-discs	Filaments		
Glial protein	51,000	Polymer	Structural protein of glial cells			
Unnamed	68,000	Polymer	Forms neurofilaments			
Tubulin (α and β)	55,000	Polymer	Structural, vesicle guide	Microtubules	25	Colchicine
Dynein	370,000	Dimer	ATPase			vinblastine

and Mg^{2+} ATP, the gel contracts. It is likely, therefore, that the meshwork actin serves two important functions. One is the elastic support of the cytoplasm by gel formation and the other is the transduction of the energy of ATP-hydrolysis into translational movements of the cell. At regions of blebs and ruffles rapid gel–sol changes must take place to allow the fluid extension of the cytoplasm. These are controlled in fibroblasts by a regulatory protein called 'gelsolin' which reversibly inhibits the formation of actin gels in the presence of micromolar concentrations of free calcium ions. It acts in very low concentration relative to actin monomer, causing a decrease in viscosity and sedimentation coefficient of the actin polymers. By cutting the actin polymers and increasing the ratio of polymer : cross-linker it effectively prevents gel formation. Another possible regulatory protein is 'profilin' which binds to monomeric actin and prevents its polymerization.

Intermediate filaments are structural polymers which link the various organelles of the cell. There are five known classes of these filaments found in different cell types which all have a similar diameter, but are built from different subunits. It is possible to get more than one type co-existing in one cell. The ability to visualize these and other cytoskeletal elements is largely due to the development of indirect immunofluorescent techniques in which fluorescent antibodies are used to stain the cells. Different types of intermediate filament do not cross-react immunologically but have a similar molecular weight, form similar-sized filaments and have a similar function in different cells. It has been suggested, therefore, that these molecules might contain a constant region responsible for formation of the polymer and a variable region associated with their specialized function.

Microtubules are the largest cytoskeletal filament, formed from two types of tubulin subunit arranged alternatively in a large hollow helix. They are involved in the formation of mitotic spindles, the transport of material, and giving structural support to the cell. Since colchicine, which binds to the tubulin subunit and prevents polymerization, inhibits cellular processes like secretion and endocytosis, it is likely that these processes also involve microtubule formation. In hepatocytes it has been shown that the microtubules are in a dynamic state, held in balance by an assembly–disassembly cycle. Conceivably, microtubules could have a functional role in secretion by attaching to a vesicle and then powering it about the cell by polymerization at one end, depolymerization at the other. In cilia microtubules are linked to each other by cross-bridges of an ATPase molecule called 'dynein'. During hydrolysis of ATP, adjacent microtubules move over each other in a sliding mechanism similar to the actomysin system of muscle. This sliding movement results in the beat of the cilia.

The energetic origin of cellular locomotion almost certainly lies in the interaction of these cytoskeletal elements with the cell surface. It is still necessary, however, to explain how the cell surface expands in localized areas where blebs and ruffles form, and for this membrane circulation remains a valid concept.

6.7.4 Patching and capping of surface antigens

When a multivalent reagent reacts with a plasma-membrane component of a mammalian cell it becomes redistributed into a cap over one pole of the cell. This may be readily observed in the light microscope if a fluorescent-labelled antibody is used to cap a membrane antigen. Two steps in the process may then be discerned. At low temperatures (0–4 °C), or in the presence of metabolic inhibitors, small patches are formed giving the cell a freckled appearance. In a second, energy-dependent, step these patches are transported to the pole of the cell to give a fluorescent crescent when viewed in the light microscope. The whole process takes between 2 and 10 min at 37 °C. Although patching and capping induced by adding antibody or lectin to a cell suspension is completely artefactual it does resemble some processes that occur *in vivo* and has therefore aroused considerable interest. Capping, for example, resembles the collection of particles over the nucleus in motile fibroblasts and has therefore been studied as a model of cell motility. Patching of surface receptors, on the other hand, has been shown to be an early event in receptor-mediated endocytosis, stimulation of exocytosis in mast cells and the activation of adenylate cyclase by cholera toxin. It is possible, therefore, that the aggregation of surface antigens into groups on the cell surface is linked to the transfer of information into the cell. Evidently, patching and capping display the characteristic dynamic property of biological membranes and are therefore of interest *per se*.

The principal requirement to produce a cap is the cross-linking of surface components. Monovalent Fab fragments of antibodies binding to a surface antigen thus do not evoke capping, but when cross-linked by addition of an anti(Fab) antibody normal capping occurs. Patching is energy independent and is the result of the lateral movement of the surface components in the bilayer until they combine with free binding sites on partially reacted, multivalent antibodies. Capping, on the other hand, is energy dependent. Considerable controversy still exists as to the mechanism of capping. The observation that capping is inhibited by combinations of cytochalasin B and colchicine has suggested that the movement of the patches to the pole of the cell is powered by the skeletal elements of the cytoplasm. Some confirmation of this theory is provided by the observation that actin-containing filaments in the cytoplasm line up beneath the fluorescent patches on the cell surface when cells are incubated with fluorescent-labelled antibodies. This observation has been quantified by measuring the amount of actin bound to membrane antigens using muscle myosin as an affinity label. As predicted the amount of actin associated with the membrane antigen is greater after capping than before. One of the effects of reacting surface antigens with multivalent antibodies or lectins is therefore the attachment of cytoskeletal elements to the membrane protein.

This theory suffers from a number of logical drawbacks. In the first place, only transmembrane proteins or components linked to a transmembrane protein could react with both an exogenous antibody and with skeletal elements in the cytoplasm. Therefore, capping of the ganglioside, GM1, in response to cross-linking by cholera toxin is difficult to explain. The ganglioside, situated in the outer half of the bilayer, would have to be coupled to a transmembrane protein in order to account for its capping by cytoskeletal elements. In fact this is probably what happens in this instance. The five B-subunits of the toxin bind to the ganglioside–protein complex in the membrane forming a channel for the entry of the A-subunit into the cell. This latter process probably occurs at low pH after the toxin particles have been capped and transported to the lysosomes in a similar manner to the entry of SFV particles. If all outer membrane components which may be capped are linked to transmembrane proteins in this manner a bewildering multiplicity of recognition sites for attachment of cytoskeletal elements would be necessary. Consequently, it has been suggested that one transmembrane protein might be common to many receptor molecules. Cross-linking of the receptors would cause the attachment of this transmembrane protein which in turn could be linked to the cytoskeleton and cause capping. The existence of such a protein has not yet been demonstrated. Finally, it is difficult to imagine the nature of the signal that cross-linking transmits across the membrane. If the signal is a conformational change in the transmembrane protein then it is difficult to see why monovalent Fab fragments of antibody do not produce a similar effect. On the other hand, if cross-linking produces a particular pattern in the cytoplasmic domains of transmembrane proteins, it is surprising that lateral diffusion of membrane proteins does not produce similar patterns by momentary proximity of membrane molecules and hence a spontaneous capping of all membrane proteins.

In view of these difficulties, it has been proposed that the capping of surface antigens occurs as a result of the relative rates of protein diffusion in the bilayer, and membrane flow. In this model it is proposed that the lipid component of the plasma membrane is continuously taken up at one region of the cell surface and inserted at the opposite pole so as to establish a directional flow of lipid. This assumption was based on the observation of particle movements during fibroblast locomotion (see Figure 6.18). Proteins the size of normal membrane components have a rate of diffusion in the bilayer greater than the rate of lipid flow and so can remain uniformly dispersed on the cell surface. After cross-linking, however, the raft of protein or lipid has a much smaller diffusion coefficient and so is swept through the membrane to the pole of the cell. Although this attractive hypothesis can account for the capping of cross-linked surface lipid as well as proteins, no direct evidence in support of its validity has been produced.

6.7.5 Flow-differentiation

In the electron microscope it is difficult to obtain evidence for a flow of material through the cell since each picture presents only a static image.

Time-course experiments in which tissue is fixed and processed for electron microscopy at various times after addition of radio-labelled amino acids have been able, however, to give a general route for the transport of newly synthesized proteins. Autoradiography of these tissue sections shows that radioactivity first appears in the endoplasmic reticulum and then in the Golgi apparatus and secretory vesicles. It is not clear during this process how the secretory protein moves from one organelle to the next. Since each organelle in the pathway has a distinct morphology and composition it is easiest to imagine them as static structures to which the secretory proteins are delivered by a

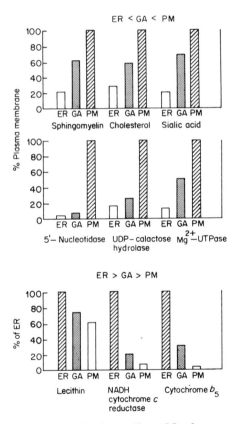

Figure 6.19   Flow-differentiation in rat liver. Membrane components synthesized in the endoplasmic reticulum move through the cell to other membranes where growth or expansion are required. In secretory cells an obvious route for this membrane flow is in the membranes surrounding secretory vesicles. Each organelle of the secretory pathway, however, has its own characteristic composition, so any flow of membrane through this pathway must be accompanied by differentiation. If this model is correct then one might anticipate a smooth change in the properties of each membrane involved in the pathway, as shown here. (Reprinted with permission from Morré, D. J. (1977) in *Cell Surface Reviews*, **Vol. 4, Poste, G., and Nicolson, G. L. eds., Elsevier, Amsterdam, pp. 1–83)**

shuttle of vesicles. The high concentration of clathrin in this region of the cell suggests that coated vesicles might be involved. Such a mechanism would preserve the identity of each organelle membrane, provided that the vesicles in the shuttle discharge their contents into the lumen but do not allow mixing of membrane components.

Alternatively the organelles of the secretory pathway could be regarded as being in a dynamic state, constantly being formed and degraded. It is well known, for instance, that the Golgi apparatus of some cells is in a dynamic state, new cisternae being continuously formed at the *cis*-face of the stack, while older ones disappear at the *trans*-face. One need only suggest that the new Golgi cisternae are formed from transitional elements of the endoplasmic reticulum (where the secretory proteins are concentrated) and that the older ones form secretory vesicles and a complete flow of membrane from endoplasmic reticulum to cell surface has been postulated. Evidently, such a flow of membrane must be coupled with a differentiation process in order to achieve the differences in composition between each organelle. Enzymes and lipids may be synthesized or inserted, while others are degraded or inhibited. One might predict, therefore, a gradual and continuous change in properties between the endoplasmic reticulum and the plasma membrane. This does indeed seem to be the case (Figure 6.19). In rapidly growing cells flow-differentiation must be very fast since large numbers of Golgi-derived vesicles are transported to the expanding cell surface. In the elongating pollen tube of the Easter lily, for instance, it has been calculated that at least 1000 secretory vesicles must be produced by the Golgi apparatus each minute in order to account for the rate of growth of the cell surface. In this case the membrane is largely synthesized *de novo* by the endoplasmic reticulum. In non-growing cells a unidirectional flow of membrane from the endoplasmic reticulum to the cell surface cannot occur. One might therefore expect to find a recycling of membrane from the cell surface back to the endoplasmic reticulum or Golgi apparatus. This would also explain the lack of radioactive label in the membranes of secretory vesicles compared with their contents when secretory cells are given radioactive amino acids.

6.7.6 Membrane recycling

We have now seen how some of the theories of cell locomotion and capping of surface antigens have embraced the concept of a dynamic cell surface membrane. This concept has been recently confirmed and extended using both morphological and biochemical techniques. The qualitative description of a cycle of membrane circulation has been derived principally from studies using the electron microscope. In the motor nerve terminals of frog sartorius muscle, for instance, a cycle of membrane circulation is used to maintain a constant cell surface area during stimulation of the nerve. During electrical stimulation, synaptic vesicles containing transmitter substances move to the presynaptic membrane and fuse so as to release their contents into the synaptic cleft. The excess plasma plasma membrane is then recovered by the pinching-off of coated vesicles. Short-term electrical stimulation (<1 min) of these nerves causes a concomitant increase in the area of plasma membrane. During long-term stimulation (>10 min) of these nerves, however, the number of synaptic vesicles decreases, but reaches a constant level despite the continuing release of transmitter. In fact the total amount of transmitter released may be many times the theoretical capacity of the synaptic vesicles seen in the nerve terminal at any one time. The obvious conclusion is that new vesicles are being continually formed. Since the cell surface membrane does not continually expand under these conditions it evidently must be recycled for the construction of new vesicles. This occurs by the pinching-off of coated vesicles which deliver cell surface membrane to an intracellular reticulum from which new synaptic vesicles are formed. Thus, when a soluble marker is included in the medium this too is taken up into the nerve terminal, and, if stimulation continues, is exported again.

A similar cycle of membrane between the plasma membrane and intracellular organelles has been observed using fluorescent-labelled antibodies to label the cell surface of epithelial cells grown in tissue culture. When grown into a confluent monolayer these cells show a strong polarity similar to that found *in vivo*, i.e. each cell surface has a characteristic morphology and contains characteristic membrane antigens. When the apical surface (that facing the medium) is labelled with fluorescent-labelled antibodies these are rapidly internalized so as to deplete the cell surface of these antigens. A small but significant fraction of these immune complexes are, however, re-expressed on the cell surface after about 40 min. They appear in the area where adjacent cells are in contact and then migrate into the apical membrane. The remaining immune complexes are degraded. Newly synthesized antigen behaves in a similar manner, being inserted in the region where adjacent cells are in contact and then migrating into the apical surface membrane. The point at which the proteins are inserted appears to be the junctional complexes (see section 1.2.1) which possibly also have a role therefore in the segregation of membrane proteins into various domains.

One of the first quantitative measurements of membrane circulation was performed in the electron microscope by counting the number of pinocytotic vesicles that contained extracellular markers from the medium. The rate of membrane uptake by these vesicles was estimated at about 3.1 per cent of the cell surface every minute in tissue culture fibroblasts. This corresponds to a complete uptake of the plasma membrane in about 30 min. Since this is considerably faster than the rate of synthesis of membrane proteins and lipids most of the internalized plasma membrane must be recycled to the cell surface. Using some elegant immunological techniques, it has been shown that cell surface antigens during their cycle through the cell are at some point exposed to the lysosomes, since by preloading the lysosomes with antibodies directed against cell surface components it is possible to trap these components in the lysosomal fraction.

In support of a circulation of cell surface membrane through the cell is the unexpected observation that a large amount of proteins, previously regarded as markers for the cell surface, are in fact found inside the cell as well. These include the enzyme, 5′-nucleotidase, the acetylcholine receptor and fucose-containing glycoproteins. As much as 50 per cent of the total 5′-nucleotidase may thus be found inside a cell, the exact distribution between the cell surface and intracellular sites depending on the cell type. The intracellular pool of 5′-nucleotidase is membrane bound but has a different buoyant density from plasma-membrane vesicles. Vesicles containing intracellular 5′-nucleotidase prepared by equilibrium density gradient centrifugation have been shown to have a similar phospholipid composition to coated vesicles, but are larger and less dense than the basket-like vesicles shown in Figure 6.14. On the other hand, these intracellular vesicles appear to be related to the plasma membrane since the 5′-nucleotidase they contain is indistinguishable from plasma membrane 5′-nucleotidase, has the same topographical location as the cell surface enzyme, and is even inhibited when whole cells are incubated with antiserum to the enzyme. These facts may be explained if it is proposed that the intracellular pool of this membrane-bound enzyme is an intermediate on the circulation of plasma membrane through the cell. Coated vesicles may act as a shuttle between these two pools, transferring the 5′-nucleotidase and some other cell surface proteins as well as molecules present in the medium surrounding the cells.

Evidence for the recycling of endocytotic vesicles has also been provided by studies on the receptor-mediated endocytosis of LDL particles by fibroblasts. LDL-particle uptake may continue for up to 6 h after synthesis of new receptor is blocked by cyclohexamide. If recycling did not occur, all the receptors for the LDL particles would be taken up in the first 10 min after exposure to LDL. In this case no measurable pool of receptors may be measured inside the cells, suggesting that the receptor remains in the coated vesicle membrane and returns immediately to the cell surface rather than exchanging with a pool of receptors in an intracellular membrane.

A large body of evidence from a variety of studies thus supports the notion that biological membranes are in a highly dynamic state. Some of the pathways that can occur are shown in Figure 6.20. The most frequently described movement is between the cell surface and lysosomes, endoplasmic reticulum, or Golgi apparatus (known as the endomembranes of the cell). It seems likely that in many cases the principal means of membrane uptake is via coated vesicles, whether this occurs at the cell surface during receptor-mediated endocytosis, or at the endoplasmic reticulum during the transport of newly synthesized secretory protein. The role of clathrin seems to be in the formation of a vesicle from a planar membrane surface, since after this the clathrin dissociates. During this process segregation of membrane proteins occurs, some proteins (e.g. the receptor proteins) being included in the vesicle membrane whilst others are excluded. This is the most likely mechanism for preserving the integrity of each organelle. Specificity of fusion must also occur

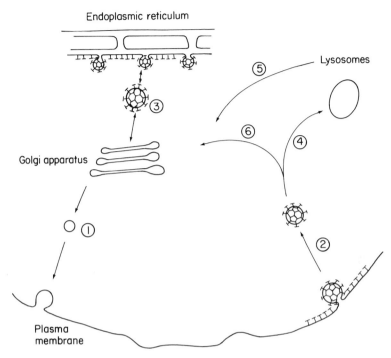

Figure 6.20 Pathways of membrane circulation. Two well-characterized pathways of vesicle movements are the export of secretory vesicles during exocytosis (1), and the endocytosis of plasma membrane via coated vesicles (2). Coated vesicles are probably also involved in a shuttle between the endoplasmic reticulum and the Golgi apparatus (3). Some vesicles taken up by receptor-mediated endocytosis are directed to the lysosomes (4) and the membrane may either be degraded or recycled (5). Other vesicles may travel direct to the Golgi apparatus or endoplasmic reticulum (6)

since proteins taken up by receptor-mediated endocytosis have different fates within the cell. This scheme of membrane movement is obviously of prime importance in the export and import of proteins. For this reason clathrin is a ubiquitous component of cells. It also appears that other functions are possible, e.g. the action of hormones at intracellular sites and the repair of the cell surface membrane. Cell locomotion and capping of surface antigens may also involve these pathways of vesicle movement. Evidently the structure and function of clathrin and other molecules involved in the uptake and sorting of intracellular vesicles will be of great importance to the understanding of these processes.

Summary

Large areas of membranes move around the cell in the form of vesicles.
Energy for these movements is supplied by a close interaction with the cytoskeleton.
This circulation plays a vital role in cell function.

FURTHER READING

Bedouelle, H., Bassford, P. J., Fowler, A. V., Zabin, I., Beckwith, J., and Hofnung, M. (1980) Mutations which alter the function of the signal sequence of the maltose binding protein of *Escherichia coli*, *Nature*, **285**, 78–81.

Blobel, G., and Dobberstein, B. (1975) Transfer of proteins across membranes, *J. Cell Biol.*, **67**, 835–851.

Bretscher, M. S. (1976) Directed lipid flow in cell membranes, *Nature*, **260**, 21–23.

Goldstein, J. L., Anderson, R. G. W., and Brown, M. S. (1979) Coated pits, coated vesicles, and receptor-mediated endocytosis, *Nature*, **279**, 679–685.

Harris, A. K. (1973) Cell surface movements related to cell locomotion, *Ciba Foundation Symp.*, **14**, 3–26,

Heuser, J. (1980) Three-dimensional visualisation of coated vesicle formation in fibroblasts, *J. Cell Biol.*, **84**, 560–583.

Higgins, J. A. (1981) Biogenesis of endoplasmic reticulum phosphatidylcholine. Translocation of intermediates across the membrane bilayer during methylation of phosphatidylethanolamine, *Biochem. Biophys. Acta*, **640**, 1–15.

Katz, F. N., Rothman, J. E., Lingappa, V. R., Blobel, G., and Lodish, H. F. (1977) Membrane assembly *in vitro*: synthesis, glycosylation, and asymmetric insertion of a transmembrane protein, *Proc. Natl. Acad. Sci.*, **74**, 3278–3282.

Lazarides, E., and Revel, J. P. (1979) The molecular basis of cell movement, *Scientific American*, **240** (May), 88–100.

Loor, F., Forni, L., and Pernis, B. (1972) The dynamic state of the lymphocyte membrane. Factors affecting the distribution and turnover of surface immunoglobulins, *Eur. J. Immunol.*, **2**, 203–212.

Michell, R. H. (1979) Inositol phospholids in membrane function, *Trends Biochem. Sci.*, **4**, 128–131.

Morré, D. J., Kartenbeck, J., and Franke, W. W. (1979) Membrane flow and interconversions among endomembranes, *Biochim. Biophys. Acta*, **559**, 71–152.

Parodi, A. J., and Leloir, L. F. (1979) The role of lipid intermediates in the glycosylation of proteins in the eucaryotic cell, *Biochim. Biophys. Acta*, **559**, 1–37.

Rothman, J. E. and Kennedy, E. P. (1977) Rapid transmembrane movement of newly synthesized phospholipids during membrane assembly, *Proc. Natl. Acad. Sci.*, **74**, 1821–1825.

Schneider, Y-J, Tulkens, P. and Trouet, A. (1977) Recycling of fibroblast plasma membrane antigens internalised during endocytosis. *Biochem. Soc. Trans.*, **5**, 1164–1167.

Stanley, K. K., Edwards, M. R. and Luzio, J. P. (1980) Subcellular distribution and movement of 5′-nucleotidase in rat cells, *Biochem. J.*, **186**, 59–69.

Steinman, R. M., Brodie, S. E., and Cohn, Z. A. (1976) Membrane flow during pinocytosis, *J. Cell Biol.*, **68**, 665–687.

Taylor, R. B., Duffus, W. P. H., Raff, M. C., and de Petris, S. (1971) Redistribution and pinocytosis of Lymphocyte surface immunoglobulin molecules induced by anti-immunoglobulin antibody, *Nature New Biol.*, **233**, 225–229.

Tweto, J., and Doyle, D. (1976) Turnover of the plasma membrane proteins of hepatoma tissue culture cells, *J. Biol. Chem.*, **251**, 872–882.

Waksman, A., Hubert, P., Crémel, G., Rendon, A., and Burgun, C. (1980) Translocation of proteins through biological membranes, *Biochim. Biophys. Acta*, **604**, 249–296.

Walker, J. H., Burgess, R. J., and Mayer, R. J. (1978) Relative rates of turnover of subunits of mitochondrial proteins, *Biochem. J.*, **176**, 927–932.

Wickner, W. (1979) The assembly of proteins into biological membranes: the membrane trigger hypothesis, *Ann. Rev. Biochem.*, **48**, 23–45.

Chapter 7

Dynamic membrane–associated processes

7.1 INTRODUCTION

We have developed the theme in preceding chapters that biological membranes are not simply inert boundaries which limit a cell, but are dynamic structures with a highly organized, asymmetric design. In this chapter we shall try and illustrate some of the dynamic functions of membranes, i.e. where the movement of molecules in or across the membrane is of physiological importance. Much of what follows is therefore concerned with membrane transport. This is tackled in an order of increasing complexity commencing with simple diffusion which is largely a property of the lipid bilayer, and progressing to the large, multifunctional proteins which can catalyse the transport of materials across the membrane against huge concentration gradients. This order might well describe the evolutionary development of such proteins. Wherever possible we have sketched the background of the physiological processes in which these transport systems are involved. We hope that this will give the reader an appreciation of the fundamental role played by membrane transport in a variety of sophisticated processes including muscle contraction, vision, intracellular metabolism and energy production.

We have deliberately avoided reproducing pages of equations which describe the kinetic models of transport systems. The reason for this is twofold. Firstly, they have been adequately described in other textbooks, but more importantly, they do not in fact define a unique physical model of a transport protein. We have been more concerned with the concept of what transport proteins actually look like and how this structure is related to their remarkable dynamic function.

A more recent, but equally exciting, example of dynamic membrane function is the association and disassociation of membrane proteins during their lateral movements through the bilayer. These are now known to be involved in the action of some hormones at the surface of cells. In this case an effect similar to transport is achieved, namely the transmission of a signal across the otherwise impermeable membrane.

7.2 PASSIVE DIFFUSION

Random Brownian movements of molecules in solution cause a solute, over a period of time, to disperse from areas of high concentration until the solution is homogeneous. This process of diffusion can also occur between compartments separated by a biological membrane. In this case the close packing and hydrophobic nature of the hydrocarbon chains in the centre of the membrane adds a constraint on the types of molecule that can diffuse into the cell at any meaningful rate. Charged molecules and very large molecules are almost entirely excluded unless a specific transport process for them exists, even small molecules like water and methanol diffuse across biological membranes at 10^{-2}–10^{-3} times the rate that is observed across an aqueous barrier. Diffusion is therefore not a rapid process, the cell membrane presenting an obstacle to diffusion rather like a sheet of synthetic rubber of the same thickness.

Passive diffusion across a biological membrane obeys Fick's first law in which the flux per unit area of a molecule through the membrane (J) depends only on the diffusion coefficient of the molecule (D) and its concentration gradient across the membrane dc/dx:

$$J = D \frac{dc}{dx}$$

No term is included for the binding of the substrate to the membrane since discrete binding sites are not involved. Consequently, passive diffusion does not saturate at high substrate concentration.

Two theories describe the mechanism by which molecules can diffuse through biological membranes. For many non-electrolytes the lattice model adequately accounts for the rate of transport. In this model diffusion occurs by the movement of the permeant molecule into a 'vacancy' in the lattice structure of the membrane (Figure 7.1a). The principal energetic barrier to diffusion in this case is the activation energy required to move the substrate from the aqueous phase into the apolar environment of the phospholipid bilayer. Thus the permeability of the membrane of *Chara ceratophylla* correlates with the hydrophobicity of the permeant molecule (Figure 7.2). The best correlation of permeability is in fact observed if it is expressed as a function of the number of hydrogen bonds that must be broken as molecules pass from aqueous solvent into the bilayer.

This still leaves the question of how the postulated 'vacancies' in the bilayer

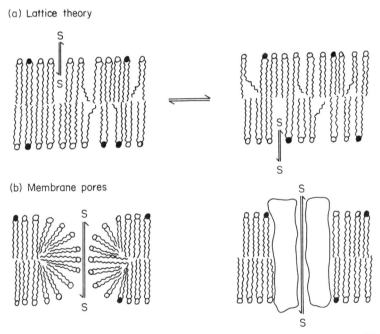

(a) Lattice theory

(b) Membrane pores

Figure 7.1 Mechanisms of membrane diffusion. Biological membranes at 37 °C are in a fluid state allowing small molecules to diffuse into, and out of, 'vacancies' in the outer leaflet of the bilayer (a). As they do so they move from an aqueous environment into the apolar interior of the membrane. If a 'vacancy' in the lattice of phospholipids in the inner leaflet becomes available the substrate can diffuse right across the membrane. Membrane pores (b) could be formed by a micellar rearrangement of the phospholipid bilayer or by proteins in the membrane. In both cases solutes pass into the cell without entering the hydrophobic core of the bilayer

lattice arise. Most simply these are envisaged as temporary discontinuities created by the lateral movement of phospholipid molecules. As expected, therefore, progressive removal of cholesterol from mammalian membranes, so as to increase the fluidity of the hydrocarbon chains of the phospholipids, causes a corresponding increase in the permeability. Anomalously high rates of diffusion are also observed at the phase transition temperature in synthetic phospholipid vesicles. This may be attributed to diffusion at the boundaries between crystalline and liquid-crystalline domains where the lattice structure of the bilayer is disturbed (see section 2.7).

This simple lattice theory cannot account for the diffusion of all molecules through biological membranes. It may be seen in Figure 7.2, for example, that small molecules like water and methanol have a greater permeability than would be predicted on the basis of their partition coefficient or degree of hydrogen bonding. Furthermore, dissolved gases such as oxygen and carbon dioxide diffuse through biological membranes at a similar rate despite their large difference in size. These observations suggest that aqueous pores must exist in the membrane for the transport of these materials. In the human erythrocyte the diameter of these pores calculated from the rate of water movement across the membrane is approximately 0.7 nm. Since this flow of water is inhibited by the sulphydryl reagent DTNB (5,5'-Dithiobis-(2-nitrobenzoate), it appears that the pore is bounded by protein although no membrane protein with this function has as yet been isolated and reconstituted. It is also possible that these substances pass in a non-specific way through the pores of transport proteins for other larger molecules.

When the pore size approaches a diameter of about 1 nm, diffusion becomes of relatively little importance. Instead molecules move across the membrane according to the hydrostatic or osmotic pressure difference by a process called 'bulk' or 'mass' transport. Diffusion through non-specific, water-filled pores in the membrane cannot, however, explain the diffusion of non-electrolytes in general since no single estimate of pore size is in agreement with the diffusion rate of all substances.

The rate of diffusion of ions across membranes is also anomalous, with no solution in terms of a simple general model. As expected, the permeability of ions in synthetic phospholipid vesicles is very small—a consequence of the large activation energy required to solvate an ion in the hydrophibic interior of the bilayer. Small ions like K^+ can penetrate phosphatidyl choline bilayers, but the rate of diffusion is very dependent on the surface charge of the bilayer and on the membrane potential. Changing the surface charge by introducing as little as 5 mol per cent of a long-chain cationic detergent, for instance, can completely abolish potassium loss from the vesicles. A residual rate of ion diffusion may also be measured in biological membranes after poisoning to prevent active transport of ions. Different membranes display different ion selectivities. In the squid axon, for example, the passive flux of K^+ ions is about 10^{-9} mol cm^{-2} s^{-1} which is comparable with the rate of water transport, whilst the diffusion of K^+ ions across the erythrocyte membrane is some five

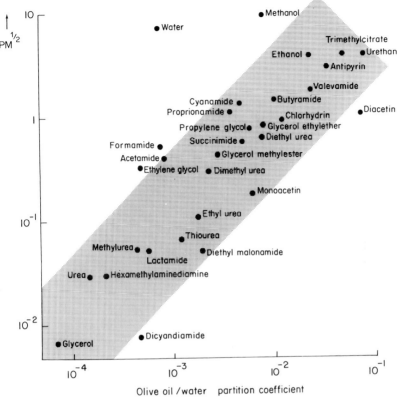

Figure 7.2 **The correlation of diffusion with hydrophobicity. On the lattice model the rate of diffusion is predicted to be inversely proportional to the square root of the molecular weight for molecules smaller than the lattice dimensions. Thus, $PM^{1/2}$ is a constant. Data taken from the alga *C. ceratophylla* show however, that the value of this constant changes with the hydrophobicity of a molecule as measured by its oil : water partition coefficient. This suggests that a second factor determining the rate of diffusion is the lipid solubility of a molecule. It may also be seen that the molecules on the lower limit (like diacetin, monoacetin and malonamide) are all large molecules, whilst those with higher permeabilities than the average (like water, methanol, formamide and ethylene glycol) are all small molecules. A third factor influencing the rate of diffusion may therefore be the presence of pores of small diameter which span the bilayer.**
(Reprinted from Collander, R. (1949) *Physiol. Plantarum*, 2, 300–311)

orders of magnitude smaller. The erythrocyte membrane, on the other hand, is capable of rapid chloride ion diffusion. A different phospholipid composition of these membranes, such that the erythrocyte contained a greater surface positive charge, could explain the high anion selectivity of this cell.

Biological membranes are in general more permeable than synthetic phospholipid vesicles, showing that some of the ion-diffusion properties of a membrane are a feature of the proteins in the membrane rather than of the phospholipids. Since the differences noted between squid axon membranes and those of the erythrocyte membrane reflect their different biological activities, it seems likely that some of the anomalous ion fluxes could also be the result of residual ionophoric activity of the ion-transport proteins in the membrane.

Summary

Diffusion occurs by several different mechanisms, neither the lattice nor the pore theory can account for the diffusion of all molecules through biological membranes.

Biological membranes represent a considerable barrier against the diffusion of all types of molecule, the fastest are retarded by a factor of 100 to 1000.

Permeability is influenced by molecular size and interactions at the surface of the membrane, i.e. hydrogen bonds that must be broken or repulsive forces between charges on the substrate and the phospholipid headgroups.

7.3 IONOPHORES

An ionophore may be defined as a substance that enhances the movement or incorporation of an ion from the aqueous phase into the hydrophobic phase. This definition encompasses mobile, ion-carrying molecules (ionophore literally means 'ion bearer'); molecules that form ion channels through membranes; and large polypeptides covalently linked to eukaryotic transport proteins that aid the movement of specific ions through the membrane.

Small molecular weight ionophores (200–2000 daltons) have been isolated from a number of bacteria and fungi and have been termed 'antibiotics' although they frequently are more active in disrupting membrane transport in higher organisms. Two classes of compound may be distinguished. The antibiotic valinomycin is typical of ionophores that are mobile within the hydrophobic core of the bilayer. It is a cyclic dodecadepsipeptide (Figure 7.3) in which the bracelet-shaped backbone of the molecule allows an accurate focus of the peptide oxygen atoms so that they can interact with a cation in the centre of the ring. The size of this ion is very critical, small cations only being able to interact with one oxygen at a time, whilst large ions require too much distortion of the polypeptide to fit. Consequently, valinomycin shows a $10,000 : 1$ preference for K^+ ions (diam. 0.27 nm) over Na^+ ions (diam. 0.19 nm). Ions in aqueous solution are solvated by water molecules, and the exchange of solvent for the ionophore oxygen system must be a concerted reaction in order to achieve a low energy of activation for transport. At the polar interface of the membrane the carboxyl oxygens of valinomycin are relatively mobile, swinging out of the plane of the molecule and facilitating the entry of a K^+ ion into the ion site. As the complex diffuses into the bilayer, however, all the oxygen atoms move into the centre of the molecule, thus aligning the non-polar side chains with the hydrocarbon tails of the phospholipids and fully complexing the bound ion.

An example of a carboxylic acid ionophore is A23187. In this molecule a closed structure is formed by head-to-tail hydrogen bonding (Figure 7.3), a

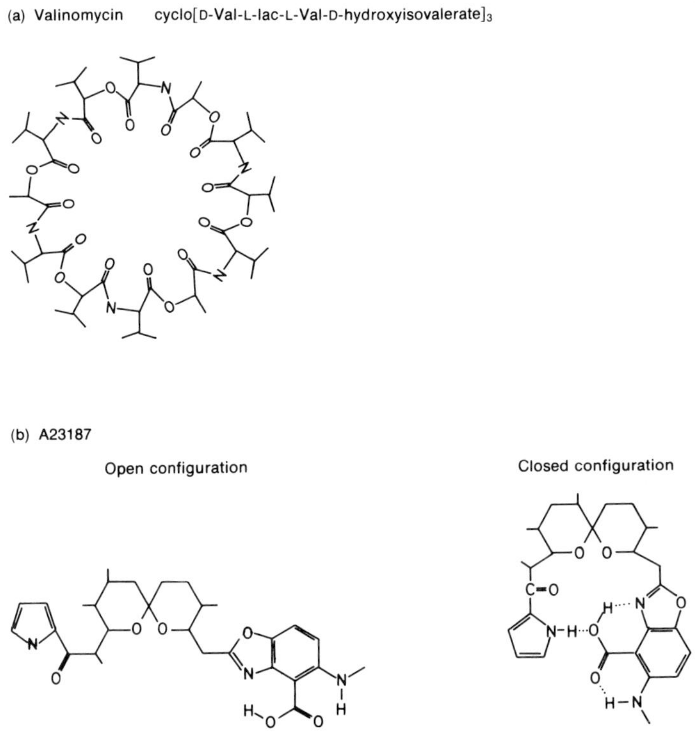

(a) Valinomycin   cyclo[D-Val-L-lac-L-Val-D-hydroxyisovalerate]₃

(b) A23187

Open configuration   Closed configuration

Figure 7.3   The structure of some mobile ionophores. Ionophores that transport ions by a shuttle across the membrane are usually ring structures such as valinomycin (a) or carboxylic acids which can form ring structures by head-to-tail hydrogen bonding as in A23187 (b). The ion (K$^+$ in the case of valinomycin and Ca^{2+} in the case of A23187) is chelated in the centre of the ring; interactions with the solvent having been replaced by interactions with electronegative atoms in the structure. The large number of apolar residues on the outside of the ring then allow the complex to diffuse across the membrane

fairly common structural feature of ionophores. The selectivity of A23187 is unique for divalent cations over monovalent cations (Table 7.1) and has therefore been extensively used to study the effects of Ca^{2+} ions on metabolism without affecting intracellular levels of K$^+$ and Na$^+$ ions. Two A23187 molecules are required to bind one Ca^{2+}ion, although it is not known which oxygen or nitrogen atoms are involved. Unlike valinomycin, A23187 catalyses an electrically neutral exchange of ions across the membrane suggesting that either a cation moves in the opposite direction across the membrane or that anions are taken across with the Ca^{2+}. The simplest model is for the ionophore to return across the membrane in the protonated form, although this does not agree with all the available data.

Nigericin is a carboxylic acid ionophore with a high degree of specificity for K$^+$ or H$^+$ ions. Unlike valinomycin, however, it cannot move across the mem-

Table 7.1 Properties of some ionophores

Ionophore	Molecular weight	Mechanism	Ion selectivity
Valinomycin	1110	Mobile carrier	Monovalent cations $(K^+ \gg Na^+)$
Nigericin	724	Mobile carrier	Monovalent cations (K^+, H^+)
A23187	523	Mobile carrier	Divalent cations (Ca^{2+})
Gramicidin A	1882	Channel formed by transverse association	Monovalent cations
Amphotericin B	924	Channel formed by lateral association	Anions > cations

brane without a bound cation. Nigericin therefore has a complementary action to valinomycin, dissipating the K^+ gradient across a membrane but maintaining the membrane potential by the exchange with H^+. These two ionophores have, therefore, been of particular importance in the study of ion gradients and membrane potential as an energy store in the mitochondrial membrane.

Each of these inonophores catalyses ion transport by a shuttle to and fro across the membrane. This is a relatively slow process and is limited by the rate of diffusion of the ionophore in the membrane.

The second class of ionophore forms hydrophilic pores across the membrane. The polyenes, amphotericin and nystatin, for example, consist of conjugated lactone rings which resemble phospholipids in their size and amphipathic nature (Figure 7.4a). One side of the collapsed ring structure is hydrophobic and interacts with cholesterol in a similar manner to phospholipid. The other side of the molecule contains many hydroxyl groups which can form a hydrophilic channel when associated in an oligomeric complex. Space-filling models suggest that about 10 molecules can be assembled with cholesterol in each half of the bilayer to give a pore of approximately 1.0 nm diameter (Figure 7.4a). This is sufficient to allow the passage of fully hydrated ions and consequently a broad specificity for anions is observed. The rate of diffusion is inversely related to the hydrated radius of the ion.

Another group of channel-forming antibiotic peptides are the gramicidins. Gramicidin A (Figure 7.4b) contains 15 alternate D- and L-amino acids arranged in a left-handed helix having 6.3 residues per turn, a length of 2.5–3.0 nm and a pore diameter of 0.4 nm. These peptides spontaneously associate as dimers in the membrane with the hydrophobic amino-acid residues facing the lipid of the membrane and a continuous file of water molecules in the centre of the helix. Gramicidin A has no charged amino-acid residues, but has the peptide carbonyl oxygen atoms lining the centre of the helix rather like the binding site of valinomycin. The pore size is also smaller than that of the multimeric polyene complexes and therefore shows more

(a) Amphotericin B
Structure

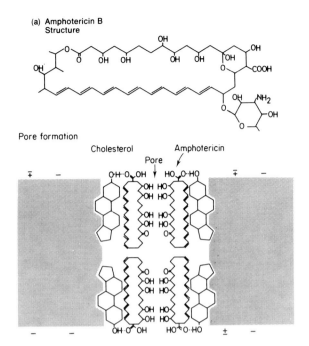

Pore formation

(b) Gramicidin A
Structure

HCO-L-Val-Gly-L-Ala-D-Leu-L-Ala-D-Val-L-Val-D-Val-[L-Trp-D-Leu]₃-Trp-NHCH₂CH₂OH

Pore formation

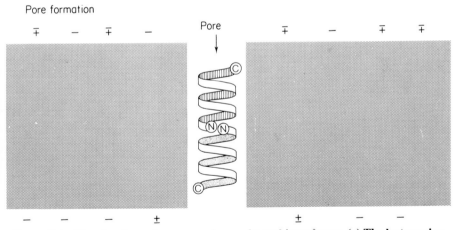

Figure 7.4 **The structure of some membrane channel ionophores. (a) The lactone ring of amphotericin B has one side composed of a hydrophobic polyene and the other containing many hydroxyl groups. These amphipathic molecules spontaneously associate in an apolar environment to form pores which span half of the bilayer. Cholesterol is also required for stable pore formation and is believed to interact with the hydrophobic face of the molecule as shown. (b) Gramicidin A is a peptide containing 15 amino acids. In the apolar environment of the phospholipid hydrocarbon chains it assumes a helical configuration which spans half of the bilayer. When these helices dimerize in a head-to-head fashion a transient channel is formed allowing a pulse of ions to enter the cell**

specificity (Table 7.1). Dimer formation is a reversible process and ion transport via gramicidin pores may be observed as short-lived pulses of conduction across the membrane. Nevertheless, the diffusion of ions through membrane channels is a more rapid process than transport via mobile ionophores (up to four orders of magnitude faster) since the diffusion of a large molecule across the lipid bilayer does not have to occur.

Although antibiotic ionophores have been widely used in biochemical research in the investigation of the effect of ions on cell metabolism, there is no evidence that they serve any transport function *in vivo* or that ionophores associated with transport enzymes behave in a similar manner. Indeed, as will be seen below, the ionophores of Ca^{2+}-dependent ATPase and Na^+/K^+ ATPase are covalently attached on one side of the membrane and may only be ascribed ionophoric function after proteolytic cleavage from the rest of the protein.

Summary

Ionophores carry ions across membranes by either a mobile carrier mechanism or by forming a fixed channel through the membrane.

Ion specificity depends on the geometry, rigidity and polarity of the ion binding site.

7.4 MEMBRANE CHANNELS AND PORES

Although no example of a shuttle of protein carrier molecules across a membrane has yet been satisfactorily established, there are examples of non-specific protein pores and channels through which diffusion or bulk transport of molecules can occur. These pores are usually formed by the assembly of aggregates of membrane proteins rather than by a single molecule with a hydrophilic channel down the centre. In *Escherichia coli* outer membrane lipoprotein, perhaps the most primitive of these pores for which structural details are known, there is no mechanism for regulating the flux of molecules through the membrane. The pores in mammalian membranes, however, are more sophisticated, having elaborate mechanisms for opening and closing.

7.4.1 E. Coli outer membrane lipoprotein

Escherichia coli, like other Gram-negative bacteria, has two distinct membranes limiting the cell. Both membranes are composed of phospholipids and proteins and have the normal trilaminar appearance by electron microscopy. Separating the two membranes is a peptidoglycan network which gives structural support to the organism. The inner membrane contains a number of specific transport systems such as the lactose permease and the dicarboxylic acid transport protein. In contrast no specific transport function has been shown for the outer membrane and it is frequently assumed that it simply acts as a barrier against penetration by antibodies and other macromolecules. Small molecules, however, require access to the transport proteins of the inner membrane so it is not surprising to find an abundant protein in the outer membrane which forms channels through which bulk transport may occur. This protein is the lipoprotein which is an intrinsic membrane protein spanning the outer membrane. Since there are about 7.5×10^5 copies of this molecule per cell it must be the most abundant protein in *E. coli*. About one-third of the total is covalently bound to the peptidoglycan network while the rest is free to diffuse in the bilayer. This protein, which contains 58 residues, has been sequenced and shown to consist almost entirely of an α-helix spanning the membrane.

The sequence of amino acids is very striking, with hydrophobic and hydrophilic amino acids arranged in a seven-residue repeat. When the polypeptide chain is folded into a helix, these hydrophobic residues fall into two adjacent straight lines on one side of the structure, giving what is known as an amphipathic helix, i.e. a helix with one face hydrophobic in nature and the other hydrophilic. Such structures spontaneously aggregate in the bilayer to form micelles with all the hydrophobic faces interacting with the membrane lipids and all the hydrophilic faces creating an aqueous pore through the centre of the complex (Figure 7.5). The hydrophobic face of each lipoprotein does not run exactly parallel to the axis of the helix but turns $166°$ round the α-helix in the overall length (7.6 nm) of the molecule. This necessitates a

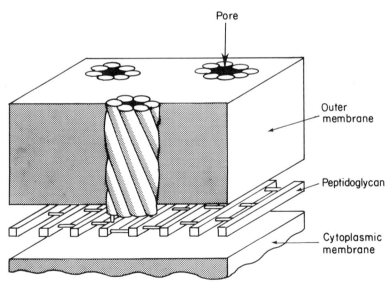

Figure 7.5 An example of a proteinaceous membrane pore is provided by the lipoprotein of the outer membrane of *E. Coli*. In this case the pore is composed of a multimeric aggregate similar in structure to amphotericin B, but which spans the whole distance of the bilayer. The pore is quite large, allowing bulk transport of all small molecules to the specific transport sites of the inner membrane, and has no known regulatory properties. (**Reprinted by permission from Inouye, M. (1974)** *Proc. Natl. Acad. Sci., U.S.A.*, **71, 2396–2400**)

supercoiled structure where the helices lie at 25° to the axis in order to achieve a pore structure with no strain on the peptide bonds of each subunit. In the most stable hexameric lipoprotein complex seven ionic interactions are also formed between adjacent molecules.

As more lipoprotein molecules are added to the complex the tilt of the subunits increases and the pore becomes larger, but less interactions are formed. On this model the pore diameter of 1.3 nm for a hexameric aggregate is consistent with the permeability of the outer membrane. As new lipoprotein molecules are synthesized and inserted into the membrane it is possible that higher aggregates are made until it is energetically more favourable to split into two smaller channels.

7.4.2 Gap junctions

An interesting example of a membrane channel in mammalian cells is the pore (connexon) formed by the protein connexin in gap junctions. Where adjacent cell membranes are apposed a variety of junctional structures may be formed. Gap junctions were first observed in electron micrographs as small areas where the normal separation of about 25 nm between membranes had narrowed to a gap of 2–3 nm. The gap can be penetrated by electron-dense heavy-metal stains such as lanthanum hydroxide and uranyl acetate and is then seen in the electron microscope as a lattice of cylindrical particles with a stain-filled core (Figure 7.6). Each particle, or connexon, is aligned with a similar unit in the adjacent cell membrane to form a channel which connects the two cells. In electrically excitable tissues these channels transmit electrical signals in the form of ion currents and achieve a much more rapid response than the release of a chemical transmitter across a synapse. They are import-ant, therefore, in giving precise synchronization of response, for example in the contraction of heart muscle cells, or the firing of excitable cells in the electric organ of some fishes. The pores also allows the diffusion of sugars, amino acids, nucleotides and other metabolites ($M_r \leqslant 1000$) and are found between adjacent cells in many tissues.

Although the channels are non-specific they are capable of opening and closing—a response perhaps to prevent loss of cytoplasmic contents on dam-age of the adjacent cell. This response is provoked by a rise of intracellular Ca^{2+} ion concentration. At about 5×10^{-5}M Ca^{2+} the pores close and the cells become uncoupled. During this process a change in the regularity and density of packing of the connexon units has also been observed. Attention has been drawn to the function of gap junctions by the fact that some malignant trans-formed cells do not form proper connections with other cells suggesting that the intracellular communication through gap junctions might be used to transmit substances that control growth. Another interesting observation is that at Ca^{2+} ion concentrations below that required for complete uncoupling, partial changes in the permeability and spacing of the connexon units occur. In this condition the gap junction acts as a filter and only allows the movement of small molecules between cells. An apparently physiological condition of partial uncoupling has been described in several types of embryonic cell. This raises the exciting possibility that a selective modulation of intracellular permeability could also play a role in cell differentiation.

Gap junctions are stable structures which may be isolated from other membranes in a cell homogenate by centrifugation techniques which exploit their characteristic density. When this junction-rich fraction is treated with detergents the morphology of the gap junctions remain unchanged, but other contaminating membranes are removed. The connexons in these detergent-treated gap junctions are organized in a crystalline array and may be examined by negative staining and low-dose electron diffraction in the electron microscope. From these diffraction patterns an averaged structure with 8.5 nm resolution has been constructed (Figure 7.6b). This structure contains a channel in the centre surrounded by six identical subunits. Although the continuity of the pore is not resolved by this technique it seems likely from permeability studies that it is a fixed continuous channel through the membrane. The six subunits are tilted at approximately $12°$ to the vertical. After dialysis in distilled water these proteins assume a second configuration which is shown by electron diffraction to contain almost vertical subunits and no pore through the centre. This change in structure is reversible on addition of detergent showing that no covalent modification of the protein has occurred. Although this change in pore size is brought about by a change in detergent, rather than Ca^{2+} ion concentration, it is tempting to speculate that a similar conformational change between connexin subunits might bring about the opening and closing of the pore *in vivo* (Figure 7.6c). Interestingly, the pore diameter of 1.2 nm predicted from this structural analysis agrees well with the permeability limit of molecules through the gap junction.

7.4.3 Ion channels

Historically, some of the earliest investigated membrane channels were the Na^+ and K^+ channels found in the plasma membrane of excitable tissues like the nerve axon, although the evidence for their existence still rests on measurements of ion conductance rather than on the isolation and reconstitution of their activity. All living cells maintain gradients of Na^+ and K^+ ions across their plasma membrane. These ions are pumped across the membrane in most cases by the Na^+/K^+-dependent ATPase enzyme, which is an active transport protein using the energy of ATP to establish a concentration gradient of ions. Sodium ions are pumped out of the cell while potassium ions are transported in. These ions also leak back across the membrane, and since K^+ ions diffuse at a faster rate than Na^+ ions, an electrical as well as a chemical gradient is established. In a resting nerve cell the electrical potential across the plasma membrane is between 60 and 90 mV with the outside of the cell positive. It is the rapid, self-propagating depolarization of this membrane potential which causes the electrical impulse or action potential to travel down a nerve axon.

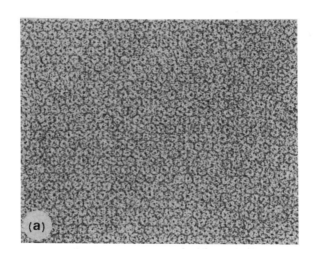

(a)

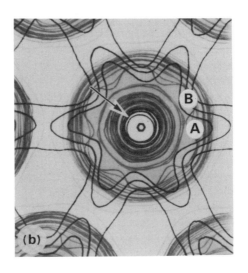

(b)

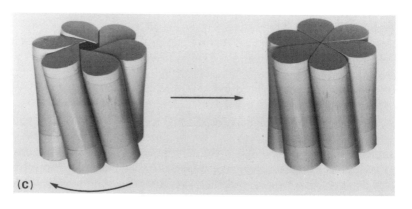

(c)

This depolarization and subsequent recovery of the membrane potential is achieved by Na^+ and K^+ channels in the membrane. These ion-channels have the special property of opening and closing in response to changes in the membrane potential. Thus, if an electrical impulse is applied to the membrane such that a certain threshold is achieved, then the Na^+ ion channels in that vicinity are opened and Na^+ ions rush into the cell until the local membrane potential is abolished or even slightly reversed. This action is sufficient to trigger the opening of adjacent Na^+ channels so that a self-propagating signal is produced (Figure 7.7). The Na^+ ion permeability is then suddenly shut off and K^+ ions moving in the opposite direction restore the resting membrane potential. While the Na^+ channel has closed but the K^+ channel is still open, the membrane potential achieves a higher value than the resting potential, thus giving a refractory period before the membrane can 'fire' again, and giving directionality to the impulse. All this takes place in a matter of milliseconds at the expense of energy stored in the Na^+ and K^+ gradients. Since the changes in Na^+ and K^+ permeability are very localized and short-lived the net change in the ion gradients are very small for each impulse and are quickly restored by the Na^+/K^+ ATPase. Even when the latter has been inhibited by depleting cytoplasmic ATP with metabolic poisons, several hundred thousand impulses may be generated before the ion gradients have collapsed.

It is clear from this description that the action potential is not caused by the conduction of an electric current along the nerve axon, but in fact relies on the high resistance and capacitance of the membrane to achieve a self-propagating local disturbance. The size of this disturbance, the steepness of the leading edge of the action potential, and the rate of transmission of the signal along the membrane all depend on the rapid response of the Na^+ channels to changes in membrane potential. Recent experiments have therefore attempted to identify the components which are activated by the electric field across the membrane. Three voltage-dependent first-order reactions have been postulated. The fastest of these is thought to be the activation of

---

Figure 7.6 **The structure of isolated gap junctions. Gap junctions stained with uranyl acetate (a) appear in the electron microscope in plan view as a collection of annular units, the connexons, which are organized on a hexagonal lattice. Each connexon has a central dark region which is interpreted as stain within the connexon pore. More detail of the connexon structure may be obtained by studying the electron diffraction pattern of these crystalline arrays of connexons, since they represent an average structure. The density contour map (b) constructed from the electron diffraction data shows that a connexon has a sixfold symmetry about the central pore region (indicated by the arrow). The pore diameter is about 2 nm at the surface but narrows within the membrane. It is also apparent that each 'subunit' is tilted with respect to the axis of the connexon, having opposite ends (A and B) displaced when seen in plan view (b). After removal of detergent a second form of connexon is obtained which has untilted subunits and no central pore (c). (Reprinted by permission from Unwin, P. N. T., and Zampighi, G. (1980) *Nature*, 283, 545–549. Copyright © 1980, Macmillan Journals Limited)**

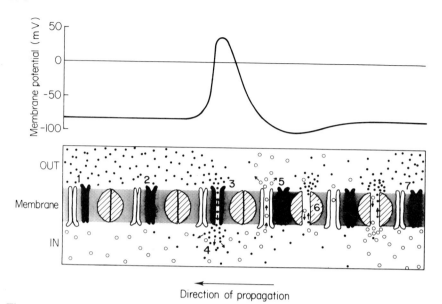

Direction of propagation

Figure 7.7 Generation of an action potential. The upper part of this figure represents the membrane potential that might be recorded along the membrane of a nerve axon shown diagramatically below. Sodium channels are represented by the solid structures, potassium channels by the open structures and the Na^+/K^+ ATPase by the shaded structure. At a point in front of the advancing membrane disturbance (1) the resting conditions is shown in which sodium ions (●) have been pumped out of the cell by the Na^+/K^+ ATPase while potassium ions (○) have been accumulated inside the cell. Although both ion channels are closed the more rapid leakage of K^+ ions across the membrane generates a potential of 60–90 mV. The flux of Na^+ ions across an open Na^+ channel (3) causes a sharp rise in the intracellular sodium concentration (4) and hence a depolarization of the membrane. This causes the voltage-sensitive Na^+ channel at (2) to open and so propagates the disturbance from right to left. In the recovery phase the potassium channel opens (5) to restore the membrane potential and even causes some overshoot. Eventually the operation of the Na^+/K^+ ATPase (6) restores the original ion distribution (7)

the inward sodium current. The slower reactions control the inactivation of the sodium current and the activation of the potassium current. In experiments in which Na^+ and K^+ ions were replaced by non-penetrating ions it was possible to measure the gating currents associated with these events. A simple interpretation of these currents would be the orientation of membrane molecules with a large dipole moment into channels which subsequently can conduct ions (Figure 7.8). There is, however, no structural data to confirm or refute this idea and no attempt has been made to reconcile the movement of these polar components in the hydrophobic core of the membrane with the thermodynamic model of membrane structure.

Ion-channels have to be capable of transferring ions at high fluxes ($> 10^6$ ions sec channel^{-1}). Such a flux can only be achieved if the electrostatic energy of an ion in the channel is similar to that of an ion in solution. One

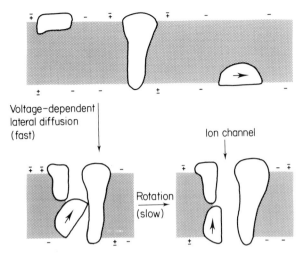

Figure 7.8 **Opening of voltage-sensitive ion-channels. Two events can be determined in the opening of voltage-sensitive ion-channels by measurement of the gating currents. The first is very rapid (time constant 10 μs) and could correspond to the association of mobile components in the plane of the bilayer. A slower process (time constant 100 μs) then results in the opening of the channel. In this diagram this second process is illustrated as the orientation of a high-dipole-moment protein in the electric field across the membrane. The relaxation time constant is consistent with the rotation of a protein of M_r = 100,000. (Reprinted by permission from Rojas, E., and Bergman, C. (1977)** *TIBS*, **2, 6–9)**

method of lowering the electrostatic energy of an ion is to provide a cage of fixed charges around it, as is used by the ionophores (see section 7.3). Ionophores are slow ion carriers, however, as they rely on diffusion through the lipid bilayer. For a proteinaceous pore a continuous lining of charges of opposite sign would be required to achieve a sufficiently rapid transfer of ions. Such a structure is most unfavourable energetically. It has therefore been suggested that ion channels could be lined with a series of rings of water molecules. These could form quite stable structures, hydrogen bonded to the surrounding amino acids of the protein, whilst providing enough dipole interactions to stabilize an ion, provided that it is in contact with every member of one ring. This model has several interesting features. Firstly, in the absence of an ion, the water dipoles may rotate so that they adopt a low electrostatic energy configuration. This overcomes the energetic problem of a channel lined with fixed charges. Ion specificity, too, could be controlled by the size of each water ring in the channel, rings with five members showing a preference for Na^+ over K^+ and rings of puckered hexagons having the opposite specificity. Most interesting is the possibility that a simple rotation of the water dipoles could control the ion translocation by virtue of the large electric field produced in the channel when the water dipoles are in the electrically ordered state. Electrical gating of water-lined channels removes the need to postulate a physical gating particle, and could be controlled by the membrane potential or the proximity of a suitably charged group.

A definitive answer to the structure of ion channels may have to wait their isolation and study *in vitro*. A particularly promising development, therefore, has been the isolation of neuropoisons which specifically inhibit sodium transport in excitable tissues. Tetrodotoxin, for example, reversibly binds to the sodium channel with a dissociation constant of approximately 10^{-9} M. It only binds to one site on the sodium channel which is accessible from the outside of the cell membrane and is believed physically to occlude the pore through which the ions translocate. Detergent solubilization of the toxin–membrane complex has led to the isolation of a protein of $M_r = 230,000$ to which the tetrodotoxin binds.

Another example of voltage-sensitive ion-chemicals is found in the visual response of retinal rod cells. Here, light falling on flattened disc membranes containing stacks of the integral membrane protein, rhodopsin, causes the rapid release of Ca^{2+} ions into the cytoplasm until the electrical potential of the disc membrane has fallen to a point where the channel closes. Increased Ca^{2+} ion concentration directly or indirectly causes the sealing of the Na^+ channel in the adjacent plasma membrane (in vertebrates), thus causing a hyperpolarization of the membrane potential which is transmitted to the brain. Thus, the opening of ion channels is used to amplify the initial response to a single quantum of light.

The primary response to absorption of a photon of light is thought to be a change in the conformation of the retinal molecule as illustrated in Figure 7.9, although this subject is still under debate. Subsequent light-independent

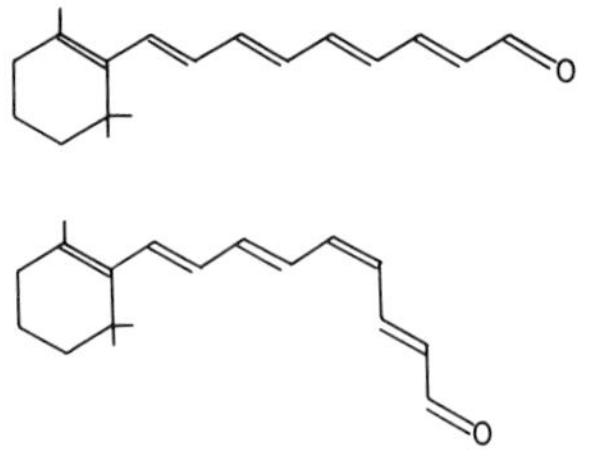

All-*trans* retinal

All-*cis* retinal

Figure 7.9 The mechanism of energy transduction in rhodopsin. Light energy absorbed by the retinal chromophore of rhodopsin gives rise to the formation of Ca^{2+} ion channels in the disc membranes of vertebrate retinas. The retinal is normally found as the 11-*cis*-isomer attached by a Schiff base to a lysine residue of the opsin protein. After exposure to light a series of intermediates are generated which eventually result in the dissociation of all-*trans* retinal from the protein. All-*trans* retinal does not itself recombine with the opsin apo-protein unless converted to 11-*cis* retinal by the enzyme retinene isomerase. It is possible, therefore, that the photoisomerization of 11-*cis* to all-*trans* retinal places some strain on the protein molecule which triggers a conformational change in the rhodopsin and leads to its aggregation into Ca^{2+} ion channels

changes occur which result in the bleaching of the retinal chromophore and conformational changes in the rhodopsin protein that are most likely responsible for the ensuing chain of events. Finally the retinal chromophore is released, although this reaction is reversible, and in the dark-adapted state all discs contain complete rhodopsin molecules. When rhodopsin is reconstituted in phospholipid vesicles loaded with a variety of small molecules, exposure to light increases the peremeability to molecules less than about 1 nm in diameter, suggesting that a non-specific transmembrane channel is formed. There is a lag of several seconds for this to happen *in vitro*, although permeability changes in the disc membrane *in vivo* may be measured in milliseconds. This observation has prompted the suggestion that the light-induced conformational change in rhodopsin causes an aggregation of this protein into a multimeric pore structure similar to that of *E. coli* lipoprotein. This would be a very interesting mechanism for regulating ion permeability through a transmembrane pore in response to an external stimulus. About 80 per cent of the rhodopsin molecule is α-helix, but most of the peptide oxygen atoms are exposed to aqueous solvent. The transmembrane pore is therefore imagined as an aqueous channel between the subunits of a multimeric aggregate.

7.4.4 Complement

When a foreign cell (e.g. a bacterium) invades a mammalian host a complex defence mechanism is activated which results in the destruction of the foreign cell. It was recognized in the nineteenth century that two soluble components of the blood were required for this, antibody and complement. The antibody binds to antigenic sites on the foreign cell with a high degree of specificity and then the antibody-labelled cells are destroyed, either by phagocytosis or by irreversible membrane damage that leads to cell lysis. Both processes are influenced by the family of proteins known collectively as complement. Phagocytosis and a variety of other cellular functions are activated by byproducts of the reaction of complement components with the immune complex on the foreign cell. These byproducts are short polypeptides released during the proteolytic activation of complement proteins. Cell lysis, on the other hand, is mediated entirely by complement proteins.

Altogether at least 18 serum proteins are involved in these reactions. They have been assigned to two pathways which have different mechanisms of activation. The 'classical' pathway refers to the model system for measuring the lysis of sheep erythrocytes using rabbit antibodies and human complement. Cell lysis is activated by IgG- or IgM-immune complexes on the erythrocyte surface. The alternate or 'properdin' pathway may be activated by aggregates of IgA or by naturally occurring polysaccharides and liposaccharides. This suggests that it may be responding to the surface charge of the foreign material. Both pathways terminate in the formation of a complex of proteins which binds to the membrane and leads to the collapse of ion gradients across the membrane and cell lysis by osmotic bursting of the cell.

300

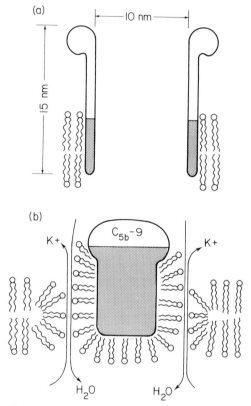

Figure 7.10 **Pore formation by the membrane attack complex of complement.**
Complement-mediated cell lysis is produced by a complex of complement components
C_{5b-9}, the so-called 'membrane attack complex' (MAC). Although each of the individual
components is a hydrophilic protein of the blood serum, when activated in the MAC a
large hydrophobic area is exposed which is capable of spontaneous association with
phospholipid bilayers. Two mechanisms have been proposed to account for the rapid
loss of ions from the cell which follows this event. In (a) the MAC is shown as forming a
huge pore in the membrane. Measurements taken from electron micrographs suggest
that the MAC does not quite traverse the bilayer but has an opening of about 10 nm
through the centre. (Reprinted with permission from Bhakdi, S., and Tranum-Jensen,
J. (1978) *Proc. Natl. Acad. Sci., U.S.A.*, 75, 5655–5659.) An alternative model (b), based
on biochemical measurements of the pore size and lipid binding of the MAC, suggests
that the hydrophobic complex causes a rearrangement of the phospholipid bilayer and
secondary changes in ion permeability. (Reprinted with permission from Podack, E. R.,
Biesecker, G., and Muller-Eberhard, H. J. (1979) *Proc. Natl. Acad. Sci., U.S.A.*, 76,
897–901)

It is the molecular mechanism of this membrane attack complex (MAC)
which is of principal interest in this context.

Initial recognition of foreign material in either pathway is followed by the
assembly of a specific, membrane-bound proteolytic enzyme from several
complement components. This enzyme (C5-convertase) cleaves a polypeptide

fragment from component C5 to generate C5b, the first component of the MAC. Self-assembly of the terminal complex is then a spontaneous process. Components C6, C7, and C8 bind to the C5b on the membrane surface, followed by three molecules of C9. Two of these aggregates form the MAC which has the structural formula $(C5b, C6, C7, C8, C9_3)_2$ and a $M_r = 1,700,000$. Although each of these complement components is a soluble protein, the complex in its active state has a high affinity for membranes, and has the properties of an integral membrane protein, i.e. it requires detergents for elution from the membrane and is an amphipathic complex which binds large quantities of detergent molecules. The molar ratio of phospholipid to protein in the terminal complex isolated from synthetic phospholipid vesicles is 1460 : 1. The MAC only lasts in the active state for < 0.1 s, secondary changes then occurring which destroy its affinity for membranes. Consequently the MAC can only be inserted in the vicinity of the membrane-bound C5-convertase and inactive forms are found circulating in the blood. The insertion of the MAC into membranes correlates well with the appearance of characteristic lesions which appear in the electron microscope as circular structures with a dark central portion surrounded by a light ring in negative stained images. This has been interpreted as a doughnut-shaped pore structure in the membrane with an internal diameter of about 10 nm (Figure 7.10a). An alternative model is shown in Figure 7.10b. This proposes that the MAC is able to form mixed protein–phospholipid micelles in the membrane due to the large exposed hydrophobic areas on the complex. Phospholipid membrane channels would thus be created in the immediate vicinity of the MAC. It is not at present clear which of these models is correct.

Summary

Fixed pore structures exist in a variety of membranes which allow the flux of ions or uncharged molecules through the membrane.

The specificity of these pores is not usually very great, offering a barrier to the diffusion of large molecules only.

Ion channels, however, are very specific.

Diffusion through pores may in some cases be regulated by restriction or closure of the pore.

7.5 FACILITATED DIFFUSION

Many substrates useful to the cell are transported more rapidly than would be predicted on the basis of diffusion through the membrane. Glucose, for example, at a concentration of 1 mM is transported across the erythrocyte membrane about 10^4 times faster at 25 °C than would be predicted on the basis of its lipid solubility or number of hydrogen bonds that form in aqueous solution (see section 7.2). Transport is independent of an energy source, however, and only proceeds as long as a concentration gradient of glucose persists. It has therefore been called facilitated diffusion. A complete set of criteria are listed in Table 7.2 on page 311. The ability to saturate the transport system at high substrate concentration and in some cases to distinguish between optical isomers of the substrate provides a clear distinction from passive diffusion and suggests that an enzyme-like protein with a substrate binding site is involved. Recent isolation of purified membrane proteins and reconstitution of facilitative transport phenomena in phospholipid vesicles confirms this view.

The similarity of a transport protein to an enzyme having a binding reaction followed by a rate-limiting first-order transport process is reflected in their similar kinetics. The flux of substrate through the membrane (J) may be described by the Michaelis–Menten equation for enzyme velocity:

$$J = \frac{S V_{max}}{K_m + S}$$

Since, however, a flux can occur in both directions the net flux is given by the equation:

$$\text{Net flux} = J_{in} - J_{out}$$

$$= \frac{S_{out} V_{max}}{K_m + S_{out}} - \frac{S_{in} V_{max}}{K_m + S_{in}}$$

A convenient way of following the movement of substrate is by using a radioactive label attached to some part of the permeant molecule. When there is no substrate inside the cell the flux measured in this way is entirely undirectional since $J_{out} = 0$. However, as the concentration of substrate increases inside the cell towards a chemical equilibrium, exchange of substrate will occur and the increase in labelled substrate inside the cell no longer represents the net flux. If isotopically labelled substrate is added at chemical equilibrium, exchange will still occur, thus giving equilibration of the label even though no net flux of substrate passes across the membrane. For the transport of glucose across the human erythrocyte membrane the net flux of glucose is some 85 times slower than the rate of exchange. The demonstration that the net flux across the membrane is not the same as the rate of exchange is an important criterion for distinguishing facilitative transport from simple diffusion.

In some cases the net flux is so small that the system may be regarded simply as an exchange reaction. An example of this type of transport is the carnitine : acylcarnitine exchange transporter of the mitochondrial inner membrane. This exchange system provides a mechanism by which the fatty acyl substrates of β-oxidation may pass from the cytoplasm to the matrix compartment of the mitochondria where they are oxidized. Although fatty acids are activated as acyl–co-enzyme A derivatives in the cytoplasm, they are converted into carnitine esters for the transport across the mitochondrial inner membrane and then back into co-enzyme A esters for the pathway of β-oxidation. This complex procedure is required because long-chain fatty acids and fatty acyl–co-enzyme A compounds cannot pass directly through the mitochondrial inner membrane.

Four modes of exchange are possible:

$$\text{carnitine}_{out} \rightleftharpoons \text{carnitine}_{in}$$
$$\text{acyl-carnitine}_{out} \rightleftharpoons \text{carnitine}_{in}$$
$$\text{carnitine}_{out} \rightleftharpoons \text{acyl-carnitine}_{in}$$
$$\text{acyl-carnitine}_{out} \rightleftharpoons \text{acyl-carnitine}_{in}$$

Only one of these (acyl-carnitine$_{out}$ $\rightleftharpoons$ carnitine$_{in}$) achieves the primary object of transporting fatty acyl groups into the mitochondria for oxidation. Since the mitochondrial matrix has a fixed concentration of carnitine, however, the apparently 'futile' export of acyl-carnitine rapidly becomes important as the intramitochondrial carnitine becomes acylated. This effect is exacerbated by the much lower K_m for acyl-carnitine compared with that for carnitine. Thus the export of acyl-carnitine from the mitochondria is preferred over the export of carnitine at a low degree of intramitochondrial carnitine acylation and the net influx of acyl groups is effectively shut off (Figure 7.11). In this way the kinetics of exchange transport allow a control of the net flux of substrate whilst not affecting the overall V_{max} of the system. This probably

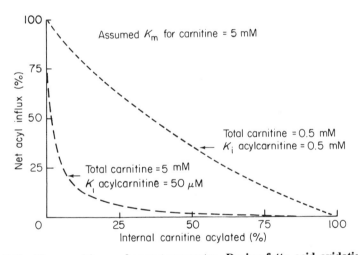

Figure 7.11 The carnitine exchange transporter. During fatty acid oxidation, fatty acyl groups are transported into the mitochondria by exchange across the mitochondrial inner membrane of cytoplasmic acyl carnitine with mitochondrial carnitine. Diffusion of carnitine across the membrane occurs at < 0.5 per cent of the exchange rate. The transporter is energy independent and entirely symmetrical, allowing exchange of carnitine and acyl carnitine in both directions. Thus the export of acyl-carnitine from the mitochondria limits the net influx of acyl groups for β-oxidation. Shown here are two theoretical curves based on the kinetic parameters of the exchange transporter in heart and liver mitochondria where the concentrations of total carnitine are approximately 5.0 and 0.5 mM respectively. (Reprinted with permission from Tubbs, P., and Ramsay, R. (1979) in *Function and Molecular Aspects of Biomembrane Transport* (Quagliarello, E., Palmieri, F., Papa, S., and Klingenberg, M., eds.) Elsevier/North Holland Biomedical Press)

serves a rather important role in β-oxidation. By maintaining a pool of unacylated carnitine and hence also of co-enzyme A within the mitochondrion, the β-oxidation of long-chain fatty acyl–co-enzyme A compounds (which require an acceptor co-enzyme A molecule for each 2-carbon cycle of oxidation) is allowed to proceed.

Another consequence of the reversible nature of facilitative diffusion is 'counter-transport'. This can occur when more than one substrate binds to the same transport protein. When erythrocytes are equilibrated with a non-metabolizable sugar, for instance, and then a high concentration of glucose is added outside the cells, the glucose analogue is transported out of the cell against a concentration gradient. The energy for this process comes from the coupled diffusion of glucose and will occur in cells which have been depleted of ATP. Counter-transport is therefore good evidence that two substrates share the same transport protein and that transport occurs by facilitative diffusion.

7.5.1 Models of membrane transport proteins

Early models of facilitated transport invoked a shuttle of carriers between the two surfaces of the membrane (Figure 7.12a), or tumbling molecules which alternately revealed a substrate binding site at each surface (Figure 7.12b). As we have seen in section 7.3, some antibiotics have a structure which allows them to transport ions across biological membranes by this type of mechanism, but no analogous structures have yet been observed in mammalian transport systems. Transport proteins are in general the largest proteins found in the membrane and contain hydrophilic regions which extend into the aqueous regions on both sides of the membrane. Whilst moving carrier models might therefore aptly describe the kinetics of transport, they are unlikely to exist *in vivo* since the movement of large hydrophilic regions of protein through the apolar core of the membrane is not thermodynamically feasible.

Direct evidence against the models in Figure 7.12a and 7.12b comes from measurements of the orientation of proteins in the membrane. Transport proteins like band 3 in the erythrocyte membrane, Ca^{2+} ATPase in sarcoplasmic reticulum, and the Na^+/K^+ ATPase are not found in random orientations as would be predicted on this model, but have all copies of the protein inserted in the same direction. Furthermore, it has been shown that the attachment of bulky antibody or lectin molecules to the surface parts of these proteins, or the cross-linking of the proteins to form large aggregates does not adversely affect their function.

The current concept of a transport protein is shown in Figure 7.12c. This model proposes that the substrate passes through hydrophilic pores formed between the subunits of a multimeric membrane transport protein. Since the pore is not a fixed structure giving permanent access of substrate across the membrane, it is quite distinct from a membrane channel, and has been described as a gated-pore. The model proposes that the transport protein can

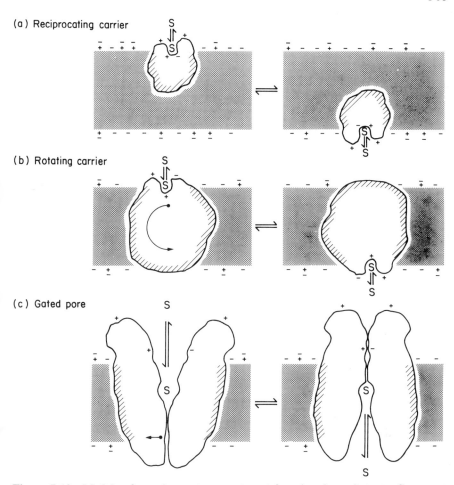

(a) Reciprocating carrier

(b) Rotating carrier

(c) Gated pore

Figure 7.12 Models of membrane transport proteins. A polar substrate, S, cannot pass directly across the hydrophobic core of the membrane (shown in stipple) because of the high inactivation energy involved. The moving carrier models depicted in (a) and (b) account for the kinetics of facilitative diffusion, but are unlike the structures of known membrane transport proteins, and are energetically unlikely since they require hydrophilic regions (shown by + and − charges) of the protein to pass across the bilayer. All characterized transport proteins have a structure more akin to that shown in (c). In this model the protein does not move across the membrane but acts as a gated pore allowing substrate to pass across the membrane following a conformational change. (Adapted from Singer, S. J. (1977) *J. Supramol. Struct.*, **6**, 313–323. Reprinted with permission from the author and Alan R. Liss, Inc.)

exist in two states which are interconvertible by a conformational change. One of the properties of a multimeric protein is that a relatively low input of energy can give rise to a large rearrangement of the subunits. It is quite conceivable, therefore, that the energy of substrate binding in the case of facilitative transport should be sufficient to achieve this. During active transport the phosphorylation of the subunits would be capable of giving profound conformational changes as the phosphate group can carry three negative charges. Theoretically, a movement of < 0.1 nm could close a pore to a small molecule like glucose, whereas the mobile carrier model requires a rotational or translational movement of a large molecule over the dimensions of the bilayer, about 7 nm. Although the substrate binds to the same binding site from each side of the membrane the topography of the surrounding protein is different. Thus the observation of binding sites with different properties on either side of the membrane is strong evidence for this model and cannot be explained by the moving carrier models of Figure 12a and 12b.

An alternative gated-pore model proposes that an 'arm' of the protein containing the substrate binding site moves across a narrow hydrophobic barrier in a fixed pore through the membrane. This is really a modification of the carrier theory except that the distance of 'carrier' movement has been reduced. In both cases the absolute asymmetry and thermodynamic stability of the membrane are maintained and kinetics identical to those of a moving carrier are generated.

One consequence of the gated-pore model is that molecules smaller than the substrate might be expected to be able to penetrate the pore volume and be transported by a conformational change, even if they have no affinity for the substrate binding site. This permeability should be proportional to the volume of distribution of the solute in the pore and should therefore decrease with increasing molecular size, decreasing to zero for molecules larger than the pore radius. Of course, a mobile carrier would also have a binding site of finite volume, but this would be expected to be considerably smaller than a pore. When the permeability of 4- and 5-carbon polyols was investigated in human erythrocytes it was found that a significant fraction of the permeability was inhibited by glucose, suggesting that they shared the same transport system. The fact that the smaller compound was transported most rapidly, however, argues against transport by competition with the glucose binding site and suggests that the rate-limiting step of transport is the diffusion of the compound into the pore of the glucose transport protein. Calculations of the pore volume from this rate of transport give rise to a hypothetical pore diameter of 0.8 nm. This compares well with the size of a glucose molecule (diam. 0.72 nm) and is more in keeping with the expected dimensions of a pore rather than the non-specific volume in a carrier molecule.

Direct experimental tests of the gated-pore hypothesis are not easy. The best evidence comes from structural investigations of transport proteins. Since these are so important for cellular function they are frequently major proteins of the membrane and are therefore relatively easy to isolate and purify. Until recently, however, it was difficult to identify which solubilized membrane protein was responsible for a particular transport process because they could only be assayed in their native environment of a sealed membrane envelope. The development of techniques of reconstitution of membrane proteins in synthetic phospholipid vesicles has now allowed solubilized membrane proteins to be assayed for transport activity *in vitro*. Ideally, the reconstituted transport protein should retain the kinetics of transport *in vivo* and remain sensitive to specific transport inhibitors before identification is made. Studying transport in reconstituted vesicles has some other experimental advantages. Measurements of substrate concentrations are facilitated by the absence of intracellular compartmentation and metabolism of the substrate, which otherwise can give misleading results. It is also possible to manipulate the surface charge of the bilayer by incorporating different phospholipids, and to impose or dissipate electrochemical gradients across the membranes using ionophores. Reconstitution of purified proteins in phospholipid vesicles has therefore been particularly fruitful in the study of membrane transport processes (see Chapter 5).

7.5.2 Anion exchange in erythrocytes

In cases where the conformational change producing the translocation of the substrate across the membrane can only occur when the substrate binding site is occupied, a strict 1 : 1 exchange of molecules is observed. Good examples of this type of facilitative diffusion are given by the carnitine transport system of heart mitochondria which has already been mentioned, and by the anion exchange protein of the erythrocyte membrane. Considerable structural details are known for the latter protein which is responsible for the exchange of bicarbonate ions formed from tissue CO_2 with chloride ions in the erythrocyte cytoplasm. This exchange reaction occurs during the passage of the cells through the tissue capillaries, and then later in the reverse direction as the cells pass through the lung. Since this latter process can take place in $< 1s$ it is not surprising to find that the exchange protein represents a major proportion (about 30 per cent) of the total membrane protein, and can exchange ions at the extremely rapid rate of 7×10^5 ions site^{-1} s^{-1}. In fact the half-time of exchange of Cl$^-$ ions is about 50 ms at body temperature. Despite this rapid exchange, no energy input has been discovered; anions appear to be at equilibrium across the erythrocyte membrane. The net flux of Cl$^-$ ions across the membrane is, however, very small, some five orders of magnitude less than the exchange rate. Anion transport in the erythrocyte may be identified with band 3 on SDS–PAGE. This band contains several proteins, but the major component, which comprises 75 per cent of band 3 is in fact the anion exchange protein ($M_r = 90,000$) and exists in the membrane as dimers. It is extremely unlikely that this molecule could rotate or shuttle across the membrane since 70 per cent of the protein consists of hydrophilic regions situated on both sides of the membrane (see section 4.4). The protein is distributed asymmetrically across the bilayer with all copies of the protein in the same orientation. We are left with the conclusion that facilitative exchange diffusion must occur in this case by the mechanism shown in Figure 7.12c.

7.5.3 Adenine nucleotide exchange protein

Another exchange protein which has been the subject of intense study is the adenine nucleotide exchange protein of mitochondria. This is the most active transport system in many eukaryotic cells despite the fact that ADP and ATP are amongst the bulkiest and most highly charged species that are moved across membranes. Although the turnover number of the protein is not very high it is frequently the most abundant protein in the mitochondrial membrane, accounting for 12 per cent of the protein in heart mitochondria. ATP is synthesized inside the mitochondria using an intramitochondrial pool of ADP. Access of cytoplasmic ATP consuming reactions to this pool of ATP relies on the 1 : 1 counter exchange of adenosine nucleotides which is catalysed by the translocator protein. In uncoupled mitochondria the ADP : ATP exchange reaction reaches an equilibrium in which ATP and ADP are transported symmetrically across the membrane. In coupled mitochondria, however, the membrane potential makes the exchange highly asymmetric. This is a result of the electrogenic nature of the exchange. At neutral pH, ATP^{4-} contains one more negative charge than ADP^{3-}. Thus, ADP uptake and ATP release by the mitochondria are favoured more than 20-fold over other modes of exchange. This results in a high ATP : ADP ratio in the cytoplasm and a low ratio inside the mitochondria—exactly suiting the requirements of each metabolic compartment. In this case, therefore, exchange reactions can achieve a movement of molecules up a concentration gradient at the expense of energy stored by ion-pumps in the membrane potential. It could therefore be classified as an electrogenic exchange transporter.

Dissection of this transport system has been possible due to the use of highly specific inhibitors such as atractylate and its derivatives isolated from plants, and bongkrekate from bacteria. These compounds interact with the translocator in different ways. Carboxyatractyloside binds directly to the cytoplasmic face of the adenine nucleotide exchange protein, while bongkrekate is transported into the mitochondrion before binding to the matrix side of the transporter protein. Thus bongkrekic acid inhibition requires a small amount of ADP to be present, and the amount of ADP carried into the mitochondria during this inhibition is equal to the amount of carrier inhibited. These two forms of the translocator have been isolated by detergent extraction of mitochondria after treatment with the appropriate inhibitor. The two isolated forms of the protein both have a subunit of similar size ($M_r = 30,000$) and exist as dimers in a reconstituted transport system. No other co-factors are required for activity. The two forms of the protein also have the same amino-acid composition, but are immunologically different and showed a different sensitivity to sulphydrl reagents and proteolytic digestion.

These observations fit well with the mechanism of transport shown in Figure 7.12c. The presence of two forms of the enzyme which may be interconverted by the translocation of the substrate across the membrane is good evidence for a two-state model. Carboxyatractyloside and bongkrekate act as probes for these two forms of the protein and their specificity for binding to only one face of the membrane illustrates the asymmetry of the binding site on each side of the bilayer. The different immunological and chemical properties of these two binding sites reinforce the conclusion that the topography of the protein surrounding the substrate binding site is different on the matrix face compared with the cytoplasmic face and effectively rules out a mobile carrier mechanism for this transport process.

7.5.4 Co-transport and energy coupling

We have already discussed in the case of counter-transport how the facilitative diffusion of two substances may be coupled by the existence of a common transport protein. When the co-substrate is an ion like Na^+ or K^+ it is possible for facilitative diffusion to move substances across the membrane against a concentration gradient using the energy stored indirectly in ion gradients by the Na^+/K^+-dependent ATPase. This is a particularly common mechanism of indirect active transport which is found in the absorption of neutral amino acids and sugars across the plasma membrane of intestinal brush border and other cells. An electrogenic influx with Na^+ ions appears to be a likely mechanism for this transport process, although exchange for K^+ or neutral exchanges are also possible (Figure 7.13). Although there are no structural data available for these transport systems, it seems likely that both substrate binding sites on a co-transport protein must be occupied before the conformational change depicted in Figure 7.12c can occur. Another indirect way of coupling energy to facilitative diffusion is to modify the substrate after transport into the cell. Glucose for example, is rapidly phosphorylated by muscle hexokinase, thus removing it from the chemical equilibrium of glucose across the membrane. In general, metabolites like these and energy-rich compounds do not have facilitative transport systems in the plasma membrane so that

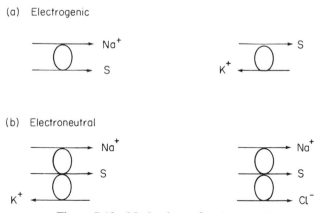

(a) Electrogenic

(b) Electroneutral

Figure 7.13   Mechanisms of co-transport

they cannot be lost once captured in this manner. In certain bacteria this metabolism of the substrate may be catalysed by the transport protein itself, but there are no examples of this among vertebrates.

Summary

Facilitative transport is faster than transport by diffusion.

Catalysis of transport is achieved by membrane proteins which bind a specific substrate and then pass it across the bilayer.

This process is energy independent, reversible and net flux stops once an electrochemical equilibrium has been attained.

Facilitative transport is better described by a two-state gated-pore mechanism rather than by moving carrier molecules.

7.6 ACTIVE TRANSPORT

During active transport the movement of a substance across the membrane is directly coupled to an energy source, which is usually ATP. This enables the cell to accumulate specific substances against a considerable concentration gradient (e.g. up to 10,000-fold for Ca^{2+} ATPase). Membrane proteins engaged in active transport are large structures, often forming multimers in the membrane and having large parts of the polypeptide chain in the aqueous medium on each side of the bilayer. This structure favours a gated-pore type of mechanism although these proteins are usually rather more complex than the structure shown in Figure 7.12c, having covalently attached polypeptides with ionophoric or ATPase activity. During each catalytic cycle the transport protein hydrolyses an ATP molecule and is itself phosphorylated in the process. It would be easy to imagine how the introduction of a highly charged phosphate group could drive the conformation change of a two-state gated-pore model, although this mechanism has not yet been proven for any specific example. Another striking feature of active transport proteins is their vectorial activity. Substrate is only passed in one direction across the membrane. This is probably a consequence of the requirement for the energy of ATP hydrolysis, rather than substrate binding, to produce translocation of the substrate. Strictly speaking, even active transport is reversible, and under extreme conditions some reverse flux of substrate may be observed. However, this is accompanied by the formation of ATP which illustrates the tight coupling of ATP hydrolysis to substrate translocation in the normal mode of operation. The distinguishing features of active transport are summarized in Table 7.2. Although a lot of structural details are known for active transport proteins like Na^+/K^+ ATPase and Ca^{2+} ATPase, little is known about the mechanism by which they catalyse transport. One of the best models for an active transport protein is bacteriorhodopsin. This is a light-driven protein pump of *Halobacterium halobium* which generates an electrochemical gradient across the bacterial membrane. Although ATP is not directly involved in the mechanism of proton transport, the energy (light in this case) is directly absorbed by the protein and used to pump protons against a concentration gradient. We may therefore classify it as an active transport protein.

Table 7.2 Properties of transport processes

Diffusion	Facilitated diffusion	Active transport
Mediated by membrane lipid	Mediated by membrane proteins	Mediated by membrane proteins
Net flux ceases at electro-chemical equilibrium	Net flux ceases at electro-chemical equilibrium	Achieves transport against electrochemical gradient
No energy coupling	May be indirectly coupled to electrochemical energy	Directly coupled to energy supply
Low specificity	High specificity	High specificity
Non-saturating	Saturates at high substrate concentration	Saturates at high substrate concentration
No counter transport	Displays counter transport	Almost irreversible

7.6.1 Bacteriorhodopsin

Bacteriorhodopsin is synthesized and inserted into the membrane of *H. halobium* when the organism is grown in the absence of oxygen, and the electrochemical gradient that it produces is used to generate ATP and keep the organism alive. Unlike other membrane proteins of *H. halobium*, the bacteriorhodopsin is segregated into patches which contain no other membrane proteins. There is, however, some lipid present (about 25 per cent of the total mass) and this is as heterogeneous as the lipid in the bulk of the membrane. Bacteriorhodopsin and rhodopsin from animal retina should not be confused as their structures are not related. They do, however, share a common retinal chromophore which gives both proteins a purple colour.

The purple membranes from *H. halobium* are simply prepared by lysing the cells in hypotonic salt solutions, washing away soluble components and then separating the purple membranes from other membranes by sucrose density gradient centrifugation. This gives a preparation of oval-shaped discs of membrane containing only one membrane protein, the bacteriorhodopsin. Curiously, this protein is present in an almost perfect crystalline array in the bilayer, thus enabling it to be studied by X-ray and electron diffraction techniques without pretreatment with detergents or other substances that might alter its structure. From these studies we have a good idea of how the protein is integrated into the bilayer.

Figure 7.14a is an electron diffraction contour map of purple membranes which shows the potential distribution in plan view. To a first approximation this is the same as the electron density distribution given by X-ray analysis and shows the main structural features to a resolution of 0.7 nm. It may be seen that there is a threefold axis of symmetry in the membrane which is created by the aggregation of bacteriorhodopsin molecules into trimers. Each molecule consists of seven α-helices which span the membrane and lie roughly perpendicular to it. The inner three helices lie close to the perpendicular and so appear as clear circular cross-sections in Figure 7.14a, while

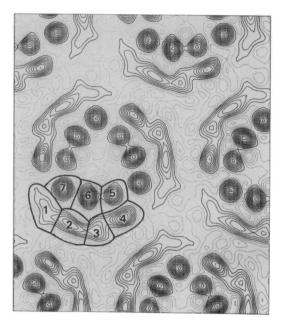

(a)

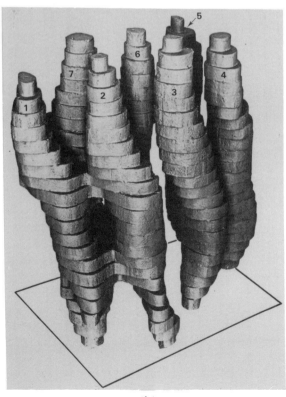

(b)

the outer four lie at an angle to the perpendicular and so have a more diffuse image in this projection. When electron diffraction patterns are collected from specimens tilted at various angles to the electron beam a three-dimensional contour map can be obtained. A model constructed from this is shown in Figure 7.14b. Here the seven rod-shaped α-helices are clearly visible. Although the helices span the membrane they do not extend into the medium on either side, but are connected to each other by short interconnecting peptides. Little can be directly inferred about the structure of these at 0.7 nm resolution, but the physical constraints of the molecule suggest that only one scheme of interconnection is possible. In this scheme the internal acidic and basic residues on the helices are bound together in salt linkages which stabilize the overall structure while the hydrophobic external faces of the helices interact with lipid in the environment. The retinal chromophore, which is bound by a Schiff base to a lysine residue is also situated between the seven α-helical segments of the molecule. This arrangement suggests that the space between the rod-like helices might form a proton pore across the membrane. After absorption of a quantum of light energy (about 45 kcal) an excited state is formed which is reversible at very low temperatures (i.e. when no conformational changes of the protein can occur). A cycle of temperature-dependent intermediates eventually lead back to the native rhodopsin structure. During this decay process protons are lost to one side of the membrane and then subsequently picked up from the other. It seems probable, therefore, that a two-state mechanism accounts for proton transport, although in this case the conformational change occurs between transmembrane segments of one polypeptide chain rather than between subunits of a multimeric protein, as shown in Figure 7.12c.

One of the curious features of bacteriorhodopsin is the presence of such an organized intermolecular structure when the translocation of protons occurs by an intramolecular process. Looking at Figure 7.14a one might have supposed that the trimeric arrangement of bacteriorhodopsin molecules enclosed an aqueous pore. This central structureless region is, however, filled with

Figure 7.14 **Structure of bacteriorhodopsin. Bacteriorhodopsin is a light-driven proton pump which is present in a crystalline array in the purple membrane of *H. halobium*. Electron-diffraction studies of these membranes have enabled a density contour map to be constructed to a resolution on 0.7 nm as shown in projection in (a). The symmetry of the structure is clearly visible. Three bacteriorhodopsin molecules, each with seven rod-shaped α-helices spanning the membrane, cluster together about the centre of symmetry to form two rings of 9 and 12 helices. The central featureless region is about 2 nm in diameter and is filled with bilayer lipid. In three dimensions the α-helices are more obvious (b). Very little protein extends out of the membrane on either side of the membrane and the light-absorbing retinal chromophore and proton 'channel' are thought to be located in the centre of each molecule. (Reprinted by permission from Henderson, R., and Unwin, P. N. T. (1975) *Nature*, 257, 28–32, copyright © 1975, Macmillan Journals Limited, and with permission from *J. Mol. Biol.*, 94, 425–440, copyright Academic Press Inc. (London) Ltd)**

lipid. Bacteriorhodopsin is quite different, therefore, from its animal counterpart.

7.6.2 Binding protein transport processes

In a number of bacteria, binding proteins specific for substrates transported across the inner membrane have been found in the periplasmic space, i.e. the space between the inner and outer membrane. These proteins are not membrane transport proteins themselves since they are not hydrophobic and could not enter the bilayer. Even if they did manage to cross the membrane they would not be able to release their bound substrate at a sufficient rate to account for the rate of transport *in vivo* since they have very low dissociation constants. One clue to their function is given by the observation that mutant strains which do not have these proteins still manage to transport material through the inner membrane, but that the K_m for transport is increased by as much as 1000 times. This suggests that they act as ancillary binding proteins which increase the affinity of the system by scavenging the periplasmic space for substrate and delivering it to the transport protein. A possible hypothesis is that the inner membrane transport protein catalyses the phosphorylation of the binding protein using cytoplasmic ATP. The binding protein then releases its bound substrate which passes down a channel through the membrane. Thus the bacterial binding proteins behave in a similar manner to the ionophore of eukaryotic active transport proteins except that they occur in free solution in the periplasmic space rather than being covalently linked to the membrane transport protein. They are also the possible forerunners of a two-site mechanism of transport, binding in this case occurring on an independent peripheral protein before a conformational change releases the substrate and allows it to pass across the membrane.

7.6.3 Electron transport

The origin of a membrane potential across the mitochondrial inner membrane is not an ion-pumping ATPase, but is derived from the energy of biological oxidation reactions inside the mitochondrion. It is generally accepted that these oxidation reactions result in the efflux of protons, and possibly other cations, from the mitochondrial matrix, although the mechanism by which this occurs is still under debate. Also situated in the mitochondrial inner membrane is a complex protein structure which can couple the return of these protons to the synthesis of ATP. The mechanism of this reaction is even less well understood. The mitochondrial inner membrane acts, therefore, as a coupling membrane between the oxidation reactions which establish the high-energy electrochemical gradient and a variety of driven reactions, one of which is ATP synthesis.

Several biological molecules may be oxidized in the mitochondrion, but the principal one of these is NADH produced in the citric acid cycle and β-oxidation. During oxidation a pair of electrons is lost which eventually are passed to oxygen:

$$NADH = NAD^+ + H^+ + 2e^-$$

$$\tfrac{1}{2}O_2 + 2H^+ + 2e^- = H_2O$$

This pair of electrons is passed down a series of membrane-bound components, comprising the electron transport chain, during which sufficient energy is stored to synthesize three molecules of ATP. As already stated, however, the coupling of these two processes is indirect and interest has focused on the nature of the high-energy intermediate or state that must be involved. Reconstitution of cytochrome components in artificial phospholipid vesicles has shown beyond reasonable doubt that the primary reaction during electron transport is a vectorial movement of protons across the membrane. In the original formulation of the 'chemiosmotic' theory this proton translocation was envisaged as a consequence of hydrogen and electron carriers arranged in redox loops (Figure 7.15a). During the first half of each cycle the pair of electrons is transported across the membrane in combination with two protons on a hydrogen carrier. Since only an electron carrier is available for the return of electrons the protons must dissociate from the electrons and pass into solution, now on the opposite side of the membrane. This theory stimu-

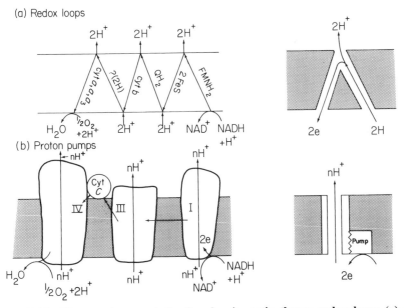

Figure 7.15 Electron transport. In the chemiosmotic theory redox loops (a) are proposed to pass the protons across the bilayer. In the proton-pump theory (b) this is achieved by pumps coupled to electron transport. The path of electrons is shown by closed arrowheads

lated a great deal of research into the formation of proton gradients and the identity of the postulated hydrogen and electron carriers. Unfortunately, there are some problems in the assignment of a hydrogen carrier in the final redos loop and complex movements of co-enzyme Q have to be postulated.

An additional problem is the number of H^+ ions ejected from the mitochondrion for each pair of electrons passed down the electron transport chain. Although this is predicted as $2H^+$/electron pair on the chemiosmotic theory, values considerably greater than this have recently been measured. In some ways this is reassuring as the energy of the electrochemical gradient set up by the ejection of $2H^+$/electron pair is barely sufficient to account for the amount of ATP synthesized. On the other hand, it is difficult to rationalize on the model of hydrogen and electron carrier loops. Consequently, it has been proposed that protons may be 'pumped' across the membrane using the energy of electron transport (Figure 7.15b). This idea has received some support from the ability to separate ion and electron transferring activities in isolated complexes of the electron transport chain.

A precedent for an indirect energy coupling is seen in bacteriorhodopsin which transports protons in response to energy input. Two separate processes can be detected. The initial reaction is a rapid energy-dependent conformational change of the protein which is followed by a much slower proton translocation activity. For bacteriorhodopsin the stoichiometry of proton transport is also irregular. This could be explained on the proton pump model by the ionization of amino-acid side chains involved in the proton transport. This has been called the membrane Bohr effect.

Cytochrome oxidase, which is the terminal complex of the electron transport chain has been shown by external labelling procedures and by electron diffraction techniques to span the membrane (see section 4.7.2). Seven subunits are involved in the complex, the three largest of which are hydrophobic and are synthesized on mitochondrial DNA. The complex is aligned flush with the matrix surface of the bilayer where reaction with oxygen occurs, but projects by 6–7 nm on the cytoplasmic face where its other substrate, reduced cytochrome c, is bound. Measurements of the transfer of fluorescence energy from cytochrome c to the two haem a groups of cytochrome oxidase shows that the electrons must 'jump' a surprisingly large distance, about 2.5 nm, and the mechanism for this process is still unclear.

Cytochrome oxidase and a number of other integral proteins of the mitochondrial inner membrane can undergo both lateral and rotational diffusion. It is now thought that the interaction between a number of the components of the electron transport chain may be dependent upon their diffusion and collision within the plane of the bilayer. One approach to investigate this problem has been to extend the distance between diffusing entities by increasing the size of the lipid pool. This can conveniently be done by fusing lipids into the membrane (see section 5.4) and observing its effects on electron transfer rates. These, and other experiments identify a diffusion-limited step between the membrane dehydrogenases and cytochromes bc_1 and indicate that the dehydrogenases, ubiquinone and cytochromes bc_1 are independent diffusible components. Indeed a specific interaction, after collision, between NADH-ubiquonone reductase and ubiquinol-cytochrome c reductase has been suggested as essential for electron transfer from NADH to cytochrome c to occur. These components apparently dissociate and associate at rates which are equal to or faster than the actual rates of electron transfer. The lipid soluble, ubiquonone, will also of course act as a mobile redox carrier, shuttling reducing equivalents between various components by diffusing in the plane of the bilayer.

The electrochemical gradient of protons is used for ATP synthesis by another membrane protein complex called the F_0, F_1 ATPase. This has a distinctive morphology in the electron microscope showing the spherical F_1 component bound by a stalk to the F_0 proteins in the membrane. Altogether at least five different types of subunit are involved. When the F_1 complex is sheared from the membrane it exhibits ATPase activity whilst the remaining F_0 proteins allows proton leakage across the bilayer, suggesting that it functions *in vivo* as a proton channel. When F_0, F_1 ATPase is reconstituted with bacteriorhodopsin in artificial phospholipid vesicles the shift of protons caused by incident light on the bacteriorhodopsin is capable of generating ATP via the separate F_0, F_1 ATPase (see section 5.5). These experiments confirm that this complex normally functions in the direction of ATP synthesis. The mechanism for this process is unknown, but current concepts favour a proton-gradient-induced conformational change of the ATPase enzyme which facilitates synthesis of ATP from ADP and inorganic phosphate. The model for this energy transduction is the muscle actinomyosin system and the reversible Ca^{2+} ATPase from sarcoplasmic reticulum.

Summary

Active transport proteins are the most complex membrane transport system, frequently involving the co-operative action of many subunits, or the covalent linkage of different functional domains in one polypeptide chain.

Energy transduction is achieved by conformational change of the transport protein.

ATP, light energy or redox potential may all serve as an energy source.

7.7 DYNAMIC MEMBRANE EVENTS IN HORMONE ACTION

During active transport, membrane proteins transduce metabolic energy into the vectorial movement of their substrates across the membrane by virtue of their asymmetric location and energy-linked conformational changes. We have seen how structural studies of membrane transport proteins have revolutionized our understanding of how they participate in the transport reaction. Another cell surface phenomenon where an understanding of the dynamic properties of membranes has shed light on their biological function occurs during the stimulation of cells by some hormones (e.g. adrenalin, glucagon, insulin and enkephalin). Direct penetration of these hormones into the cytoplasm does not occur. Instead the membrane provides a relay service which measures hormone concentration in the environment and transmits the information in the form of a 'second messenger' molecule. The two essential components of this relay system are the receptor proteins, which are transmembrane proteins carrying specific binding sites for various hormones on the external surface of the membrane, and the catalytic units which are the proteins responsible for producing the cytoplasmic second messenger. A particularly widespread catalytic unit is the enzyme adenylate cyclase which is integrated into the plasma membrane with its active site on the cytoplasmic surface, but which probably does not extend across the bilayer.

This enzyme catalyses the conversion of ATP into cyclic AMP which has been shown to be an important regulatory factor in glycogen metabolism, glycolysis, lipolysis and DNA transcription. Extensive regulation of intracellular processes may therefore be carried out via this enzyme. Once again it is the asymmetry of these membrane components which ensures vectorial flow of information across the membrane. Most exciting of the properties of this system, however, is the use of the lateral mobility of membrane proteins to achieve an economy of catalytic units, provide a unique amplification mechanism, and give a means for regulating the response of the cell to chronic stimulation by hormone.

7.7.1 Lateral mobility of membrane components

Two models have been suggested to account for the stimulation of adenylate cyclase on the cytoplasmic face of the plasma membrane by hormone binding to the receptor binding site on the extracytoplasmic face of the membrane (Figure 7.16). In the static model one or more receptor molecules are permanently associated with adenylate cyclase and transmit the information of hormone binding by a conformational change occurring between the protein components. The alternative, mobile receptor, model proposes that the receptors and catalytic units normally move independently in the two-dimensional plane of the membrane. Only after hormone binding does a transmembrane complex between receptor and adenylate cyclase occur in which the enzyme becomes activated.

In the rat epididymal fat cell up to seven different hormones bind to distinct receptor proteins, but activate a common pool of adenylate cyclase. If all seven types of receptor were permanently coupled to a single adenylate cyclase then a macromolecular complex of over 10^6 daltons would be observed. This does not seem to be the case.

Measurements of the molecular size of the adenylate cyclase have been made by the technique of irradiation inactivation. A beam of 12 MeV electrons is 'fired' from a linear accelerator at the membrane. Destruction of a protein molecule depends on the target area it presents for absorption of the energy from an accelerated electron. Consequently the changes in molecular size of adenylate cyclase during hormone stimulation may be determined by measuring the change in the susceptibility of the enzyme to inactivation.

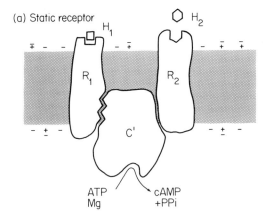

(a) Static receptor

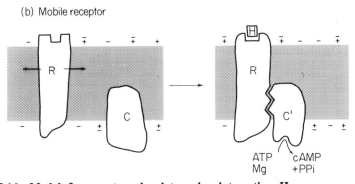

(b) Mobile receptor

Figure 7.16 Models for receptor-adenylate cyclase interaction. Hormone receptors (R) are, in general, transmembrane proteins containing high-affinity binding sites for a single type of hormone (H). The receptor–hormone complex is capable of activating the enzyme adenylate cyclase (C → C′) which is also an intrinsic membrane protein. Cytoplasmic Mg.ATP is converted by this enzyme into cyclic AMP which acts as a 'second messenger', spreading the information of extracellular hormone concentration to intracellular metabolic processes. Since different types of receptor share a common pool of adenylate cyclase molecules, an array of receptors must be clustered around each enzyme molecule in the static model (a). Alternatively, mobile receptors moving in the plane of the membrane may develop a high affinity for adenylate cyclase molecules after binding hormone, as shown in (b). The information of hormone binding is shown here as being transferred by a conformational change in the protein, illustrated by the zig-zag interface. The GTP-binding protein is not shown

Similarly, the size of the receptor may be determined from the loss of hormone binding. By using these techniques it may be shown that the size of the adenylate cyclase of rat liver plasma membranes is increased on the addition of the hormone glucagon, and that the increase in size is equal to the size of the glucagon receptor. This implies that glucagon causes the receptor and catalytic unit to associate into a complex presenting a single target. Furthermore, if the hormone is washed off the membrane after partial inactivation, the remaining functional receptors are able to recombine in a random fashion with the catalytic units in the membrane showing that receptor–catalytic unit binding is reversible and dependent on the presence of hormone.

A different experimental approach, which developed out of the elegant fusion experiments of Frye and Edidin (see section 2.12), has led to similar conclusions. The idea was to fuse a cell containing adenylate cyclase but no receptors in its membrane, with a cell containing receptors but no cyclase and see if any active receptor–cyclase complexes could be formed by lateral diffusion of the membrane components from each type of cell within the common bilayer of the heterokaryon. Turkey erythrocytes were chosen as a source of β-receptors which bind the catecholamines adrenalin and noradrenalin and various β-agonists. The adenylate cyclase of these cells was destroyed by heat treatment or by using the thiol reagent, N-ethyl maleimide. Fusion with Freund erythroleukaemia cells which contain a functional adenylate cyclase but no β-receptors was then achieved using Sendai virus. The result was a population of heterokaryon cells with a β-agonist-sensitive cyclase activity. This experiment has since been repeated with many other cell types in which a variety of receptor species known to activate adenylate cyclase have formed active complexes with adenylate cyclase from different sources. From this it may be inferred that there must be a high degree of conservation of the coupling interface between the receptor molecules and adenylate cyclase. This class of receptor protein might therefore be expected to be multifunctional membrane proteins with a highly conserved binding site for adenylate cyclase and a separate domain with specificity for a particular hormone.

A third membrane protein which is mobile in the plane of the bilayer and interacts with the hormone–receptor–cyclase complex is the GTP regulatory protein. In the presence of hormone and GTP, the receptor, regulatory protein and adenylate cyclase form a transient complex that rapidly dissociates to release the adenylate cyclase in an activated form; GTP is hydrolysed during this reaction. Both the receptor–hormone complex and the activated adenylate cyclase released in this reaction only have a finite lifetime. Owing to the rapid lateral diffusion of protein through the bilayer, however, it is possible for one receptor–hormone complex to activate several catalytic units by this collision–coupling mechanism before the activated catalytic unit returns to its basal state. Rapid diffusion in the plane of the bilayer thus gives a unique amplification mechanism. This allows the hormone to trigger a large stimulation under conditions when only a small fraction of the receptors are occupied. As already noted for the fat cell, it is possible for several types of receptor molecule to compete for a single type of catalytic unit. This not only achieves an economy in membrane protein synthesis but also allows one type of receptor to vary in amount (e.g. by down-regulation, see section 6.7) without changing the number of receptors or catalytic units for other agonists.

7.7.2 Entry of hormones into cells

Following the binding of a hormone to its membrane receptor two processes can occur. Rapid lateral diffusion of the receptor–hormone complex allows (in some cases) a rapid initial response to be communicated by interaction with catalytic units in the membrane which have an enzyme-active site facing the cytoplasmic compartment of the cell. Binding of the hormone to cell surface receptors also triggers the internalization of the hormone by absorptive pinocytosis (see section 7.6). This may result in the degradation of the hormone (down-regulation) or in some cases, longer-term action within the cell.

Insulin binds with extremely high affinity to receptor glycoproteins exposed at the cell surface. These insulin receptors cluster and are internalized upon the addition of insulin. This internalization is of importance in regulating the number of insulin receptors and in certain cellular responses to this hormone, namely the short-term anti-lipolytic effect of insulin on adipocytes and its long-term effects on lipoprotein lipase. Similar responses can be obtained by cross-linking the insulin receptors using bivalent anti-(insulin receptor) antibody. In contrast, univalent fragments of these antibodies, whilst still binding to the receptor, were ineffective unless they themselves were cross-linked to form a bivalent reagent capable of aggregating the receptors. It would appear, therefore, that some, but certainly not all, of the effects of insulin can be mimicked by cross-linking the receptors. As to whether cross-linking is the only trigger, or whether internalization and processing of insulin are also required for some responses, remains to be seen.

The mitogenic (growth-promoting) peptides, epidermal growth factor (EGF) and nerve growth factor (NGF) also cluster and trigger the internalization of their receptors. This appears to be important for their biological function, as derivatives of EGF which are unable to trigger clustering yet are still capable of binding with high affinity to their receptors appear to be biologically inactive. However, by cross-linking these derivatives with anti-(EGF) antibody a bivalent reagent capable of specifically cross-linking the receptors is produced which restores biological activity. As with insulin, the second messengers of the action of these two mitogenic peptides is unknown. However, all three species can affect the phosphorylation of plasma membrane and other proteins, suggesting that this might hold the key to the initial responses that occur at the plasma membrane upon receptor occupancy.

Summary

Lateral mobility of membrane proteins allows a transitory interaction of hormone receptor, G-protein and catalytic unit of adenylate cyclase.

Transient association and modulation of proteins in the lipid bilayer can achieve a unique flexibility of biological function.

Receptor-mediated internalization can provide a means of down-regulation of receptors and a means of expressing the biological action of hormones.

7.8 LATERAL DIFFUSION OF CYTOCHROME b_5 IN MICROSOMAL ELECTRON TRANSPORT

In rat liver microsomes, cytochrome b_5 and cytochrome b_5-reductase provide an electron transport chain capable of cytoplasmic NADH oxidation. Both proteins have a globular hydrophilic portion containing the catalytic centre and a hydrophobic pedicle which attaches the protein to the membrane. Cytochrome b_5 is a small haemoprotein of $M_r = 16,700$, while the reductase is a flavoprotein enzyme of $M_r = 43,000$. Both may be purified to homogeneity and are readily reconstituted into artificial phospholipid vesicles. *In vivo* they function as a source of reducing equivalents for the membrane-bound enzyme stearoyl CoA desaturase and also in NADH-linked cytochrome P_{450} catalysed reactions.

Some elegant experiments using reconstituted components have shown that the cytochrome b_5 and cytochrome b_5-reductase can intereact as a result of collisions between the proteins whilst they are diffusing laterally through the bilayer. The electron transport pathway was assayed in these experiments in two different ways, as shown in Figure 7.17. When ferricyanide was used as an electron acceptor only the electron transport through the reductase enzyme was measured, but when oxidized cytochrome c was added the flux of electrons through the whole pathway was obtained. Table 7.3 shows how the flux of the pathway varies with different ratios of cytochrome b_5 to cytochrome b_5-reductase. In all cases the molar amount of the reductase enzyme was kept constant at 0.5 mmol mol^{-1} of phospholipid. Although cytochrome b_5 was always in a molar excess it is clear that the rate of electron transport increases with cytochrome b_5 concentration, suggesting that effective interaction between the molecules is increased at higher protein concentration. The nature of this interaction was demonstrated by measuring the temperature

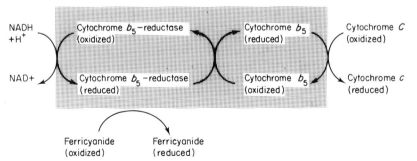

Figure 7.17 **Microsomal electron transport. Cytochrome b_5-reductase and cytochrome b_5 together constitute an electron transport chain of microsomol membranes which normally provides reducing equivalents for the enzyme stearoyl CaA desaturase and some P_{450} catalysed reactions. This figure shows how the whole electron transport pathway may be assayed using oxidized cytochrome c as an electron acceptor, or simply the cytochrome b_5-reductase if ferricyanide is used as substrate**

Table 7.3   The effect of cytochrome b_5 concentration on reconstituted electron transport

Experiment	Ratio cytochrome $b_5 : b_5$-reductase	Calculated distance between molecules (nm)	NADH : cytochrome c-reductase activity (arbitrary units)
1	2 : 1	10	1
2	5 : 1	5	3
3	15 : 1	0.5	10

Purified cytochrome b_5 and cytochrome b_5-reductase were incorporated into dimyristoyl phosphotidyl choline vesicles as indicated. In each case the reductase concentration was 0.5 mmol/mol lipid. NADH : cytochrome c-reductase activity measured at 25 °C was sensitive to cytochrome b_5 concentration despite its molar excess in each case. Close packing of the molecules therefore appears to be important for catalytically productive interactions between the reductase and the haem protein.
Reprinted with permission from Strittmatter, P. and Rogers, M. J. (1975) *Proc. Natl. Acad. Sci.* (USA) **72**, 2658–2661.

dependence of electron transport. Cytochrome b_5-reductase itself (measured using ferricyanide) does not show any break in the Arrhenius plot (Figure 7.18), which is not surprising since only a small portion of this protein is embedded in the bilayer. The whole electron transport pathway, on the other hand, is extremely sensitive to phase transitions in the artificial phospholipid bilayer (Figure 7.18). At low cytochrome b_5 concentration the pathway shows a tenfold decrease in rate at temperatures below the lipid phase transition of the phosphatidyl choline vesicles. This strongly suggests that the bilayer itself acts as the coupling between the reductase enzyme and its cytochrome substrate. As the phospholipid mobility decreases so does the rate of electron transport since the two reactive components are frozen in the bilayer at intervals of 10 nm on average (Table 7.3). At high cytochrome b_5 concentration the effect of the phospholipid phase transition is abolished. This can be attributed to the close spacing (0.5 nm) of protein molecules in these vesicles (Table 7.3) which allows interaction to occur without lateral diffusion through the bilayer. Since cytochrome b_5 and cytochrome b_5-reductase do not associate to form a static complex in reconstituted vesicles it is unlikely that one exists in normal membranes. This mechanism of collision coupling probably therefore reflects the mechanism of electron transfer *in vivo*.

We might tentatively conclude that lateral diffusion of protein molecules in the bilayer may be a general mechanism of membrane protein function. Diffusion in the two-dimensional surface of a membrane is a much more rapid process than diffusion in the three dimensions of the aqueous phase. Thus, collision–coupling is a feasible mechanism for membrane protein interaction. Furthermore, the transient nature of the interaction offers a unique opportun-

324

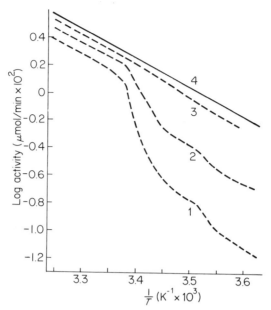

Figure 7.18 **Arrhenius plots of microsomal electron transport. Each line of this figure shows the temperature dependence of electron transport under different conditions. Line 4 shows the linear plot obtained when ferricyanide is used as electron acceptor. Curves 1–3 refer to the complete electron transport pathway, using different concentrations of cytochrome b_5 as listed in Table 7.3. Only at very high concentrations of cytochrome b_5 is a linear plot obtained. The break temperature occurs close to the phase transition temperature of the dimyristoyl phosphatidyl choline used in this reconstituted system and has been interpreted as evidence for the collision–coupling of cytochrome b_5-reductase and cytochrome b_5. (Reprinted with permission from Strittmatter, P., and Rogers, M. J. (1975) *Proc. Natl. Acad. Sci., U.S.A.,* 72, 2658–2661)**

ity for flexibility in response and regulation. What does seem abundantly clear from the studies reported in this chapter is that a structural understanding of membrane components has had far-reaching consequences in our understanding of a number of perplexing biological problems and that in many of these it is the asymmetry and dynamic properties of membranes which are used by the cell to achieve its ends.

Summary

Cytochrome b_5 and cytochrome b_5-reductase interact by lateral diffusion in the plane of the membrane.

This may be a general mechanism of membrane protein function.

FURTHER READING

Green, D. E. (1981) A critique of the chemosmotic model of energy coupling, *Proc. Natl. Acad. Sci.,* **78,** 2240–2243.

Hinkle, P. C. and McCarty, R. E. (1978) How cells make ATP, *Scientific American*, **238** (March), 104–123.

Houslay, M. D. (1981) Mobile receptor and collision coupling mechanisms for the activation of adenylate cyclase by glucagon, *Advances in Cyclic Nucleotide Research*, **14**, 111–119.

Houslay, M. D. (1981) Membrane phosphorylation : a crucial role in the action of insulin, EGF and pp. 60^{src}? *Bioscience Reports*, **1**, 19–34.

Klingenberg, M. (1981) Membrane protein oligomeric structure and transport function, *Nature*, **290**, 449–454.

Limbird, L. E. (1981) Activation and attenuation of adenylate cyclase, *Biochem. J.*, **195**, 1–13.

Mayer, M. M. (1972) Mechanism of cytolysis by complement, *Proc. Natl. Acad. Sci.*, **69**, 2954–2958.

Peracchia, C. (1977) Gap junction structure and function, *Trends Biochem. Sci.*, **2**, 26–30.

Pressman, B. C. (1976) Biological applications of ionophores, *Ann. Rev. Biochem.*, **45**, 501–530.

Singer, S. J. (1977) Thermodynamics, the structure of integral membrane proteins, and transport, *J. Supramol. Struct.*, **6**, 313–323.

Wilson, D. B. (1978) Cellular transport mechanisms, *Ann. Rev. Biochem.*, **47**, 933–965.

Index